DATE DUE

NO 19 03			

DEMCO 38-296

WORKFORCE READINESS

Competencies and Assessment

WORKFORCE READINESS

Competencies and Assessment

Edited by

Harold F. O'Neil, Jr.

University of Southern California
and
National Center for Research on
Evaluation, Standards, and
Student Testing (CRESST)

LAWRENCE ERBLAUM ASSOCIATES, PUBLISHERS
1997   Mahwah, New Jersey   London

Lawrence Erlbaum Associates, Inc., Publishers
10 Industrial Avenue
Mahwah, New Jersey 07430

Library of Congress Cataloging-in-Publication Data

Workforce readiness : competencies and assessment / edited by Harold
F. O'Neil, Jr.
      p. cm.
   Includes bibliographical references and index.
   ISBN 0-8058-2149-X (cloth : alk. paper). -- ISBN 0-8058-2150-3
(pbk. : alk. paper)
    1. Workplace literacy--United States--Evaluation. 2. Educational
tests and measurements--United States. 3. School-to-work
transition--United States.   I. O'Neil, Harold F., Jr. 1943-
LC149.7.W68   1997
370.ll'3--dc21
                                          97-13791
                                             CIP

Books published by Lawrence Erlbaum Associates are printed on acid-free paper,
and their bindings are chosen for strength and durability.

Printed in the United States of America
10 9 8 7 6 5 4 3 2 1

Contents

Preface

Workforce Readiness: Competencies and Assessment is designed for professionals in the assessment/evaluation/measurement and the vocational, technical, and educational psychology communities. It explores the state of the art in the specification of competencies (skills) and their assessment for students entering the world of work from both high school and college. Both individual and team competencies are explored.

Many high school and college graduates lack the necessary knowledge and skills to be productive members of a workforce that focuses on high-performance/high-paying jobs. By "high performance" we mean work settings committed to excellence, product quality, and customer satisfaction. Lack of skills for such a workplace in an entry-level workforce may be a major reason for potential U.S. economic noncompetitiveness. This development means that much more is expected of even entry-level members of the American workforce. Thus, even more is expected of our high schools and colleges to provide this type of workforce. Such issues are a common concern in the United States and other countries (e.g., Australia).

The most promising intellectual framework to deal with these issues is provided by the Secretary's Commission on Achieving Necessary Skills (SCANS; U.S. Department of Labor, 1992). This framework of competencies is updated in this book and is focused on a high-skill, high-wage, and high-performance workplace. Competencies were defined by SCANS as part of "workplace know-how" (U.S. Department of Labor, 1992, p. 6) or the skills that young people need to succeed in the world of work. This framework is common to many of the chapters of this book.

Another reality of the workplace is the increase in technology, which results in an increase in cognitive complexity. For example, instead of performing simple procedural and predictable tasks, a worker becomes responsible for inferences, diagnosis, judgment, and decision making, often under severe time pressure. Trends of increasing requirements of both knowledge and skills of workers coupled with an increase in technology in the workplace are made worse by the increased influence of international markets. In the future, one will compete worldwide or not at all. In summary, there is a potential skill gap for the high-skill, high-wage, high-productivity jobs.

The need of American management for workers with greater skills who can take on greater responsibility has spawned many commissions, task forces, and studies. All of them have contributed to the vast evidence documenting the need for a more highly skilled workforce. These studies are summarized and synthesized in this book in the initial set of chapters on "Specifications of Workforce Competencies." However, what remains largely undone is the development of methods to assess the necessary competencies that have been identified. The final set of chapters on "Assessment of Workforce Competencies" deal with such assessment issues. The measurement issues to assess these competencies are conceptualized as (a) what to measure (e.g., cognitive processes, tasks, or characteristics of jobs and the setting or context); (b) performance assessment approaches (e.g., portfolios or simulations); (c) criteria (e.g., validity, fairness, transfer and generalizability, cost, and efficiency); and (d) type of technology (e.g., paper-and-pencil, computer). Various chapters will deal with either individual or team competency assessment.

In summary, this book is characterized in the following montage of factors: (a) a focus on competencies of teams or groups as well as individuals; (b) a focus on the assessment of competencies; (c) contexts in training as well as education; (d) a focus on the transition from high school to the world of work as well as college to the world of work; (e) it adapts a modified version of the SCANS framework as its conceptual model of the competencies needed for this world of work; (f) it focuses on entry-level positions; (g) it uses a broad variety of methodological approaches from qualitative to instructional design to quantitative (structural equation modeling); and (h) it is multidisciplinary. The data that will be reported were collected in various settings, for instance, schools, laboratories, and industrial settings. The authors are the researchers that, in many cases, have defined by their work the state of the art in competency identification and assessment in both this country and overseas.

This book could not have come into existence without the help and encouragement of many people. Our thanks to our editor, Ray O'Connell of Lawrence Erlbaum Associates, for his support and guidance in the publication process. We thank Katharine Fry for her excellent assistance in preparing the manuscript.

ACKNOWLEDGMENTS

The work reported herein was supported under the Educational Research and Development Center Program cooperative agreement R117G10027 and CFDA catalog number 84.117G as administered by the Office of Educational Research and Improvement, U.S. Department of Education. The findings and opinions expressed in this report do not reflect the position or policies of the Office of Educational Research and Improvement or the U.S. Department of Education.

REFERENCES

U.S. Department of Labor. (1991, June). *What work requires of schools: A SCANS report for America 2000.* Washington, DC: U.S. Department of Labor, Secretary's Commission on Achieving Necessary Skills.

U.S. Department of Labor. (1992). *Skills and tasks for jobs: A SCANS report for America 2000.* Washington, DC: U.S. Department of Labor, Secretary's Commission on Achieving Necessary Skills.

Part I

SPECIFICATIONS OF
WORKFORCE COMPETENCIES

Chapter 1

Review of Workforce Readiness Theoretical Frameworks

Harold F. O'Neil, Jr.
CRESST/University of Southern California

Keith Allred
Eva L. Baker
CRESST/University of California, Los Angeles

At the National Center for Research on Evaluation, Standards, and Student Testing (CRESST), we have been working to develop the needed assessment measures of identified workforce readiness skills.

Our development has been in the context of the Secretary's Commission on Achieving Necessary Skills (SCANS). The purpose of this chapter is to identify and categorize workforce skills identified in five major studies (O'Neil, Allred, & Baker, 1992). The five studies we examined are described in the following reports: (a) *What Work Requires of Schools*, conducted by the Secretary's Commission on Achieving Necessary Skills (SCANS) for the U.S. Department of Labor (U.S. Department of Labor, 1991); (b) *Workplace Basics: The Essential Skills Employers Want*, conducted by the American Society for Training and Development (ASTD) with the support of the Department of Labor (Carnevale, Gainer, & Meltzer, 1990); (c) the *Michigan Employability Skills Employer Survey*, conducted by the Michigan Employability Skills Task Force (Employability Skills Task Force, 1988, 1989; Mehrens, 1989); (d) *Basic and Expanded Basic Skills*, conducted by the New York State Education Department (1990); and (e) *High Schools and the Changing Workplace: The Employers' View*, conducted by the National Academy of Sciences (NAS; 1984). Also, for the purpose of this chapter, we updated our 1992 analysis.

These studies have their source in a common concern. Current economic difficulties and the challenge of competing in the world market have necessitated a rethinking of American approaches to the utilization of people in organizations (Office of Technology Assessment, 1990, 1995; Packer & Pines, 1996; Stasz, 1995). The early part of this century witnessed a burgeoning concern about how to organize the efforts of individuals in the large organizations that resulted from the industrial revolution. Previously, productivity was regarded as a function of technology and access to raw materials. Increasingly, however, the human factor came to be recognized as critical to productivity. Frederick W. Taylor (1916), considered the father of the field of American management, recognized the limitations in skills of most workers in these organizations. His answer to the deficiency in worker skills was to emphasize the need for an extensive managerial structure to organize and supervise the workers, leaving as little responsibility and discretion as possible to the common worker. Although it remained dominant in the American workforce for decades, this emphasis has been fundamentally rethought in the last decade, and especially in the last several years.

In contrast to the traditional approach in the United States, management now recognizes a need to have workers take on more responsibility at the points of production, of sales, or of service rendered, if we are to compete in rapidly changing world markets. In order to adapt to the need to introduce new products and services quickly with high quality, new directions in management emphasize participative management, flatter organizational structure, just-in-time management, and team work. This development means that much more is expected of even entry-level members of the American workforce. The cry of American management for workers with greater skills and who can take on greater responsibility has spawned many commissions, task forces and studies, including the five studies previously mentioned. All of them have contributed to the vast evidence documenting the need for a more highly skilled workforce. What remains largely undone is the development of methods to assess the necessary skills that have been identified.

In general, the five studies examined all began with a similar first step. Experts, generally educators, business people, scholars, and policy makers, were assembled to identify skills necessary for the workforce. The experts generated a framework of skills based on their own knowledge and experience, in addition to various informal investigations of the workforce and its requirements. All but the NAS study also included a second, validation phase. In this phase, employers and/or employees were asked how necessary each of the identified skills was for the world of work.

The five major frameworks of skills which these studies identified are first reviewed individually. Subsequently, the frameworks are compared and summarized.

SCANS

SCANS was charged by the U.S. Secretary of Labor to investigate what is required in today's and tomorrow's workplace and to determine the extent to which high school students are able to meet those requirements. Specifically, SCANS was directed by the Secretary of Labor to (a) define the skills needed for employment, (b) propose acceptable levels of proficiency, (c) suggest effective ways to assess proficiency, and (d) develop a dissemination strategy for the nation's schools, businesses, and homes. In June 1991, the Commission issued a report (U.S. Department of Labor, 1991) concerning the first two directives. The Commission, based on its discussions and meetings with business owners, public employers, unions, and workers and supervisors in shops, plants, and stores, identified five competencies in accordance with the first directive: the ability to efficiently use (a) resources, (b) interpersonal skills, (c) information, (d) systems, and (e) technology (see Table 1.1). Additionally, the Commission found that these five competencies are based on a three-part foundation: (a) basic skills, (b) thinking skills, and (c) personal qualities (see Table 1.2).

ASTD

The American Society for Training and Development (ASTD), a nonprofit professional association representing approximately 50,000 practitioners, managers, administrators, educators, and researchers in the field of human resource development, also undertook an extensive study of skills required in America's workforce. Supported by a grant from the Department of Labor, ASTD reviewed relevant research and conducted extensive on-site studies and telephone interviews to find out what employers thought that their workers needed to be successful. Again, the emphasis was on skills that are needed across all jobs. ASTD also consulted with more than 400 experts in interpreting their findings. The result was the identification of the 16 skills within the seven skill groups listed in Table 1.3 (Carnevale et al., 1990).

MICHIGAN

In 1987, the Employability Skills Task Force was convened by Governor Blanchard's Commission on Jobs and Economic Development to identify the generic skills employers believed to be important in jobs across all sectors of the economy. Composed of leaders from business, education, and labor across the state, the Task Force identified 26 skills falling into the three domains of academic, personal management, and teamwork skills. The skills are presented in Table 1.4.

TABLE 1.1
Five Competencies

Resources: Identifies, organizes, plans, and allocates resources

- A. *Time*—Selects goal-relevant activities, ranks them, allocates time, and prepares and follows schedules
- B. *Money*—Uses or prepares budgets, makes forecasts, keeps records, and makes adjustments to meet objectives
- C. *Material and Facilities*—Acquires, stores, allocates, and uses materials or space efficiently
- D. *Human Resources*—Assesses skills and distributes work accordingly, evaluates performance and provides feedback

Interpersonal: Works with others

- A. *Participates as Member of a Team*—Contributes to group effort
- B. *Teaches Others New Skills*
- C. *Serves Clients/Customers*—Works to satisfy customers' expectations
- D. *Exercises Leadership*—Communicates ideas to justify position, persuades and convinces others, responsibly challenges existing procedures and policies
- E. *Negotiates*—Works toward agreements involving exchange of resources, resolves divergent interests
- F. *Works With Diversity*—Works well with men and women from diverse backgrounds

Information: Acquires and uses information

- A. *Acquires and Evaluates Information*
- B. *Organizes and Maintains Information*
- C. *Interprets and Communicates Information*
- D. *Uses Computers to Process Information*

Systems: Understands complex interrelationships

- A. *Understands Systems*—Knows how social, organizational, and technological systems work and operates effectively in them
- B. *Monitors and Corrects Performance*—Distinguishes trends, predicts impacts on system operations, diagnoses deviations in systems' performance and corrects malfunctions
- C. *Improves or Designs Systems*—Suggests modifications to existing systems and develops new or alternative systems to improve performance

Technology: Works with a variety of technologies

- A. *Selects Technology*—Chooses procedures, tools, or equipment, including computers and related technologies
- B. *Applies Technology to Task*—Understands overall intent and proper procedures for setup and operation of equipment
- C. *Maintains and Troubleshoots Equipment*—Prevents, identifies, or solves problems with equipment, including computers and other technologies

---

Note. From U.S. Department of Labor, 1991, p. 12.

TABLE 1.2
A Three-Part Foundation

Basic Skills: Reads, writes, performs arithmetic and mathematical operations, listens and speaks

A.   *Reading*—Locates, understands, and interprets written information in prose and in documents such as manuals, graphs, and schedules
B.   *Writing*—Communicates thoughts, ideas, information, and messages in writing; and creates documents such as letters, directions, manuals, reports, graphs, and flow charts
C.   *Arithmetic/Mathematics*—Performs basic computations and approaches practical problems by choosing appropriately from a variety of mathematical techniques
D.   *Listening*—Receives, attends to, interprets, and responds to verbal messages and other cues
E.   *Speaking*—Organizes ideas and communicates orally

Thinking skills: Thinks creatively, makes decisions, solves problems, visualizes, knows how to learn, and reasons

A.   *Creative Thinking*—Generates new ideas
B.   *Decision Making*—Specifies goals and constraints, generates alternatives, considers risks, and evaluates and chooses best alternative
C.   *Problem Solving*—Recognizes problems and devises and implements plan of action
D.   *Seeing Things in the Mind's Eye*—Organizes and processes symbols, pictures, graphs, objects, and other information
E.   *Knowing How to Learn*—Uses efficient learning techniques to acquire and apply new knowledge and skills
F.   *Reasoning*—Discovers a rule or principle underlying the relationship between two or more objects and applies it when solving a problem

Personal Qualities: Displays responsibility, self-esteem, sociability, self-management, and integrity and honesty

A.   *Responsibility*—Exerts a high level of effort and perseveres toward goal attainment
B.   *Self-esteem*—Believes in own self-worth and maintains a positive view of self
C.   *Sociability*—Demonstrates understanding, friendliness, adaptability, empathy, and politeness in group settings
D.   *Self-Management*—Assesses self accurately, sets personal goals, monitors progress, and exhibits self-control
E.   *Integrity/Honesty*—Chooses ethical courses of action

Note. For U.S. Department of Labor, 1991, p. 16.

These 26 skills were further refined into 86 subskills and then validated through a survey in which over 2,500 Michigan employers rated the importance of the skills identified. In general, all the skills were rated as very important. The average overall rating corresponded to "Highly Needed" on the response scales. However, personal management and teamwork skills were more highly valued than academic skills (Mehrens, 1989).

TABLE 1.3
ASTD Skills

I. The Foundation:
    1. Learning How to Learn

II. Basic Competency Skills:
    2. Reading
    3. Writing
    4. Computation

III. Communication Skills:
    5. Speaking
    6. Listening

IV. Adaptability Skills:
    7. Problem Solving
    8. Thinking Creatively

V. Developmental Skills:
    9. Self-Esteem
    10. Motivation and Goal Setting
    11. Career Development (Planning)

VI. Group Effectiveness Skills:
    12. Interpersonal Skills
    13. Teamwork
    14. Negotiation

VII. Influencing Skills:
    15. Understanding Organizational Culture
    16. Sharing Leadership

NEW YORK

The New York State Department of Education convened a blue-ribbon committee of educators, scholars, and business people who developed a framework of skills necessary for the workforce. The result, as seen in Table 1.5, was two categories of Basic Skills—Language Arts and Mathematics—and eight Expanded Basic Skills.

This framework (Anderson Committee, no date; New York State Education Department, 1990) was subsequently validated in a study in which 1,400 workers in jobs that did not require a college degree were interviewed and observed with regard to what was required of them at work. Based on these interviews and observations, the investigators rated the level of a given skill necessary for that job. The New York study is currently the only one to have examined this critical issue of what *levels* of the skills identified are required in the workforce.

TABLE 1.4
Michigan Skills

Academic Skills	Personal Management Skills	Teamwork Skills
• Understand spoken language and speak in the language in which business is conducted.	• Identify personal job-related interests, strengths, options, and opportunities.	• Identify with the goals, norms, values, customs, and culture of the group.
• Read written materials (including graphs, charts, and displays).	• Demonstrate personal values and ethics in the workplace (e.g., honesty, fairness, and respect for others).	• Communicate with all members of a group.
• Write in the language in which business is conducted.	• Exercise a sense of responsibility.	• Show sensitivity to the thoughts and opinions of others in a group.
• Understand and solve problems involving basic arithmetic and use the results.	• Demonstrate self-control.	• Use a team approach to identify problems and devise solutions to get a job one.
• Use the tools and equipment necessary to get a job done.	• Show pride in one's work.	• Exercise "give and take" to achieve group results.
• Access and use specialized knowledge when necessary (e.g., the sciences or skilled trades) to get a job done.	• Be enthusiastic about the work to be done.	• Function in changing work settings and in changing groups.
• Think and act logically by using the steps of the scientific method (i.e., identify problems, collect information, form opinions, and draw conclusions).	• Follow written or verbal directions.	• Determine when to be a leader or a follower depending on what is necessary to get a job done.
	• Learn new skills and ways of doing things.	• Show sensitivity to the needs of women and ethnic and racial minorities.
	• Identify and suggest new ideas for getting a job done.	• Be loyal to a group.
	• Be a leader or a follower depending upon what is necessary to get a job done.	

Note. Adapted from Employability Skills Task Force, 1989, p. 2.

TABLE 1.5
New York Skills

Basic skills

1. Language arts
    Listening and speaking for:
    • personal response
    • social interaction
    • information and understanding
    • critical analysis and evaluation
    Reading for
    • aesthetic and personal response
    • acquisition, interpretation, and application of information
    • critical analysis and evaluation
    Writing for
    • personal expression
    • social interaction
    • information and understanding
    • critical analysis and evaluation

2. Mathematics
    Basic operations
    Logic
    Probability
    Statistics
    Measurement
    Algebra/geometry

    Expanded basic skills
        1. Reasoning
        2. Interpersonal
        3. Working as a member of a team
        4. Using information systems
        5. Setting priorities
        6. Personal work skills and behaviors
        7. Personal and civic responsibility
        8. Manual dexterity

Note. From New York State Education Department, *Basic and expanded basic skills. Scales for validation study,* pp. 45-46, Albany, NY, July 1990

No precise theoretical or conceptual framework was offered to explain the discrete differences between skill levels; however, Table 1.6 lists the dimensions that underlie the general progression from skill level 1 to skill level 6. An extensive report was generated to provide examples for each level of skill for all of the basic and expanded basic skills. Table 1.7 includes the examples for each skill level with respect to interpersonal skills.

Table 1.8 summarizes the results by showing the percentage of jobs that require competency at each skill level for each of the identified skills (Anderson Committee, no date). In general, the results were similar to those of the Michigan study in that interpersonal skills, working as a member of a team, and personal work skills and behavior were more highly valued than traditional skills taught in high schools.

TABLE 1.6
Dimensions Underlying New York Skill Levels

Low	→	High
Simple	→	Complex
Routine	→	Variable
Concrete	→	Abstract
Structured	→	Unstructured
Recall of knowledge	→	Evaluation of knowledge
Directed	→	Independent
Conventional	→	Innovative

Note. From New York State Education Department, 1990, p. 1.

NAS

In 1984 the National Academy of Sciences convened a panel of employers, labor union representatives, scholars and educators to examine the skills required of high school graduates who enter the workforce after graduation. The panel identified 10 "core competencies" required of all workers,

TABLE 1.7
Skill Level Examples for Interpersonal Skills

The ability to interact effectively professionally and socially.

Descriptors

1. • tolerates others
   • refrains from causing conflict when faced with potentially difficult interactions
2. • conducts oneself with courtesy and tact
   • provides positive feedback to others
3. • recognizes and handles own routine interpersonal problems while respecting cultural, gender, and ethnic diversity
4. • is empathic when providing negative criticism
   • reacts constructively to positive and negative criticism
   • diffuses anxiety in others when dealing with controversial information or unsettling events
5. • is sought out often by others for advice in interpersonal conflict situations
   • uses a variety of approaches for handling conflict
   • deals with personal conflicts objectively
6. • helps others to resolve interpersonal differences
   • fosters an atmosphere of harmony among individuals
   • serves as a role model in formal and informal interactions

Note. From New York State Education Department, *Basic and Expanded Basic Skills. Scales for Validation Study,* pp. 45-46, Albany, NY, July 1990.

TABLE 1.8

Summary of Validation Data Results: Sample of Employee Observations/Interviews—New York

Competency Scale	Percent of Jobs at Competency Level						
	N/A	One	Two	Three	Four	Five	Six
Reading for personal response	71%	8%	9%	6%	4%	2%	0%
Reading for information	4%	5%	13%	27%	30%	17%	5%
Reading for critical analysis and evaluation	20%	6%	10%	24%	26%	12%	3%
Writing for personal expression	66%	11%	9%	8%	3%	2%	0%
Writing for social interaction	48%	17%	15%	10%	7%	2%	1%
Writing for information	7%	13%	27%	25%	16%	8%	3%
Writing for critical analysis and evaluation	26%	11%	22%	21%	12%	7%	2%
Listening/speaking personal response	40%	7%	19%	16%	9%	7%	2%
Listening/speaking social interaction	12%	6%	25%	28%	17%	8%	4%
Listening/speaking for information	1%	4%	14%	25%	33%	16%	7%
Listening/speaking for critical analysis and evaluation	8%	8%	18%	23%	28%	11%	3%
Math–basic operations	4%	10%	13%	25%	26%	11%	12%
Math–logic	7%	10%	15%	28%	24%	10%	6%
Math–probability	22%	19%	12%	23%	17%	6%	2%
Math–statistics	46%	16%	12%	11%	6%	5%	3%
Math–measurement	21%	7%	10%	17%	23%	14%	9%
Math–algebra and geometry	78%	7%	5%	2%	2%	4%	2%
Manual dexterity	1%	3%	17%	30%	24%	21%	4%
Reasoning	1%	5%	11%	19%	32%	28%	5%
Interpersonal skills	1%	3%	12%	21%	27%	20%	16%
Working as member of team	1%	9%	11%	21%	22%	23%	13%
Using information systems	22%	10%	16%	18%	15%	15%	5%
Setting priorities	4%	6%	15%	20%	27%	20%	9%
Personal work skills and behaviors	1%	7%	9%	18%	27%	25%	14%
Personal and civil responsibility	20%	25%	7%	7%	19%	16%	6%

Note. From Anderson Committee, Report to the Board of Regents on Career Preparation Validation Study, Appendix V, Albany, NY, no date.

regardless of education or specialty (see Table 1.9). The panel summarized their results into three basic findings: (a) The major asset required in the workforce is the ability to learn new knowledge and skills to adapt to the rapidly changing workplace; (b) the core competencies are required at all levels in the workforce; and (c) positive attitude and sound work habits are especially valued by employers (National Academy of Sciences, 1984, p. xi).

SUMMARY

Several commonalities in the findings of the five studies are apparent. Specifically, four major categories of job-readiness skills can be seen running through the five frameworks. First, each study identified the need for basic academic skills. As seen in Table 1.10, these include the three R's as well as speaking and listening skills. Study participants judged job-related speaking and listening skills to be particularly important in both the Michigan and New York studies. Of the four common categories found, this one exhibited the greatest similarity across studies. This is not surprising, given that the basic skills have received the most attention and elaboration in the past.

Second, all studies identified the need for higher order thinking skills, as seen in Table 1.11. In general, these skills were deemed necessary because of the rapidity of change in the workforce. The most common higher order thinking skills identified can be seen as skills in adapting to these changes. Although the New York framework includes reasoning as an expanded basic skill, it clearly did not identify higher order thinking skills to the extent that the other studies did. In the SCANS, ASTD, Michigan and NAS studies, problem-solving skills were identified as important higher order thinking skills. Creativity also emerged in one form or another in those studies, although the skill was not always referred to as "creativity" per se. Decision making was also identified as an important higher order thinking skill in the SCANS, Michigan, New York, and NAS studies, but not in the ASTD study. In the SCANS and ASTD studies, learning how to learn was identified as an important higher order thinking skill. Indeed, for the ASTD study, it was identified as *the* foundation skill. The ability to learn was also identified as one of the three basic findings in the NAS study, although it is not identified in the framework of core competencies. The Michigan study also identified the need to learn new skills, but did not identify the ability to learn new skills as a job-readiness skill itself.

Third, within all five frameworks interpersonal and teamwork skills were judged to be essential, as indicated in Table 1.12. These skills have become important, the studies emphasize, because as responsibility is shifted further down the management hierarchy to groups of workers, the average worker needs to communicate and cooperate with other members of the organi-

TABLE 1.9
National Academy of Sciences Skills

Command of English language

Reasoning and problem solving, including:

>Identify problems
>Consider and evaluate possible alternative solutions, weighing their risks and
>   benefits
>Formulate and reach decisions logically
>Separate fact from opinion
>Adjust to unanticipated situations by applying established rules and facts
>Work out new ways of handling recurring problems
>Determine what is needed to accomplish work assignments

Reading, including:

>Understand the purpose of written material
>Note details and facts
>Identify and summarize principal and subsidiary ideas
>Be aware of inconsistency in written material
>Verify information and evaluate the worth and objectivity of sources
>Interpret quantitative information; for example, in tables, charts, and graphs

Writing, including:

>Gather information suitable for the purpose
>Organize information in a logical and coherent manner
>Use standard English syntax
>Apply the rules of correct spelling, punctuation, and capitalization
>Attribute references correctly
>Use reference books such as a dictionary, a thesaurus, and an encyclopedia
>Write legibly

Computation, including:

>Add, subtract, multiply, and divide whole numbers, decimals, and fractions
>   accurately
>Calculate distance, weight, area, volume, and time
>Convert from one measurement system to another, for example, from English to
>   metric
>Determine the costs, time, or resources necessary for a task
>Calculate simple interest
>Compute costs and make change
>Understand simple probability and statistics
>Calculate using information obtained from charts, graphs and tables
>Use ratios, proportions, percentages, and algebraic equations with a single unknown
>Estimate results and judge their accuracy

Science and technology

Oral communication, including:
>Communicate in standard English
>Understand the intent and details of oral communications

(Continued)

TABLE 1.9
(Continued)

Understand and give instructions
Identify and summarize correctly principal and subsidiary ideas in discussions
Obtain, clarify, and verify information through questioning
Participate effectively in discussions

Interpersonal relationships, including:

Interact in a socially appropriate manner
Demonstrate respect for the opinions, customs, and individual differences of
    others
Appreciate the importance and value of humor
Offer and accept criticism constructively
Handle conflict maturely
Participate in reaching group decisions

Social and economic studies, including:

The history of present-day American society
The political, economic, and social systems of the United States and other
    countries
The fundamentals of economics, including a basic understanding of the roles of
    money, capital investment, product pricing, cost, profit, and productivity,
    and market forces such as supply and demand
The concept of "trade-offs" and the differences between economic principles,
    facts, and value judgments
The forms and functions of local, state, and federal governments
The rights and responsibilities of citizens
Civil rights and justice in a free society

Personal work habits, including:

A realistic, positive attitude toward one's self
A positive attitude toward work and pride in accomplishment
A willingness to learn
Self-discipline, including regular and punctual attendance and dependability
The ability to set goals and allocate time to achieve them
The capacity to accept responsibility
The ability to work with or without supervision
Appropriate dress and grooming
An understanding of the need for organization, supervision, rules, policies, and
    procedures
Freedom from substance abuse
Appropriate personal hygiene

Note. Adapted from National Academy of Sciences, 1984, pp. 20-22, 24-27.

TABLE 1.10
Major Skill Category 1: Basic Skills

SCANS	ASTD	Michigan	New York	NAS
Basic Skills Foundation Reading Writing Arithmetic/ Mathematics Listening Speaking	Basic Competency Skills Reading Writing Computation Communication Skills Speaking Listening	Academic Skills (see Table 1.4)	Basic Skills Language Arts Listening and speaking Reading Writing Mathematics	Reading, Writing Computation Oral Communication

zation to an increasing degree. Relative to the other major categories of skills identified, this category was identified as being especially important. On average, employers in the Michigan survey rated skills in working in groups and working with others between the "Critical" and "Highly Needed" points on the response scale. In the New York study, 84% of the jobs investigated were judged to require interpersonal skills at level 3 or higher, and 79% required team work skills at these levels (see Table 1.8).

Although all five studies identified interpersonal and team work skills and emphasized their importance, this category exhibited the greatest diversity in terms of the specific subskills that constitute it. It would seem that, at least at this point, these skills, although uniformly recognized as critical, are the most difficult to define and identify. Despite the differences, three common sets of subskills are apparent.

The single set of subskills in the interpersonal and teamwork skills category that all five studies identified were negotiation/conflict-resolution skills. Again, however, there was some diversity among the studies in how these skills were defined. The SCANS study defined negotiation skills as the ability to work toward agreements involving exchange of resources and resolution of divergent interests (see Table 1.1). ASTD reviewed definitions of negotiation skills found in the negotiation literature and emphasized the "principled" negotiation skills identified by Fisher and Ury (1981; see Carnevale et al., 1990, pp. 330–350). The Michigan study operationalized negotiation skills simply as willingness to compromise (Mehrens, 1989, p. 10). Compromise is viewed quite differently in the negotiation literature from the notion of "resolving divergent interests" identified by SCANS (see O'Neil et al., 1992, for a summary). The NAS study does not define negotiation/conflict-resolution skills except to state that it is necessary for workers to realize that conflict is inherent but can be handled through "constructive means" (National Academy of Sciences, 1984, p. 25). In the

TABLE 1.11
Major Skill Category 2: Higher Order Thinking Skills

SCANS	ASTD	Michigan	New York	NAS
Thinking Skills Foundation	*The Foundation*	*Academic Skills*	*Reasoning*	*Reasoning and Problem Solving*
Creative Thinking Decision Making Problem Solving Seeing Things in Mind's Eye Knowing How to Learn Reasoning	Learning How to Learn *Adaptability Skills* Problem Solving Thinking Creatively	Including: Understand and solve problems involving basic arithmetic and use the results *Personal Management Skills* Including: Identify and suggest new ideas for getting a job done	The ability to draw conclusions through the use of rational processes	Identify problems Consider alternative solutions Formulate and reach decisions logically Separate fact from opinion Adjust to unanticipated situations by applying established rules and facts Work out new way of handling recurring problems Determine what is needed to accomplish work assignments

TABLE 1.12

Major Skill Category 3: Interpersonal and Teamwork Skills

SCANS	ASTD	Michigan	New York	NAS
Interpersonal competency Participates as Member of a Team Teaches Others New Skills Serves Clients/Customers Exercises Leadership Negotiates Works with Diversity	*Group effectiveness skills* Interpersonal Skills Teamwork Negotiation *Influencing skills* Understanding Organizational Culture Sharing Leadership	Teamwork skills (see Table 1.4)	*Interpersonal* The ability to interact effectively, professionally, and socially *Working as a member of a team* The ability to conduct oneself according to the expressed or unexpressed norms of a group and to participate according to one's talents	*Interpersonal relationships* Interact in socially appropriate manner Demonstrate respect for the opinions, customs, and individual differences of others Appreciate the importance and value of humor Offer and accept criticism constructively Handle conflict maturely Participate in reaching group decisions

New York study, although negotiation and conflict resolution skills are not explicitly identified within the definition of interpersonal and team work skills, several aspects of conflict resolution skills are virtually the only examples of interpersonal skills offered for levels 5 and 6 of the interpersonal skills (New York State Education Department, 1990, pp. 45–46).

Second, leadership skills were identified as a category of interpersonal and teamwork skills in three studies. The SCANS, ASTD, and Michigan studies identified leadership skills as important, but again, there was substantial diversity in *how* these skills were defined and identified. The SCANS study spoke of persuasion (see Table 1.1). The ASTD study emphasized the skill of sharing leadership (see Table 1.3) and reviewed a number of current theories of leadership (Carnevale et al., 1990, pp. 377–398). The Michigan study emphasized the skill of recognizing when to be a leader and when to be a follower (see Table 1.4).

Third, the ability to work with others from diverse backgrounds was a category of interpersonal and teamwork skills identified by four studies. The SCANS, Michigan, New York, and NAS studies identified skills in being sensitive and responsive to the ethnic, cultural and gender differences that exist between workers.

The fourth major category of workforce competency common to the five reviewed studies focused on personal characteristics and attitudes rather than particular skills. As seen in Table 1.13, the important themes in this category were self-esteem, motivation, and responsibility. These types of worker qualities were generally rated as more critical than workforce "skills" in the Michigan survey. Eighty-four percent of the jobs investigated in the New York study were judged to require the personal work skills and behaviors competency at level 3 or higher (see Table 1.8).

When persons (e.g., employees, graduates, or students) are asked to rate the relative importance of such skills, generic skills such as thinking or

TABLE 1.13
Major Skill Category 4: Personal Characteristics and Attitudes

SCANS	ASTD	Michigan	New York	NAS
Personal Qualities Foundation Responsibility Self-Esteem Sociability Self-Management Integrity/Honesty	*Developmental Skills* Self-Esteem Motivation and Goal Setting Career Development (Planning)	*Personal Management Skills* (see Table 1.4)	*Personal Work Skills and Behavior* *Personal and Civic Responsibility*	*Personal Work Habits* (see Table 1.9)

TABLE 1.14
Comparison of Competencies: High School Versus College

	High School (SCANS)	College (Sinclair, in press)	College (NCES)
Basic skills	Basic skills foundation (e.g., reading, writing, listening, speaking)	Communication skills (e.g., writing, speaking)	Communication skills (e.g., speaking, reading, interpretation)
Higher order thinking skills	Thinking skills foundation (e.g., decision making, problem solving, knowing how to learn)	Thinking/decision-making skills (e.g., solve problems)	Critical thinking skills (e.g., verbal reasoning skills, decision-making and problem-solving skills)
	Interpersonal competency (e.g., participates as member of a team; exercises leadership; negotiates)	Capacity for cooperation and teamwork	Willingness to plan, self-correct, and be mindful (e.g., metacognition)
Personal characteristics, attitudes	Personal qualities foundation (e.g., responsibility, honesty)	Standards of personal and business conduct (e.g., work with minimum supervision)	Disposition to think critically (e.g., willingness to engage in and persist at a complex task)
Other competencies or knowledge	Resources, information, systems and technology	Professional skills, general academic subjects	Citizenship skills

TABLE 1.15
Workplace Readiness Framework

A. Personal Management

Develop and maintain personal characteristics and behaviors necessary for success in the workplace to:

1. act responsibly, dependably, and conscientiously
2. behave with integrity
3. refrain from substance abuse
4. work safely
5. demonstrate initiative, motivation, and perseverance
6. demonstrate promptness
7. adapt to change
8. manage personal resources
9. improve personal fitness/health
10. avoid absenteeism

B. Academic Foundations

Develop and improve applied academic skills necessary for the workplace in:

1. mathematics
2. communication skills
3. science and technology
4. social sciences
5. health and physical education
6. the arts

C. Career Development

Plan and prepare for current and future career options, based on personal qualities and interests:

1. evaluate own interests, strengths, and weaknesses
2. identify appropriate occupational choices
3. select personal career path(s)
4. take steps to achieve career goals
5. demonstrate self-motivated learning

D. Interpersonal

Develop and maintain effective and productive groups by demonstrating the ability to:

(Continued)

21

TABLE 1.15
(Continued)

1. provide leadership and followership as appropriate
2. build consensus
3. deal with conflict effectively
4. negotiate agreements
5. work with all members of the workforce
6. listen attentively
7. actively participate in work-related discussions
8. respect the dignity of others
9. understand differences of opinion
10. meet the needs of others, such as clients or customers
11. respect the dignity of work

E. Thinking/Problem Solving Skills

Demonstrate the ability to generate innovative and practical solutions to real-world problems:

1. define the problem
2. analyze the problem and/or situation
3. evaluate available information
4. develop and analyze potential solutions or options
5. incorporate creativity, intuition, hunches
6. allocate necessary resources
7. make defensible decisions
8. monitor progress toward goals
9. repeat steps 1 through 8 as necessary

F. Technology

Select, apply, and maintain tools and technologies

1. learn about current and emerging technologies
2. apply thinking/problem solving skills to technology situations
3. apply technology solutions to problem situations
4. evaluate and improve technologies

G. Communication

Receive, process, and convey information using a variety of sources (such as written, verbal, nonverbal, and symbolic; technological, multimedia; abstract as well as concrete) to:

1. gather information efficiently
2. organize and maintain information
3. interpret information
4. share information
5. receive and use both positive and negative feedback

(Continued)

TABLE 1.15
(Continued)

H. Workplace Systems

Determine how an individual job fits into the overall organization, how the organization fits into the industry, and how the industry fits into the overall economy, in order to:

1. identify the subparts of the system
2. know how the parts fit together
3. understanding how the work flows through the system

I. Participate in the Work Organization

Contribute to the accomplishment of the organization's purpose by working to:

1. assist the organization to set goals as well as the procedures to implement goals
2. work to help achieve organizational goals
3. assist in continuous improvement
4. initiate suggestions for improving the organization
5. demonstrate loyalty to the organization and its goals
6. communicate responsibly with coworkers
7. teach and learn from others on the job
8. carry out assigned duties

Note. From Consensus Framework for Workplace Readiness. 1995 Revision (pp. 8-10). CCSSO Workplace Readiness Assessment Consortium, July 1995. Council of Chief State School Officers, One Massachusetts Avenue, NW, Suite 700, Washington, DC 20001.

decision making, communications skills, and skills in cooperation and teamwork are rated the highest (Bikson & Law, 1994; Moore & Shaffer, 1985; National Board of Employment, Education and Training, 1992; Sinclair, chapter 5, this volume; Stasz, Ramsey, Eden, Melamid, & Kaganoff, 1995, 1996). Further, the competencies required of college graduates for high-performance/high-paying jobs also tend to be of the higher order thinking nature (e.g., Sinclair, chapter 5, this volume). Thus, we believe that students work-bound may need to learn the same competencies whether in high school or in college. What differs is their expected performance levels (see Table 1.14).

The need of American management for workers who have greater skills and who can take on greater responsibility has spawned many commissions, task forces and studies (see O'Neil et al., 1992, for a review of these commissions). All of them have contributed to the vast evidence documenting the need for a more highly skilled workforce. The most recent list of competencies can be found in the 1995 *Consensus Framework for Workplace Readiness* (Council of Chief State School Officers, 1995; Table 1.15). However, what remains largely undone is the development of methods to assess the necessary skills that have been identified and, further, the teaching of such skills, that is, their integration in some manner into the curriculum.

ACKNOWLEDGMENTS

The work reported herein was supported under the Educational Research and Development Center Program cooperative agreement R117G10027 and CFDA catalog number 84.117G as administered by the Office of Educational Research and Improvement, U.S. Department of Education. The findings and opinions expressed in this report do not reflect the position or policies of the Office of Educational Research and Improvement or the U.S. Department of Education.

REFERENCES

Anderson Committee. (no date). *Report to the Board of Regents on Career Preparation Validation Study.* Albany, NY: Author.

Bikson, T. K., & Law, S. A. (1994). *Global preparedness and human resources. College and corporate perspectives* (MR-326-CPC). Santa Monica, CA: RAND Corporation, Institute on Education and Training.

Carnevale, A. P., Gainer, L. J., & Meltzer, A. S. (1990). *Workplace basics.* San Francisco, CA: Jossey-Bass.

Council of Chief State School Officers. (1995, July). *Consensus framework for workplace readiness. 1995 revision.* Washington, DC: Author.

Employability Skills Task Force. (1988, April). *Jobs. A Michigan employability profile. Report to the Governor's Commission on jobs and economic development.* Lansing, MI: State of Michigan, Office of the Governor.

Employability Skills Task Force. (1989, October). *Employability Skills Task Force progress report to the Governor's Commission on Jobs and Economic Development and the Michigan State Board of Education.* Lansing, MI: State of Michigan, Office of the Governor.

Fisher, R., & Ury, W. (1981). *Getting to yes.* New York: Penguin.

Mehrens, W. (1989). *Michigan Employability Skills Employer Survey. Technical report.* East Lansing, MI: Michigan State University.

Moore, N. K., & Shaffer, M. T. (1985). Basic skill requirements for selected Army occupational training courses. *Contemporary Educational Psychology, 10,* 83–92.

National Academy of Sciences. (1984). *High schools and the changing workplace. The employers' view* (NTIS Report PB84-240191). Washington, DC: National Academy Press.

National Board of Employment, Education and Training. (1992, December). *Skills sought by employers of graduates* (Commissioned Report No. 20). Canberra, Australia: Australian Government Publishing Service.

New York State Education Department. (1990, July). *Basic and expanded basic skills. Scales for validation study.* Albany, NY: Author.

Office of Technology Assessment. (1990, September). *Worker training: Competing in the new international economy* (OTA-ITE-457). Washington, DC: U.S. Government Printing Office.

Office of Technology Assessment. (1995, September). *Learning to work: Making the transition from school to work* (OTA-EHR-637). Washington, DC: U.S. Government Printing Office.

O'Neil, H. F., Jr., Allred, K., & Baker, E. L. (1992). *Measurement of workforce readiness competencies: Design of prototype measures* (CSE Tech. Rep. No. 344). Los Angeles: University of California, Center for Research on Evaluation, Standards, and Student Testing (CRESST).

Packer, A. H., & Pines, M. W. (1996). *School-to-work.* Princeton, NJ: Eye on Education.

Stasz, C. (1995). *The economic imperative behind school reform: A review of the literature* (DRU-1064-NCRVE/UCB). Santa Monica, CA: RAND, Institute on Education and Training.

Stasz, C., Ramsey, K., Eden, R. A., Melamid, E., & Kaganoff, T. (1995, August). *Generic skills and attitudes in the workplace: Case studies of technical work* (DRU-1180-NCRVE/UCB). Santa Monica, CA: RAND, Institute on Education and Training.

Stasz, C., Ramsey, K., Eden, R. A., Melamid, E., & Kaganoff, T. (1996). *Workplace skills in practice* (MR-722-NCRVE/UCB). Santa Monica, CA: RAND, Institute on Education and Training.

Taylor, F. W. (1916). The principles of scientific management. *Bulletin of the Taylor Society,* December, 13–23.

U.S. Department of Labor. (1991, June). *What work requires of schools. A SCANS report for America 2000.* Washington, DC: The Secretary's Commission on Achieving Necessary Skills (SCANS), U.S. Department of Labor.

Chapter 2

Changes in the Nature of Work: Implications for Skills and Assessment

Thomas Bailey
Institute on Education and the Economy
Teachers College, Columbia University

Concerns about the condition of the economy and the skill requirements of the modern workplace have driven education reform for at least a decade and a half. As the U.S. economy lost its dominant economic position in the world, many began to blame the country's education system. The public perception grew that young Americans were not as well educated and prepared for work as their European and Asian peers and that those deficiencies accounted for growing trade deficits and a loss of jobs to foreign competitors (Commission on Skills in the American Workforce, 1990; U.S. Commission on Excellence in Education, 1983).

The role of education in determining the competitive economic position of the United States remains controversial, but international competition is not the only cause for concern. Real wages and the average standard of living has barely risen since the early 1970s and inequality has been growing in a variety of dimensions since the early 1980s. In 1972, the median earnings for 30-year-old male high school graduates (with no additional schooling) was just under $30,000 (in 1992 dollars). By 1982 it had risen to $30,000, but by 1992 it had fallen below its 1972 level. In contrast, the earnings of the median 30-year-old college graduate rose from about $36,000 (in 1992 dollars) in 1972 to almost $42,000 in 1982. But even the earnings of college graduates barely grew between 1982 and 1992 (Levy, 1995).

For the last decade there has been a growing volume of research seeking to explain these developments. During the 1980s many commentators argued that the problem was that the country's education system had deteriorated (U.S. Commission on Excellence in Education, 1993) and education reformers sought to recapture a bygone golden age (Bennet, 1984). By the early 1990s, the discussion had broadened considerably to include many factors in addition to educational problems, but to the extent that education continued to be seen as an issue, a common view was that the demands of the economy had outstripped the schools. Although there was little evidence that the schools had deteriorated, a stronger case could be made that a system that had been adequate during the two or three decades after World War II was no longer effective in preparing the workforce for the technologies and markets of the early 1990s. In order to return to an era of faster productivity growth and higher standards of living, the United States would have to combine education reform with significant additional investments in education or human capital.

As concern about the quality of U.S. education has grown, more attention has also been focussed on the assessment of skills needed in the workplace. If complex human capital has become a more important component of economic performance and individual well-being, then it becomes that much more important to develop approaches to measuring and assessing skills and capabilities. And the assessment movement has been further encouraged by a widespread intellectual shift that seeks to de-emphasize a process-oriented approach to the regulation and improvement of education and to focus instead on the nature of educational outcomes. If it is possible to measure desired outcomes, then schools can be held accountable for achieving those outcomes rather than for adhering to regulations about educational inputs. Thus both the changing nature of economic activity and changing views on how education should be regulated and improved have led policy makers to focus increasingly on assessment.

The purpose of this chapter is to examine the relationships between changes in the economy and the country's educational needs, and then to begin to develop the implications of those changes for assessment. Subsequent chapters go into this issue in great detail. I first discuss direct empirical evidence that the skill demands of the economy have risen. The chapter then turns to an analysis of the changing nature of work focussing on two broad models of work. These are referred to as the traditional and high performance models. I review information on the spread of the high performance model and the skill and educational implications of the model. Finally, I conclude with the broad implications of the analysis for education reform in general and for our understanding of competencies and assessment in specific. The changing nature of the economy has increased the importance of assessment, but ironically, it has made useful assessment more difficult.

EMPIRICAL EVIDENCE ON CHANGING SKILLS

Over the last several decades, the controversy about the changing skill requirements of the economy has ebbed and flowed. In the early 1970s, Daniel Bell (1973) argued that the *Coming of the Post Industrial Society* would end the need for lower skilled work as the production of knowledge would eclipse the production of goods. But the general notion that a more modern and technologically advanced economy needs higher skilled workers was challenged by those who argued that automation and even microelectronics had the potential to incorporate skills into the technology itself. According to this view, articulated by Braverman (1974), society would need a cadre of highly skilled engineers, planners, and managers, but the bulk of workers would simply be machine tenders and watchers. Some case study evidence seemed to provide support for this deskilling thesis. For example, numerical controlled machining equipment reduced skilled machinists to the status of machine loaders (Braverman, 1973). According to this view, to the extent that skills were still required for these processes, they were needed by a small group of programmers who were removed from the factory floor. But other examples suggested that the same technology was often used in different ways. While some employers used numerical controlled lathes to deskill the machining function, others taught their machinists how to program the new equipment (Bailey, 1989). Thus while they might need somewhat less manual skill, they needed different conceptual skills. The issue was not so much what technology was used but how it was used. For example, the introduction of personal computers could simplify the job of a typist, but it also gave the typist access to a variety of tools requiring a wide range of skills. Whether or not the individual typist needed those skills depended on how the employer chose to organize the work and the division of labor.

Braverman (1974) had argued that the logic of capitalism demanded a deskilling strategy, but analysts in the 1980s began to argue that emerging economic and competitive forces were having precisely the opposite effect. Deskilling required a well planned and orderly production process that was most effective in producing large quantities of standardized goods. The United States, with its huge domestic market, had excelled at mass production technologies designed precisely to produce such goods. But as the century came to an end, consumers became more interested in variety and ever-changing styles. Competitive advantages shifted from firms that could produce low cost commodity products to those that could produce a variety of ever-changing, more innovative goods and services (Piore & Sabel, 1984). This new environment, characterized by increased international competition, faster technological change, and a more varied and volatile product market, implied that decisions and planning had to be decentralized and carried out by so-called "front line" workers—those in manufacturing who were actually building the products and those in services who delivered the service. Organizations that relied on unskilled workers to carry out the orders of a small leadership group would be much too rigid and cumbersome to operate in fast changing contemporary markets. Flexibility and innovation were necessary to compete, and they required skilled workers at all levels.

Much of the evidence that related to this controversy was developed in case studies of particular workplaces, and these have yielded a rich lode of insights and examples. Unfortunately, proponents of both sides of the controversy can present case study evidence. In an extensive review published over 10 years ago of the case study evidence, Spenner (1985) concluded that there was not strong support for either position. Research over the subsequent decade has perhaps given some more support for the notion that skill requirements are rising, but problems of generalizability still thwart definitive conclusions, and in any case, many examples of low skilled jobs can still be found (Bailey, 1989).

Furthermore, direct measurement of skills has proved to be extremely difficult. Observation of work tasks is costly and may focus on the skills that are available rather than those that are optimal for the effective performance of the tasks. Moreover, there is no accepted conceptualization of skills or widely used methodology for measuring them.

An alternative approach has been to use broad-based data on characteristics that do not necessarily measure skills directly, but might reflect skill changes. Two approaches have been used. In one, analysts have measured the relative earnings of workers with various levels of education (Levy & Murnane, 1992). If workers with higher levels of education begin to earn more relative to those with less education, then this suggests that employers are increasingly looking for more educated workers and are therefore willing to pay more to get them. The second approach depends on forecasts of the growth of different occupations (Silvestri & Lukasiewicz, 1991). If those occupations that typically are filled with more highly educated workers grow faster than those filled by less educated workers, we can conclude that the economy requires more educated workers and, assuming that education reflects skills, more skilled workers. We now look at the evidence from wage and occupational trends.

Changing Relative Wages

Changes in wages earned by different groups of workers do suggest a steady increase in the demand for workers with higher levels of education. In 1979, college graduates earned 29 percent more than high school graduates with no post-secondary education. By 1993, that college wage premium had grown to 45 percent (Mishel & Bernstein, 1994, p. 137). This increasing educational wage gap suggests that the demand for more highly educated workers is rising relative to the demand for workers with less education. If the jobs that employers have to offer now "require" more skills, then employers would be willing to offer more money to attract adequately skilled workers.

Changing demand is not the only influence on wages. Changes in supply will also affect earnings levels. For example, in the 1980s there was a decline in the growth rate of college graduates. And a sharp drop in the supply of college graduates, by making graduates more difficult to find, could have driven up their wages even without any changes in demand. But this supply factor cannot explain the entire wage gap. This is especially true in manufacturing. If the rising wage gap had been caused primarily by a shrinking of the relative supply of college graduates, then manufacturing employers would have hired fewer college graduates and more high school graduates, but the data show a relative shift towards the higher-educated group. The employment of full-time 25- to 34-year-old college graduates rose by 10% between 1979 and 1987, while their earnings rose by one third. In contrast, the employment of high school graduates in the same age group in manufacturing rose by only 6%, while their earnings fell by 11% (Levy & Murnane, 1992, table 9). Thus in their review of the large and growing body of research on the shifting patterns of relative wages, Levy and Murnane conclude that during the 1980s there was a steady rise in the relative demand for more highly educated workers. And subsequent work by Murnane, Willett, and Levy (1995) have shown that cognitive skills are increasingly important determinants of earnings even after controlling for education levels and other personal characteristics, although it is important to note that the measure of cognitive skill that they used was test scores of math skills usually taught no later than eighth grade in the United States.

But there continues to be controversy about the causes of the growing wage gap. Several analysts argue that a variety of factors may be as important or even more important than a growing demand for higher skills. Borjas (1995) emphasizes the importance of growing international trade, arguing that trade is concentrated in industries that have many lower skilled jobs, although Krugman and Lawrence (1993) contend that while this argument may be plausible, empirical evidence suggests that trade has little to do with shifting wages. Immigration may be another factor that has lowered relative wages for unskilled workers (Borjas, 1995).

It is also important to emphasize that much of this gap has been caused by the extreme deterioration of lower skilled workers. While the earnings of college graduates have risen, that increase has not been large (Mishel & Texeira, 1991). Strong income gains have only been enjoyed by a small percentage of workers at the top of the distribution. Institutional factors such as the declining value of the minimum wage and the weakening of unions also appear to explain perhaps a quarter of the increase in the gap between the earnings of high school and college graduates (Blackburn, Bloom, & Freeman, 1990; DiNardo, Fortin, & Lemieux, 1995). Nevertheless, despite the variety of causes for the growing wage gap, economists continue to believe that a growing demand for skilled workers and an increasing importance of cognitive skills are central explanations of the trends in the distribution of earnings (see, e.g., Bound & Johnson, 1995; Juhn & Murphy, 1995).

Shifting Occupational Structure

Past and projected occupational changes have frequently been used in the controversy about changing skill requirements. Indeed, they have attracted a great deal of attention probably because, unlike wages or abstract measures of skills, different occupations are easily associated with different skill levels. A doctor or an engineer needs more skills than a fast-food worker. Long term occupation changes over the last century provide convincing evidence of a secular rise in skills. In 1900, farm workers, laborers, and operatives accounted for 63% of the labor force, but by the end of the 1980s, these groups accounted for only 19% of the labor force. Managerial, professional, and technical workers grew from 10% to 28% of the labor force (Barley, 1992).

Shorter term shifts are more ambiguous. Indeed, the discussion of occupational projections has often been misleading, and given the methodologies used, does not in the end shed much light on the nature of changes in required skills. (For a detailed discussion of the occupational analysis see Berryman & Bailey, 1992, chapter 2, and Bailey 1991).

One problem has been that the data seem to support both sides of the controversy. Those who argue that required skills are not changing point out that most of the ten occupations that are projected to add the most jobs to the economy by the year 2005 require few skills. With the exception of registered nurses and general managers, all generally require low skills. On the other hand, those who argue that skills are rising, point out that the ten fastest growing occupations are characterized by middle-level skills. Thus paralegals, system analysts and computer scientists, physical therapists, medical assistants, operations research analysts, human services workers, radiologic technologists, and medical secretaries are all among the top ten.

Both of these types of "top ten" analyses are misleading and unnecessary. Data are available to study the whole occupational distribution and this analysis suggests that overall, the new jobs that will be added over the next decade are disproportionately those that currently have more highly educated incumbents. For example, while occupations in which more than a majority of the incumbents in 1990 had some post-secondary education accounted for 38% of all employment, those occupations were projected to account for 55% of all jobs created by 2005.

On the other hand, using the current projection methodologies, if the overall distribution of occupations in the early 1990s is compared to the projected distribution of occupations in 2005, the changes in occupations of different educational levels are not dramatic. While 22% of employed workers in 1990 had four or more years of postsecondary education, the occupational projections suggested that this share would only rise to 24% by 2005. This can be easily understood by thinking about adding water to a bathtub full of water. Even if the added water is much hotter, unless the flow of new water is very large compared to the stock of existing water, it will take a long time to have a strong effect on the overall temperature of the water in the bathtub.

This illustrates the weaknesses of the methodology used to make these projections. The approach assumes that the skill and educational require-ments within occupations does not change. All change is driven by chang-ing proportions of occupations. Thus skill increases would occur because engineers replace operators or analysts replace secretaries. The method-ology assumes that the jobs of operators and secretaries have not changed. The change in the overall distribution is therefore sensitive to the projected growth rate of jobs—the volume of water entering the tub relative to the water already there. Since job growth over the next decade is projected to be only moderate, then, given this methodology, a forecast for a large change in the overall distribution would be highly unlikely.

What can be concluded from the wage and occupational data? At the least, they suggest a steady but moderate increase in the demand for more highly educated (and therefore presumably more skilled) workers. On the other hand, they are not strong support for the argument that there is a revolutionary or dramatic rise in skill requirements, although the meth-odology used for the occupational projections precludes such a conclusion. Moreover, these analyses provide meager guidance to educators because they reveal little about the types of skills increasingly needed. In order to gain a better grasp of those skills and to develop more concrete knowledge of future trends in skill requirements, we need a deeper understanding of change in the economy and the resulting shifts in the nature of work.

TWO MODELS OF THE CHANGING NATURE OF WORK

For much of the last half-century, schools evolved around an economy which provided large numbers of good jobs for semiskilled workers. Pro-duction was organized efficiently to produce large numbers of standardized goods and services. U.S. producers excelled at this mass production system, taking advantage of their access to huge and growing U.S. markets. But growing international competition, increasing demands by consumers for variety, quality, and constant style changes, and the acceleration of the pace of technological change undermined the stable markets for standardized products on which the mass production system was based.

These shifts began to create advantages for what has come to be called the high performance workplace. The shift from traditional to more innovative work organization is described in great detail by several authors (Appelbaum & Batt, 1994; Bailey, 1993; Kochan & Osterman, 1994). What is important for present purposes is that workers at all levels in the high performance workplace are expected to be actively and intellectually engaged in their work. In more traditional or mass production settings, the jobs of production or "front-line" workers are limited, well defined, and passive. Workers are expected to perform a set of tasks, and anything out of the ordinary is referred to managers or specialized support personnel. Little initiative is expected. In contrast, in high performance systems, workers are engaged in less well defined activities and are expected to be much more actively involved with their jobs, contributing their ideas and initiative to furthering the goals and objectives of their work group and organization. Rather than simply carrying out specific tasks and following well defined instructions, workers are expected to solve problems, seek ways to improve the methods that they use, and to engage actively with their coworkers. This is, therefore, much more than simply increasing the tasks that a worker can perform; rather it involves a new type of behavior and orientation towards the job. Indeed it sounds very much like the behavior that is already expected from professional and technical personnel.

HOW WIDESPREAD ARE HIGH PERFORMANCE WORKPLACES?

If innovative work organizations require new and different skills, the overall change in the demand for skills depends on the extent to which those innovations have been adopted. Although there has been a great deal of discussion of the spread of the high performance workplace, the extent to which a shift has actually occurred remains controversial. *America's Choice,* the influential report by the Commission on Skills in the American Workforce (1990), attracted a great deal of attention in 1990 when it claimed that only about 5 % of American firms had made any significant transformation.

Although the 5% figure is often cited, the methodology used by the Commission to arrive at this conclusion has never been made clear. But over the last 15 years, there have been a handful of more or less broadly based surveys designed to identify the diffusion of various work reform and employee involvement practices. The most widely cited include a 1982 survey of members of the New York Stock Exchange (Freund & Epstein, 1984), a 1987 survey by the General Accounting Office (Eaton & Voos, 1992; Lawler, Ledford, & Mohrman, 1989), and a 1990 followup to that survey (Lawler, Mohrman, & Ledford, 1992). In addition there have been a variety of more focussed surveys of particular industries or regions (or states).

In general these surveys find that often a majority of respondents (82% of the respondents to the GAO survey of the 1986 Fortune "1000")[1] have reported adopting some work reform or employee involvement technique. The extent of adoption has also grown since the 1970s. Nevertheless, at least by the mid-1980s relatively few organizations could be said to have made a significant organizational transformation. Results from the GAO survey (Lawler, Ledford, & Mohrman, 1989) suggest that "between 20% and 30% [of the respondents] have a substantial effort in place that uses a number of different kinds of approaches with a large percentage of employees" (p. 59). Manufacturing industries appear to rely more on employee involvement.

A 1992 survey by Osterman (1994) is also more or less consistent with these general findings. In a national survey of firms with 50 or more employees, Osterman also found that a quarter or a third of the firms had made significant changes. Perhaps the most comprehensive survey of work practices so far was carried out by the Bureau of Labor Statistics (BLS) in 1993 as part of their Survey of Employer Provided Training. This survey was mailed to 12,000 establishments. Almost 8,000 usable surveys were returned, and about 6,000 contained information about the organization of work. This survey identifies six employment practices generally associated with innovative work organization such as the use of teams, job rotation, quality circles, and total quality management techniques. Less than 2% of the establishments with 50 or more employees used all six practices, while 37% of such establishments used none of them. About one quarter used three out of six of the practices for at least some of their workers. While these results indicate an incidence of innovation somewhat lower than Osterman's analysis, the results are within the one quarter to one third range that has emerged as a consensus estimate (Gittelman, Horrigan, & Joyce, 1995).

But the use of innovative strategies at a given time provides only an incomplete picture of the prospects for eventual spread of those strategies. Many programs that get started often do not last long. In an aptly named article, "Quality Circles after the Fad," Lawler and Mohrman (1986) state that "few QC programs turn into other kinds of programs, more commonly, decline sets in" (p. 69). Few programs last more than 4 years (Griffin, 1988; Bailey, 1993; Schuster, 1985). Often after initial success, interest

---

[1]The survey had a 51% response rate.

begins to wane and as the agenda of easily solved problems grows smaller, meetings become less frequent (Walton, 1985; Griffin, 1988).

Furthermore, while many firms are experimenting with new forms of human resource management, others are pursuing a more ominous strategy. In many cases, employers appear to be searching for greater efficiency by cutting staff, relying where possible on temporary employees, and in other ways reducing their commitment to their labor force. The two trends are not completely inconsistent. Firms may make a commitment to a small "core" of workers while they rely on a large group of "peripheral" workers to absorb fluctuations. But it is still hard to imagine a high performance workplace in which firms have no commitment to a significant share of their workers.

Thus the precise nature and extent of trends in work organization remain obscure. Discussions in the business press and case study evidence do suggest a trend towards innovative work organization, and a substantial number of firms, although still a minority, have already made significant changes. Furthermore, economic forces are providing increased incentives for firms to adopt many of the practices associated with high performance work organization. But this change is difficult and there are counter forces. In a time of greater uncertainty, the incentive to shed workers and reduce commitments is also strong. On balance though, the most likely trend is one in which there will be a steady increase in the adoption and use of innovative human resource techniques. Furthermore, an increased supply of adequately skilled workers will itself facilitate the introduction of innovations by reducing the training and education that innovative firms often need to provide for their incumbent workers as well as their recent hires.

EDUCATIONAL IMPLICATIONS

What are the educational implications of the spread of innovative human resource practices? I shall argue that the case studies in particular suggest that such practices increasingly need what have been referred to as "advanced generic skills" (Stasz, Ramsey, and Eden, 1995) or skills outlined by the Secretary's Commission on Achieving Necessary Skills (SCANS, 1991). These are referred to as SCANS competencies and include such things as problem solving and working cooperatively with others. According to this perspective, schools need to make sure that their students emerge with those skills. But the educational lessons from the changing workplace go beyond the content of education. Innovative work systems suggest a very different educational process, one in which the integration of academic and vocational studies and of theoretical and practical learning must play a central role.

Skills

What skills are necessary for workers in high performance workplaces? First, case studies of innovative workplaces suggest that many production workers have an increasing need for the types of skills traditionally learned in school—literacy, arithmetic, and at a higher level, specific technical knowledge. But these studies also suggest that current changes in work call for more than simply an increase in traditional education. On the job, diverse tasks have been combined in new ways and even low-level workers have been given new responsibilities. Thus educational reform should look beyond the quantity of education. The content of education needs to be brought more in line with the types of activities students will be engaged in after they leave school.

Workers increasingly need to be able to operate more independently of their supervision and to work in a less well-defined environment. This requires a greater facility for creative thinking, decision making, reasoning, and problem solving. Workers need to have a broader understanding of the systems in which they operate (SCANS, 1991). Without this, they are much less able to make decisions about their own activities. Nor will they be able to monitor and correct the performance of those systems or to participate in the improvement of their design. This was not an issue when they were simply expected to follow instructions; their role within the broader operations of their organization was the concern of their supervisors and managers.

But even more than a broader knowledge of their context, they need a more abstract or conceptual understanding of what they are doing. This is what allows them to carry out tasks or solve problems that they have not encountered before or that they have not been shown specifically how to carry out or solve. Thus, more than in the past, individuals will need to be able to acquire, organize, and interpret information. Workers will also have more direct interaction with their coworkers, and therefore will need more experience in general social skills such as group problem solving and negotiation. These changes clearly involve more than an accumulation of the type of knowledge traditionally learned in schools. These skills increasingly required by the high performance workplace have been referred to as advanced generic skills (Stasz, Ramsey, & Eden, 1995).

Education Reform

But how should these skills be taught? With respect to this question, the innovative workplace has further lessons for educational reformers. There are strong parallels between traditional approaches to learning and teaching that have dominated education for decades, which have increasingly been questioned, and the characteristics of the mass production system. Similarly, conceptions of the pedagogic systems within which students learn most effectively have important parallels to innovative production systems (Berryman & Bailey, 1992).

Teaching in typical American schools and classrooms have the following five characteristics, all of which have been challenged and questioned in the last decade.

1. Knowledge is fragmented. Material is broken up into small segments that can presumably be absorbed one at a time. Thus material is broken up into the disciplines and then further fragmented and the school day is divided into many short periods.

2. Learners are passive. Teaching is seen as transmission of knowledge from the teacher to the student. This reduces exploration, initiative, and learning from mistakes; it creates dependence on the teacher which undermines the development of cognitive skills which are key to knowing how to learn; and it creates "crowd control" problems as bored and unengaged students look for something else to do.

3. Learning is based on teaching students the correct responses to given stimuli. This creates problems when students or adults encounter "stimuli" for which they have not learned the correct response. Thus this does not encourage imagination and creativity.

4. There is an emphasis on getting the right answer. Thus there is little attempt to get behind the answer or to learn from mistakes. And there is little emphasis on the process of solving problems.

5. Learning is decontextualized. In the past, many educators believed students would not be able to generalize material taught in a specific context—generalization required decontextualization. More recently researchers have argued that learning in context gives greater meaning to the material and is therefore more effective.

These general principles did not simply appear in the abstract, but reflect deeper elements of the way that society has been, and in many ways still is, organized. For example, there are close analogies to each of these five principles in the way that work is traditionally organized.[2]

1. Jobs and tasks are narrowly defined.
2. Workers are passive order takers in an hierarchical work organization.
3. Work is designed to be a set of routinized responses to limited and pre-specified problems.

---

[2]This list has been adapted from descriptions of "traditional" work organization. See, for example, Dertouzos, Lester, and Solow (1989).

4. The focus of work is on getting the task done rather than on systematically improving the process through which the tasks are carried out.

5. Most workers do their tasks without knowledge of or orientation towards the functions and purposes of the organizations in which they worked (their work was decontextualized).

Over the last decade, traditional pedagogic strategies based on those five principles have been increasingly challenged by a different approach based on five contrasting principles (Berryman & Bailey, 1992):

1. Knowledge and curriculum are integrated. This approach challenges a series of traditional and strongly held dualities: head and hand, knowing and doing, academic and vocational, and conception and execution. Traditional disciplinary boundaries and the typical organization of the school day have all been questioned.

2. Active learning (rather than passive), in which students are engaged in a process of discovery rather than being given information.

3. Encouragement of a deeper understanding that allows multiple responses to stimuli that the learner has not already encountered. This approach tries to connect with the natural tendency of students to want to learn and make sense of their surroundings.

4. New approaches focus on the thought processes that generate learning rather than on the "right answer." Indeed the possibilities of multiple responses to realistic and therefore often ambiguous situations are emphasized.

5. New strategies call for learning in context.

Fundamentally, this approach is based on the type of experience that learners have, as much as or more than the material or content to which they are exposed.

This approach to teaching and learning is a direct response to the perceived ineffectiveness of traditional pedagogy. But it is also consistent with the emerging needs of the economy. Indeed, the high performance work organization can also be characterized by five principles that correspond to the five principles of effective learning:[3]

1. Tasks and jobs (such as design, planning, production, repair, and maintenance) are integrated through either broader job definitions, or cross-functional teams.

---

[3]This list is adapted from descriptions of innovative work organization. (See, e.g., Kochan & Osterman, 1994.)

2. Workers are given more initiative and take more responsibility in a flatter organizational structure based more on trust and teamwork.

3. Employees are expected to be able to solve problems in nonroutine situations.

4. There is much greater emphasis on continuous improvement of the process of producing goods or services rather than simply on getting something produced.

5. Production workers as well as managers are expected to be able to understand their functions within the context of the broader functions and purposes of their organizations.

Given the similarities between the characteristics of high performance workplaces and innovative schools, students in these schools will experience an environment much like the environment that they would find in an innovative workplace.

The high performance workplace is based on a rejection of many traditional distinctions such as conception and execution, theory and practice, and supervision and subordination. New thinking about teaching and learning challenges the same distinctions, although somewhat different terms are used such as head and hand, academic and vocational, or learning and work. The new integrated workplace therefore calls for an integrated education that brings together academic and vocational studies and school and work.

IMPLICATIONS FOR ASSESSMENT

The changing role of the worker, especially in a high performance work organization, has profound implications for assessment. A shift from a traditional to a high performance work organization involves much more than a simple increase in the number of required skills. It involves a new type of behavior and orientation toward the job. As I have pointed out, this has many similarities to the type of behavior that is already expected of professional workers. Given that, there are important insights that can be gained about assessing "workplace competencies" from assessments of professional and managerial work.

There is a fundamental distinction between the traditional conceptualizations of professional and nonprofessional skills (Bailey & Merritt, 1995). The performance and responsibilities of nonprofessional workers can be characterized by dividing their jobs into a list of discrete tasks or skills and then adding up the mastered tasks. Since the focus is on the separate tasks that must be performed by the worker, this perspective can be referred to as a *skill components* model. In contrast, the nuances of the roles and responsibilities of professionals make a narrowly defined listing of their skills difficult to produce. To make sense of professional work, it is necessary to examine their performance as a whole, to study how they *combine* the many components of their skill and behavior. But the nonprofessional worker in traditional organizations works under the direction of supervisors and managers, and it is their responsibility, not the workers', to determine how to combine the tasks of subordinates into a coherent whole. Thus for the nonprofessional worker, the focus is on the separate tasks that make up the whole. In such circumstances, a worker can be thought of as a tool box—a collection of tasks that can be competently carried out. It is the responsibility of the manger to decide which tool should be used and in what combination. Assessments should provide information on the effectiveness of each separate tool. Certainly the professional worker must be able to carry out specific tasks, they must have the "tools," but competence in specific and separate tasks is not adequate. That worker must be able to combine those tasks to carry out the broad functions of the profession. The focus is on the whole, not its component parts.

Thus assessment of professional work is much more difficult than assessment of work based on a skill components model. For example, the skills of a pilot can be thought of as a list of competencies—taking off, climbing, turning, descending, and landing. It is easy to devise a test to determine whether a student has mastered these tasks. But the pilot must also be able to decide when to land or when it might make more sense not to take off. These matters of judgment are more difficult and costly to assess. Similarly, many students who do well in statistics courses find it much more difficult to come up with a question that can be appropriately analyzed using statistical techniques. Assessing a graduate student's ability to conceive of, design, and carry out a major research project is best done by observing the student's experience in their dissertation work. But in this case, the assessment must wait until the project has been carried out. That is a process that takes many months.

Most professional occupations do have tests and assessments for purposes of certification, but these tests usually include essay questions or exercises intended to simulate complex problems encountered in the performance of the occupation (Bailey & Merritt, 1995). In most cases, the certification cannot be separated form the nature of the training itself. Many professions require extensive guided practical experience to achieve various levels of certification. Students in professional training programs (such as law and business) know that they can increase their labor market prospects by finding summer internships and post-training experience that give some indication of how they will perform in realistic situations. And many graduates students find that impressive scores on comprehensive exams are soon forgotten if they cannot produce a good dissertation—that is, if they cannot perform the core research activities of their profession. And mid-career professionals, when they market themselves, emphasize their concrete accomplishments and when individual performance is difficult to measure, as it is in many professional situations, they emphasize the nature of their experience.

In the previous section I argued that there is a strong parallel between the nature of work in innovative work organizations and new thinking about effective pedagogy. The movement towards active-learning pedagogy has been accompanied by an increase in efforts to develop "authentic assessments" or assessments based on measuring effectiveness in the performance of complex functions. The changing nature of work for many lower- and middle-level positions also implies a shift away from assessments that focus on competence in performing discrete tasks and towards assessments of performance in realistic complex environments.

CONCLUSION

How have the skill requirements of the workplace changed and what are the implications of those changes for education and for assessment of workplace competencies? Wage and occupational data and the case study information all suggest that there is at least a steady rise in the demand for more skilled workers. Recent research on the nature of work in high performance work organizations and on teaching and learning also suggest that the traditional pedagogy that has dominated American education may not be effective in preparing workers for many human resource practices that appear to be successful in supplying today's markets using modern technology.

Critics argue, though, that the evidence does not suggest that the country is experiencing extremely rapid or revolutionary changes. Common business practices, labor market institutions, the structure of financial markets, the legal framework within which firms operate, and other factors all create barriers to change and innovation, and many jobs remain that require only minimal skills. Nevertheless, empirical evidence on the spread of the type of innovative work reorganization that would create a demand for new and different types of skills throughout the employment hierarchy suggests that between one quarter and one third of all firms have made some significant progress. Since these tend to be larger firms, a somewhat higher share of the workforce is employed in such firms.

Although clearly education is not the only barrier to more innovative approaches to production, a larger supply of workers with more complex skills will have its own positive impact on work organization. Education reform is not a sufficient condition for faster economic change, but it is a necessary condition.

The emergence of new forms of work and new demands on workers are also having important effects on the assessment of performance or potential performance on the job. Although this represents a significant change for many workers, the accumulated experience with the assessment of professional skills should already provide the basis for the design of new assessments for skills, for occupations, and for jobs that in the past have been well characterized by the skill components perspective. Thus new assessments will almost certainly involve a combination of complex written material, simulations, monitored performance, and evaluations of past work experience. As is the case with the assessment of professional occupations today, it seems unlikely that one assessment tool will be able to give an adequate picture of the skills and competencies of tomorrow's production and service jobs.

ACKNOWLEDGMENT

Much of the research reported in this chapter was funded by the U.S. Department of Education through the National Center for Research in Vocational Education at the University of California at Berkeley.

REFERENCES

Appelbaum, E., & Batt, R. (1994). *The new American workplace: Transforming work: Systems in the United States.* Ithaca, NY: ILR Press.

Bailey, T. (1989). *Changes in the nature and structure of work: Implications for skill requirements and skill formation.* Berkeley, CA: National Center for Research in Vocational Education, University of California at Berkeley.

Bailey, T. (1991). Jobs of the future and the education they will require: Evidence from occupational forecasts. *Educational Researcher, 20*(2), 11–20.

Bailey, T. (1993). *Discretionary effort and the organization of work: Employee involvement and work reform since Hawthorne* (report prepared for the Sloan Foundation). New York: Institute on Education and the Economy, Teachers College, Columbia University.

Bailey, T., & Merritt, D. (1995). *Making sense of industry-based skill standards.* Berkeley, CA: National Center for Research in Vocational Education, University of California at Berkeley.

Barley, S. (1992). *The new crafts: The rise of the technical labor force and its implication for the organization of work* (Working Paper WP05). Philadelphia: National Center for the Educational Quality of the Workforce, University of Pennsylvania.

Bell, D. (1973). *The coming of the post-industrial society.* New York: Basic Books.

Bennet, W. (1984). *To reclaim a legacy: A report on the humanities in higher education.* Washington, DC: National Endowment for the Humanities.

Berryman, S., & Bailey, T. (1992). *The double helix of education and the economy.* New York: Institute on Education and the Economy, Teachers College, Columbia University.

Blackburn, M., Bloom, D., & Freeman, R. (1990). The declining position of less skilled American men. In Gary Burtless (Ed.), *A future of lousy jobs? The changing structure of US wages*. Washington, DC: The Brookings Institution.

Borjas, G. (1995, January). The internationalization of the US labor market and wage structure. *Economic Policy Review 1*, 3–8.

Bound, J., & Johnson, G. (1995, January) What are the causes of rising wage inequality in the United States. *Economic Policy Review 1*, 9–17.

Braverman, H. (1974). *Labor and monopoly capital: The degradation of work in the twentieth century*. New York: Monthly Review Press.

Commission on the Skills of the American Workforce (CSAW). (1990). *America's choice: High skills or low wages!* Rochester, NY: National Center on Education and the Economy.

Dertouzos, M., Lester, L., & Solow, R. (1989). *Made in America: Regaining the Competitive Edge*. Cambridge, MA: MIT Press.

DiNardo, J., Fortin, N., & Lemieux, T. (1995, April). *Labor market institutions and the distribution of wages, 1973–1992: A semiparametric approach* (Working Paper No. 5093). Cambridge, MA: National Bureau of Economic Research.

Eaton, A. E., & Voos, P. B. (1992). Union and contemporary innovations in work organization, compensation, and employee participation. In L. Mishel & P. Voos. (Eds.), *Unions and economic competitiveness* (pp. 173–211). Armonk, NY: M. E. Sharpe.

Freund, W. C., & Epstein, E. (1984). *People and productivity: The New York Stock Exchange guide to financial incentives and the quality of work life*. Homewood, IL: Dow-Jones-Irwin.

Gittelman, M., Horrigan, M., & Joyce, M. (1995). *"Flexible" work organization: Evidence from the survey of employer provided training*. Unpublished. Washington, DC: Bureau of Labor Statistics.

Griffin, R. (1988). The consequences of quality circles in an industrial setting: A longitudinal assessment. *American Management Journal, 31*, 338–358.

Juhn, C., & Murphy, K. (1995, January). Inequality in labor market outcomes. *Economic Policy Review, 1*, 26–34.

Kochan, T. A., & Osterman, P. (1994). *The mutual gains enterprise*. Boston: Harvard Business School Press.

Krugman, P., & Lawrence, R. (1993, September). *Trade, jobs, and wages* (Working Paper No. 4478). Cambridge, MA: National Bureau of Economic Research.

Lawler, E., Ledford, G., Jr., & Mohrman, S. (1989). *Employee involvement in America: A study of contemporary practice*. Houston, TX: American Productivity and Quality Center.

Lawler, E. E., & Mohrman, S. (1985). Quality circles after the fad. *Harvard Business Review, 63*, 65–71.

Lawler, E., Mohrman, S., & Ledford, G., Jr. (1992). *Employee involvement and TQM: Practice and results in Fortune 500 companies*. San Francisco, CA: Jossey-Bass.

Levy, F. (1995, January). The future path and consequences of the U.S. earnings/education gap. *Economic Policy Review, 1*, 35–41.

Levy, F., & Murnane, R. (1992). U.S. earnings levels and earnings inequality: A review of recent trends and proposed explanations. *Journal of Economic Literature, 30*(3), 1333–1382.

Mishel, L., & Bernstein, J. (1994). *The state of working America: 1994–1995*. Armonk, NY: M. E. Sharpe.

Mishel, L., & Texeira, R. (1991). *The myth of the coming labor shortage: Jobs, skills, and incomes of America's Workforce 2000*. Washington, DC: Economic Policy Institute.

Murnane, R., Willett, J., & Levy, F. (1995, March). *The growing importance of cognitive skills in wage determination* (Working Paper No. 50-76). Cambridge, MA: National Bureau of Economic Research.

Osterman, P. (1994, January). How common is workplace transformation and who adopts it? *Industrial and Labor Relations Review 47*(2), 173–188.

Piore, M., & Sabel, C. (1984). *The second industrial divide*. New York: Basic Books.

Schuster, F. E. (1985). *Human resource management: Concepts, cases, and readings.* Reston, VA: Reston Publishing Company, Inc.

Secretary's Commission on Achieving Necessary Skills (1991). *What work requires of schools.* Washington: U.S. Department of Labor.

Silvestri, G. T., & Lukasiewicz, J. M. (1991, November). Occupational employment projections. *Monthly Labor Review, 114*(1), 64–94.

Spenner, K. (1985). Upgrading and downgrading of occupations. *Review of Educational Research 55* (Summer), 125–154.

Stasz, C., Ramsey, K., & Eden, R. (1995). Teaching generic skills. In N. Grubb (Ed.), *Education through occupations in American high schools* (Vol. 1, pp. 169–191). New York: Teachers College Press.

U.S. Commission on Excellence in Education (1983). *A nation at risk.* Washington, DC: U.S. Government Printing Office.

Walton, R. (1985). From control to commitment in the workplace. *Harvard Business Review.* March–April: 77–84.

Chapter 3

Skills for Success in Maryland: Beyond Workplace Readiness

Katharine M. Oliver
Maryland State Department of Education

Christine Russell
Berkshire Consulting Group

Lynne M. Gilli
Rita A. Hughes
Ted Schuder
Maryland State Department of Education

John L. Brown
Prince George's County Public Schools

Wayne Towers
Xerox Corporation

In an environment characterized by an intensely competitive, dynamic economy and tight public resources, demands for quality and accountability from the public schools grow more insistent every year. Legislators and taxpayers want proof that public dollars are spent judiciously; employers want workers who demonstrate initiative, work well with others, solve problems routinely, and use technology effectively; postsecondary schools want high school graduates who are ready for college rather than remediation; parents want to know that their children are being prepared for successful lives; and communities want responsible, productive citizens. These are enormous demands on public education systems—demands that reflect dramatic changes in economic and social conditions at the end of the 20th century.

The State of Maryland has faced the rising expectations for educational results head-on by developing a school reform program centered on high standards and accountability for student and school performance. The most recent component of that reform is a new High School Assessment Program scheduled for implementation with the graduating class of 2004.

47

A defining feature of the assessment program is its dual focus on essential academic knowledge and general, cross-disciplinary skills. "Skills for Success" (SFS) is the title Maryland has given to the latter.

There are five categories of SFS: Learning, Thinking, Communication, Technology, and Interpersonal Skills. They set high standards and bridge the gaps between school, workplace, postsecondary education, and social environments. The SFS are generic; they take on concrete representation in the specific domains and situations that lend them substance. Given the rapidly evolving social and economic environment at the end of the 20th century, the SFS are intended to help prepare students for success in learning, living, and work situations in the 21st century.

This chapter explores some of the unique features of the SFS, including the vision that drove them; the collaborative development process they required; the coherence, scope, and utility of the skill set; and their integration with the State's overall educational reform movement. Similarities and differences with nationally recognized workplace readiness skills are reviewed, as are the challenges that Maryland faces as it attempts to design a meaningful and enduring role for the SFS in the instruction and assessment of academic and other subjects in high school.

EDUCATIONAL REFORM IN MARYLAND

The Vision

It is difficult to understand the SFS in isolation from the context in which they were developed—an on-going educational reform movement in the State of Maryland. Maryland's current reform began with the recommendations of the Governor's Commission on School Performance (State of Maryland, 1989). In its report, the Commission challenged the educational community with three credos to guide all reform activities:

1. All children can learn.
2. All children have the right to attend schools in which they can progress and learn.
3. All children shall have a real opportunity to learn equally rigorous content.

Given these principles, the State established the vision for student success in the 21st century (State of Maryland, 1992). Students who graduate from high school in Maryland should be able to (a) succeed in postsecondary education and lifelong learning opportunities, (b) participate in a world economy and job market that is more competitive than ever before, (c) function as responsible citizens in a democratic society, and (d) achieve a productive life.

Maryland School Performance Program

The driving mechanism of Maryland's reform effort is the Maryland School Performance Program. It holds each school accountable for providing an educational program that results in success for every student by assessing school performance, measuring school progress toward State standards, reporting results publicly, requiring documented school improvement plans throughout the State, and if necessary, reconstituting schools.

Prior to 1989, Maryland assessed student attainment of basic skills in elementary and middle school through the administration of norm-referenced achievement tests in Grades 3, 5, and 8. At the high school level, students were required to pass the Maryland Functional Tests, designed to ensure minimal competence in reading, writing, mathematics, and citizenship. The Governor's Commission (State of Maryland, 1989) criticized these assessments for not establishing sufficiently high expectations and for the lack of school accountability for student performance. To address this criticism, the State developed a new elementary and middle school testing program to assess performance on learning outcomes for Grades 3, 5, and 8 and called it the Maryland School Performance Assessment Program (Maryland State Department of Education, 1995). This program uses performance assessment tasks that require students to write extensively, solve complex problems cooperatively and individually, apply what they have learned to real world problems, and relate and use knowledge from different subject areas. 1996 marks the sixth census administration across the State.

To extend its high expectations and accountability programs into high school, the Maryland State Board of Education created a series of task forces: The Maryland Task Force on Outcome-Based Graduation Requirements (1992), the Performance-Based High School Graduation Requirements Task Force (1993), and the High School Assessment Task Force (1994). The High School Assessment Task Force submitted core learning goals in English, mathematics, science, social studies, and SFS to the State Board in June of 1995. They also recommended end-of-course tests for the academic subjects and that the SFS be assessed in those tests.

Table 3.1 places these proposed high school assessments in the context of Maryland's on-going accountability program. The essential point is that SFS never were intended to be limited to basic skills, minimal competence, or workplace readiness skills. Maryland's accountability system already includes norm-referenced tests of basic skills and statewide tests of minimal competence. SFS subsume workplace readiness but serve a much broader

TABLE 3.1
Educational Accountability Instruments in Maryland

Performance Level	Domain	Grade	Instrument	Measurement Focus	Status
Basic skills	Reading, writing, mathematics	3, 5, 8 (2, 4, 6)[a]	Norm-referenced achievement test battery	Statewide achievement (sample of students)	Administered since 1970s
Minimal competence	Reading, writing, mathematics, citizenship	9 (8)[b]	Maryland Functional Tests	Individual students	Phased in during 1980s
High expectations	Reading, writing, language arts, mathematics, science, social studies	3, 5, 8	Maryland School Performance Assessment Program	School programs (census administration)	Administered since 1990
High expectations[c]	English, mathematics, science, social studies, Skills for Success	9-11	High School Assessments	Individual students	Under development: anticipated implementation with graduating class of 2004

[a]Adminstration moves to Grades 2, 4, and 6 in 1997.
[b]Change to administration in Grade 8 or earlier in 1996.
[c]Competence accumulates as performance level rises. For example, high expectations require both basic skills and minimal competence, but go beyond those performance levels.

vision than that—preparation for postsecondary education and responsible citizenship, too.

At the time of this writing, Maryland is continuing to seek widespread public and expert comment on the proposed core learning goals for the SFS and the four academic subject areas, and is beginning the design of the assessment system. Once learning goals and a testing framework are accepted by the State Board, components of the assessment system will be pilot tested, with the full system expected to be operational as a graduation requirement for students graduating in the year 2004. With this background in mind, the remainder of the chapter describes the collaborative development of the SFS, an analysis of their content, and the implications for instruction and assessment.

CRAFTING THE SKILLS FOR SUCCESS

Diverse Participants

To build SFS that would serve students and multiple constituencies, the State Board of Education turned to a partnership with the Maryland Business Roundtable for Education, a coalition of 64 companies that have made a 10-year commitment to support educational reform and improve student achievement in Maryland. The Roundtable provided a cochair (Christine Russell, one author of this chapter) and nominated representatives from business (large and small, national and local), industry, labor, and government to help define the desired performance attributes of the high school graduate of the 21st century. These representatives constituted a majority of the 41-member SFS Team.

The educational cochair of the SFS Team was the Assistant State Superintendent for Career Technology and Adult Learning with the Maryland State Department of Education (Katharine M. Oliver, first author of this chapter). Other team members included representatives of parents, teachers, schools, central offices, and the State education agency, as well as community colleges, universities, and professional organizations. Members included a local Superintendent of one school system and the Chair of the High School Assessment Task Force. The quality and diversity of the membership was critical to the credibility of the SFS in educational, employment, and political arenas.

Collaborative Development

The Team met regularly, once or twice a month, over a 9-month period. Meetings typically lasted between 3 and 6 hours and were generally held at facilities provided by the Maryland Business Roundtable for Education. The development process went through several stages: (a) define the context and the charge, (b) review and evaluate similar work locally and nationally, (c) draft Maryland's SFS, (d) obtain constituent and expert feedback, (e) revise and fine-tune as necessary, and (f) present the draft to the High School Assessment Task Force. The process was recursive rather than linear.

The project began with an overview of the purpose and context of the work provided by the Director of the High School Assessment Program. Given the group's diverse composition, it was important that everyone understand Maryland's reform program and the role the High School Assessment would play in it.

The SFS Team reviewed and evaluated similar work done at national, state, and local levels. A resource notebook provided each member with relevant national reports and summaries, including *America's Choice: High Skills or Low Wages* (National Center on Education and the Economy, 1990), *What Work Requires of Schools—A Secretary's Commission on Achieving Necessary Skills (SCANS) Report for America 2000* (U.S. Department of Labor, 1991), and *Workplace Basics: The Essential Skills Employers Want* (Carnevale, Gainer, & Meltzer, 1990). A wide range of state and local reports were also included. A concise review of five major studies in *Measurement of Workforce Readiness: Review of Theoretical Frameworks* (O'Neil, Allred, & Baker, 1992) provided a useful perspective. (See chapter by O'Neil, Allred, & Baker, chapter 1, this volume, for an update of this review.) The Team worked in groups and as a whole to ferret out skills, purposes, and perspectives from existing reports that were applicable to the SFS charge, and to identify areas where gaps existed.

Invited presenters provided a broad, critical perspective. For example, a local assistant superintendent of instruction presented a model of one school system's efforts to establish workplace/life skill competencies as overarching exit outcomes for high school graduates. These skills were used to shape curriculum, assessment, and instruction from kindergarten through graduation. During another meeting, an expert review panel was invited to comment on the draft SFS. The panel included a Director of Admissions at a local college; the Director of the Center for Learning and Competitiveness at the University of Maryland College Park, who provided an international perspective; and a Senior Fellow at the Institute for Policy Studies at the Johns Hopkins University in Baltimore, who had also been the Executive Director of the SCANS Commission. These discussions helped broaden the Team's understanding of related national and international efforts, as well as the linkages to workplace and postsecondary arenas.

The perspectives of local, state, and national leaders helped to shape the final draft. However, finding no existing skill set adequate to Maryland's purpose, the SFS Team elected to design a Maryland version that encompassed but expanded, simplified, and synthesized existing skill sets. The

Team often divided into small workgroups to develop, refine, and revise the skills. This cycle was repeated several times, assuring the engagement of every member of the Team at each meeting. Debate was open and extensive. Using a consensus-building process, the workgroups honed the SFS into an acceptable draft for review by external stakeholder groups.

External and Statewide Review

Expanding the scope of its review, the SFS Team sought input on the draft from numerous experts and stakeholders at national, state, and local levels. The State Board of Education and the four High School Assessment academic content teams (e.g., mathematics) were also consulted regularly. Surveys were administered to obtain comments on the SFS draft. The Team identified five criteria for external evaluation of their work:

1. Are the SFS significant? (Are they essential skills for students to have as they leave high school?)
2. Are the SFS comprehensive? (Has anything been left out? Are they vital to the variety of settings in which students must be able to function—work, education/training, community?)
3. Are the SFS coherent? (Are the categories logical? Does the organization make sense?)
4. Do the SFS represent high expectations? (Are they challenging? Will they help prepare all students for the demands of the 21st century?)
5. Are the SFS clear and understandable? (Will the general public understand them?)

A detailed questionnaire was sent to experts around the nation who were asked to review and evaluate the SFS draft. Experts were selected by virtue of their knowledge and expertise in national and state educational reform efforts and/or workplace skill development. They included employers known for their excellence in planning, training, and management, as well as representatives from political offices, higher education, and a national curriculum consortium. Each reviewer was mailed a package that contained background information on Maryland's SFS and a questionnaire. A total of 17 experts returned completed surveys, including representatives of the American Federation of Teachers, which is currently reviewing graduation requirements around the world, the Council of Chief State School Officers, the Center for Workforce Development, the International Center for Leadership in Education, and the National Center for Research on Evaluation, Standards, and Student Testing (CRESST).

A total of 17 stakeholder groups within Maryland was also asked to respond to a questionnaire on the draft skills. The majority of the responses were obtained from four groups: students, teachers, local supervisors, and administrators. The items on the stakeholders' survey repeated selected items in the experts' survey so that a comparative analysis could be conducted.

Based on the responses of the experts and the stakeholders, the SFS Team received strong endorsement to continue in the direction they had taken with the SFS. Overall, the respondents indicated that the draft SFS were clear, understandable, significant, comprehensive, and well organized. Several of the national experts commented that Maryland is well ahead of many other states in their quest to establish comparable skills.

The Team carefully reviewed cautionary notes and recommendations made by respondents. Several comments were made relative to the clarity and specificity of the SFS. Others dealt with how the SFS would be taught and assessed. A larger proportion of the stakeholders, compared to experts, thought that the public would have difficulty understanding concepts in the SFS. Several respondents thought there were too many indicators; however, others encouraged the Team to ensure sufficient specificity and linkages to the content areas (i.e., more indicators) to enable assessment of students on the SFS. Several national experts warned against teaching the SFS in abstract settings and encouraged connections with work and work-based learning. With several additions and modifications, the outcomes and categories were highly endorsed.

The Maryland State Board of Education was consulted frequently as the learning goals for the High School Assessment were being developed. At Board meetings and retreats, the State Board received updates on the status of skills identification, and discussed some of the major issues related to the learning goals and the possibilities for assessment. The academic content teams also reviewed the SFS draft and identified areas of overlap. In June, 1995, the core learning goals in SFS and the four content areas were formally presented to the State Board of Education. The Board advised the Task Force to obtain further public comment through public hearings and focus groups with teachers in selected local school systems and to gather feedback from every high school teacher in the State before the learning goals were adopted. Those processes were completed in January, 1996, and results reported to the State Board of Education. A summary of questions on the teacher questionnaire follows: (a) Are the SFS clear and understandable? (b) Are they comprehensive? (c) Are they important to the success of students in postsecondary education, the world of work, and in their personal/social lives in the 21st century? With more than three fourths of the high schools responding at the time of this writing, affirmative responses across these questions ranged between 84% and 100%, counting one consensus response per school. In January, 1996, the State Board of Education directed the SFS Team to proceed to the development of descriptions of performance levels, to define the relationship between the SFS and the content learning goals, and to begin to align the SFS with K–12 curricula.

DEFINING THE SKILLS FOR SUCCESS

What are the characteristics of the SFS that emerged from this collaborative process? In its charge to the SFS Team, the State Board of Education specified that learner goals should communicate high expectations for students and be benchmarked to national and international performance standards. This charge represented a major shift from the previous focus on minimum competence in the Maryland Functional Testing program. The State Board further challenged the Team to recommend a readily understandable set of skills with broadly applicable utility for students' future success in three learning environments: postsecondary education, the workplace, and the community. They were dubbed *skills* to distinguish them from the knowledge-centered core learning goals in English, mathematics, science, and social studies. That is, the SFS are process-centered; they are intended to be applicable in most knowledge domains. Both the SFS and the content learning goals could be described as learning outcomes, descriptions of what students should know and be able to do at the end of a course of study.

The development process just described allowed Maryland to benefit from established work on workplace readiness skills. It was important to the SFS Team that Maryland's SFS be a comprehensive yet concise set of skills that would provide a frame or context for gaining and using academic knowledge, as well as the tools for learning in a wide array of subjects, skill areas, and environments. In essence, the SFS Team wanted to be sure its work would challenge Maryland's schools to ensure that students not only learn in school, but also know how to learn throughout their lives. The difference is parallel with the old metaphor, giving them a fish versus teaching them how to fish. Preparing students for the next century means empowering them, developing independence of mind, helping them become lifelong learners.

Content and Organization

Table 3.2 provides a complete set of the SFS. They are organized into five core learning goals stated as broad learning outcomes. Under each core learning goal, there are three to five expectations, the level at which proficiency can be described. The indicators are clustered under each expectation. They represent or will represent observable, measurable behavior.

TABLE 3.2
Skills for Success in Maryland

Learning Skills *The student will plan, monitor, and evaluate his or her own learning experiences.*	Thinking Skills *The student will think creatively, critically, and strategically to achieve goals, make effective decisions, and solve problems.*	Thinking Skills *(continued).*
The student will establish and pursue clear, challenging goals for learning by: • developing short- and long-range goals for learning • using personal goals to guide learning experiences The student will plan, monitor, and evaluate his or her own learning experiences by: • planning learning experiences before acting • selecting and using appropriate resources and learning strategies • monitoring progress when following a learning plan • identifying and addressing problems that impede learning • evaluate learning experiences The student will adapt, as necessary, to changing needs and situations by: • reassessing learning goals based on performance or changes in situations • adjusting goals/strategies as necessary • accepting and using constructive criticism • taking advantage of new learning opportunity	The student will generate creative ideas in a variety of situations by: • brainstorming alternative perspectives or ways of thinking and acting in a variety of complex situations • representing his or her creative ideas in a verbal or nonverbal form appropriate to his or her purpose • applying and testing new ideas/solutions before adopting them The student will evaluate information, issues, and positions critically by: • identifying key issues in complex situations • evaluating relevance and utility of information for specific purposes • examining basic concepts and assumptions underlying an issue or position • recognizing bias, vested interests, stereotyping, manipulation, and misuse of information • using evidence and/or reason to support or refute an issue or position	The student will solve problems systematically and rationally by: • understanding the situation or context within which the problem is embedded • defining a problem • identifying and evaluating alternative solutions to the problem • selecting and using appropriate strategies to solve the problem • evaluating the solution and the strategies used to solve the problem

(Continued)

The student will persevere, when appropriate, in difficult learning situations by:
- Taking personal responsibility for learning in difficult situations
- allocating time and effort as needed
- seeking assistance or change strategies when necessary

The student will apply acquired knowledge and skills effectively in new learning situations by:
- identifying similarities and differences between old and new learning situations
- identifying knowledge, strategies, or skills that could be useful in new learning situations
- adjusting strategies as necessary to work in new situations
- evaluating the usefulness of acquired knowledge, strategies, and skills in new learning situations

Communication Skills
The student will plan, participate in, monitor, and evaluate communication experiences in a variety of situations. .

The student will plan for successful communication experiences by:
- identifying audiences and purposes for communicating
- identifying appropriate means for constructing and delivering messages for a variety of purposes, audiences, and situations

The student will demonstrate strategic thinking in a variety of situations to make effective decisions and achieve goals by:
- demonstrating an awareness of his or her own thinking and that of others
- framing questions, problems, and issues in an appropriate context
- identifying goals appropriate to available resources skills, and situations
- identifying alternative strategies to achieve goals
- planning and following steps to achieve goals and make decisions
- monitoring, evaluating, and making necessary adjustments in goals, plans, or actions

Technology Skills
The student will understand, apply, and evaluate technologies as labor-enhancing and problem-solving tools.

The student will demonstrate knowledge of current technologies appropriate for a variety of purposes and situations by:
- identifying and using resources and procedures for keeping abreast of advances in technology
- identifying appropriate and current

Interpersonal Skills
The student will work effectively with others and participate responsibly in a variety of situations.

The student will demonstrate effective interaction strategies by:
- accepting responsibility for self and actions
- showing respect for the opinions, rights, cultural differences, and contributions of others

(Continued)

TABLE 3.2
(*Continued*)

Communication Skills	Technology Skills	Interpersonal Skills
The student will plan, participate in, monitor, and evaluate communication experiences in a variety of situations.	*The student will understand, apply, and evaluate technologies as labor-enhancing and problem-solving tools.*	*The student will work effectively with others and participate responsibly in a variety of situations.*
• practicing before attempting to communicate • identifying potential problems and planning to prevent or solve them The student will acquire, manage, and convey information using a variety of strategies and technologies by: • gathering information from a variety of sources, using appropriate technologies and processes • using listening skills to help interpret and evaluate spoken messages • evaluating the utility of information gained for specific purposes • organizing and maintaining information in appropriate written, graphic, electronic, or other form • conveying formation using strategies and	technologies for assessing and managing information, communicating, performing work, and solving problems in a variety of situations • evaluating the use of current technology in specific situations and suggest appropriate changes • identifying future needs for technology for specific purposes in a variety of situations The student will use technology effectively for a variety of purposes and situations by: • developing computer literacy skills, including concepts and applications • using appropriate technologies to access, store, manage, analyze, and	• providing, accepting, and using constructive feedback to adjust behavior The student will work cooperatively with others in a variety of situations by: • participating in developing goals for group activities • supporting group consensus and respecting dissenting positions • participating in developing group rules or procedures and following them • demonstrating understanding of and assuming various roles in groups • contributing personal resources to the group

(*continued*)

- means appropriate to the audience, purposes, and situation

The student will monitor communication processes and make necessary adjustments to solve problems by:
- monitoring ongoing communication processes using identified purposes and plans as guidelines
- identifying problems and making adjustments to solve them as necessary

The student will evaluate communication experiences in a variety of situations by:
- evaluating success in achieving purposes and using audience feedback and other sources of information
- evaluating the effectiveness of communication strategies and technologies for the audience, purposes, and situation
- identifying problems in style, content, form, means, situation, or feedback, and planing to prevent or solve problems in the future

- communicate information
- using appropriate technologies to solve problems in complex situations
- demonstrating safe, effective, and creative use of labor-saving or -enhancing technologies in a variety of situations

The student will demonstrate an understanding of the impact of technology on the environment, society, and individuals by:
- developing criteria for evaluating the effects of technology
- evaluating the effects of technology on the environment, society, and individuals in a variety of situations
- designing technological systems with the most positive and least negative effects in specific situations

- helping resolve conflicts and bringing the group to consensus when appropriate
- developing and using criteria to evaluate individual and group performance

The student will function as a responsible citizen by:
- participating in democratic processes in a variety of situations
- demonstrating an understanding of cultural heritages and multicultural groups in the nation and the world
- demonstrating an understanding of important environmental, social, and economic issues
- demonstrating ability to make reasoned consumer decisions in a variety of situations
- demonstrating ability to manage financial resources responsibly
- planning and acting for the well being of the community

Note. This table reflects the Skills for Success at the time of writing, March, 1996.

Each of the five categories of skills—Learning, Thinking, Communication, Technology, and Interpersonal Skills—is stated as a single core learning goal. For example, Learning Skills are represented by the following goal: *The student will plan, monitor, and evaluate his or her own learning experiences.* The intent of this goal is to get students to take responsibility for their own learning by learning how to learn—how to plan for success, how to anticipate difficulties and avoid or resolve them, how to evaluate their success in learning, and how to build on success or avoid making the same mistakes in the future. This is the lifeblood of learning in any subject and in any situation.

Each core learning goal is elaborated by three to five expectations that specify what students will be able to do if they have achieved the goal. Under Thinking Skills, for example, students are expected to be able to think critically: *The student will evaluate information, issues, and positions critically.* The ability to judge the relevance, usefulness, and validity of information for any purpose is vital to success in school, on the job, and in life as a responsible citizen. Indicators of skill attainment provide further elaboration on the expectations. Using the earlier noted expectation for critical thinking, the indicators identify, in this instance, the critical competencies a student must demonstrate: identifying key issues in complex situations; evaluating relevance and utility of information for specific purposes; examining basic concepts and assumptions underlying an issue or position; establishing clear criteria for evaluating an issue or position; recognizing bias and vested interests, stereotyping, manipulation, and misuse of information; and using evidence and/or reason to support or refute an issue or position. In many instances, the indicators comprise the learning sequences needed to achieve the expectation (i.e., set goals, plan, perform, monitor, and evaluate). In a few areas, the indicators function solely as elaborations on the expectation.

The indicators only begin to flesh out the level of detail necessary for curriculum designers to integrate the SFS into the classroom. To provide more concrete representations of what was intended by the SFS, the Team illustrated each expectation with an example of a high school learning activity. The examples show what might occur in a school where students are learning and using Maryland's SFS. Examples include how a project in a specific subject such as earth science could be used to teach and assess Technology Skills; how a student, through planning a high school program and managing challenges over the four years of high school, develops and demonstrates Learning Skills; and how a multidisciplinary approach could be used to teach and assess Communication Skills. A sample activity follows:

Sample Learning Activity (Goal 4: Technology Skills, Expectation 1): Students in science classes across Maryland—such as earth science, biology, environmental science, and agricultural science—are participating jointly in environmental investigations aimed at studying the relationships among living resources, pollution, and water quality in the Chesapeake Bay. Students learn how to use various technological instruments to collect and analyze data samples. Students learn to use computer software programs that help them manage and analyze the data. Students also work in groups to understand how agricultural and fishing technologies affect living resources and their impact on the pollution of water.

The examples deliberately portray the role of the SFS in a wide variety of short- and long-range learning activities in school. These are illustrative learning activities. It is important to note, however, that the learning activities are simply *examples* of what the SFS might look like in high schools. They are not lesson plans; they are not curriculum guides; they are not "models"; and they do not tell teachers how to teach. The State has no intention of prescribing local curricula.

Scope and Applicability

The SFS clearly represent a focus different from the traditional academic skills taught in high school. First they are very general rather than subject-specific; they apply equally well to any subject. One needs to communicate effectively, for example, with or about mathematics as well as science or English literature or athletics or automobile repair. As workplace skills, the SFS represent real-world applications of academic knowledge and skills. They define the kinds of things employers want high school graduates to be able to do with academic knowledge and skills. For instance, mathematics is used to solve complex problems; English is a tool for thinking and communicating; Technology is a problem-solving as well as a labor-enhancing tool; and so forth. These skills provide an essential frame for the kinds of applications of academic knowledge and skills valued by employers. Moreover, they build on the knowledge and skill sets being assessed by Maryland's performance assessments in Grades 3, 5, and 8. For example, the following mathematics learning outcome for Grades 3, 5, and 8 (Maryland State Department of Education, 1990) clearly maps onto the SFS learning goals in Thinking, Technology, and Interpersonal Skills:

> Students will demonstrate their ability to solve problems in mathematics, including problems with open-ended answers, problems which are solved in a cooperative atmosphere, and problems which are solved with the use of technology.

Besides providing a frame or context for the use of academic knowledge and skills, the SFS also provide the tools for learning in any of the subject areas. Knowing how to learn is vital to learning in any new or challenging situation. Technology, an important workplace tool, is also a tool for learning. It provides access to information, helps to manage it, and makes it easier to analyze, represent, and convey that information.

The sense of the SFS as a foundation for learning carries them far beyond the purview of workplace readiness skills. These skills are just as essential in any postsecondary learning experience, whether it is formal, on-the-job training, or higher education. Standards will surely rise in colleges and universities in the 21st century in a continuous effort to keep America competitive in an international economy (Wingspread Group on Higher Education, 1993). High school graduates will compete for fewer positions in colleges and universities. Once there, they will find course work more demanding. And they must have the learning, thinking, and communication tools to succeed not only in higher education, but in their private and social lives as well.

The SFS are also community skills, skills necessary to function as responsible citizens in a complex, diverse, rapidly evolving, democratic society. Thinking Skills—the ability to think creatively, critically, and strategically—are important for making effective decisions, whether voting on an issue or a candidate for public office or making consumer decisions. Communication Skills provide the capacity to be informed citizens; Technology Skills provide labor-enhancing and problem-solving tools; and Interpersonal Skills are critical to working effectively with others and participating constructively in a variety of community situations.

Comparisons with Other Skill Sets

Table 3.3 contains a comparison of the content of the SFS with similar competencies developed by various national groups, including SCANS by the Department of Labor (1991), Workplace Basics by the American Society for Training and Development (Carnevale, Gainer, & Meltzer, 1990), and the Core Competencies by the National Academy of Sciences (1984). Because of their popularity, the similarities and difference between SCANS and the SFS were closely examined. Although a skill-for-skill comparison yields the appearance of many content matches as illustrated in Table 3.3, there are significant differences between Maryland's SFS and the SCANS skills. To begin, SCANS is much more workplace-oriented, while the SFS are designed to be merged with academic outcomes in a high school learning environment that prepares students for success in diverse arenas. The SFS cast a wider net than SCANS.

The SFS also present a very different principle of organization or coherence than SCANS. They are organized at several levels. The Learning Skills and Thinking Skills are core competencies that underlie, make possible, and indeed engender all other competencies. For example, the pattern of plan, perform, monitor, and evaluate recurs in other learning goals

TABLE 3.3
Comparison of Content of Skills for Success With SCANS, Workplace Basics, and Core Competencies

Skills for Success	SCANS Competencies	SCANS Foundation	Workplace Basics	Core Competencies
Learning Skills • Establish goals • Plan, monitor, and evaluate learning • Adapt to changes when appropriate • Persevere • Transfer and apply	Resources • Manage human resources	Thinking Skills • Knowing how to learn Personal Qualities • Self-management	The Foundation • Knowing how to learn Developmental skills • Self-esteem, motivation, career planning	Personal Work Habits • Self-concept, motivation, self-discipline, self-management
Thinking Skills • Creative thinking • Critical thinking • strategic thinking • Rational problem solving		Thinking Skills • Creativity • Problem solving • Decision making • Reasoning	Adaptability • Problem solving • Creativity	Reasoning and Problem Solving • Logical procedures • Decision making • Patterns of problems
Communication Skills • Plan for communication • Acquire, manage, and convey information • Monitor communication and solve problems	Information Management • acquire, evaluate, organize, maintain, interpret, and communicate information; use computers to process information	Basic Skills • Reading, writing • Listening, speaking	Basic Competency • Reading, writing Communication • Listening, speaking	Reading Writing Oral Communication

(Continued)

TABLE 3.3
(Continued)

• Evaluate communication in a variety of situations		
Technology Skills • Demonstrate knowledge of technology for a variety of purposes • use technology effectively for a variety of purposes • Demonstrate understanding of the impact of technology	**Systems** • Understand use; monitor, correct, and design technological systems **Technology** • Select, apply, maintain, and troubleshoot technology **Information** • Use computers to process information	
Interpersonal Skills • Demonstrate effective interaction strategies • Work cooperatively in a variety of situations • Function as a responsible citizen in a democratic, diverse society	**Systems** • Understand social systems and operates within them **Interpersonal** • Work with others • Teach, serve, lead, negotiate, work with diversity	**Personal Qualities** • Responsibility • Sociability • Integrity/honesty **Group Effectiveness** • Interpersonal skills • Teamwork skills • Negotiation skills **Influencing Skills** • Organizational effectiveness • Leadership **Interpersonal Relations** • Respect, humor, constructive criticism, conflict resolution, group decision making

in the SFS. Without Learning and Thinking Skills, nothing else is possible. In a world where the rate of change only accelerates, learning and higher order thinking must be continuous. Communication and Technology Skills form bridges between the individual, society, and the environment. These skills allow individuals to access other resources and to interact with other people, institutions, and the natural environment. The Interpersonal Skills make it possible for people to function effectively in the complex, diverse arenas where they will spend most of their lives. And all the preceding skills clusters accumulate and interact in the social context. This cumulative, interactive relationship is a series of nested spheres of influence, expanding their scope of application from a focus on internal conditions (Learning, Thinking) of individuals to external states (Technology) or social situations (Interpersonal Skills). The levels of organization not only speak to the interrelationships among the SFS, but also facilitate the integration of the SFS in instruction and assessment. Meaningful, recurring patterns make implementation easier.

Their organization and conceptual coherence distinguish the SFS from comparable skill sets. SCANS, for example, is more discursive and additive, separating its competencies into two sets of skills (Competencies and Foundation) which vary in level of specificity from the very general to the very specific: for example, "works to satisfy customers' expectations" (under Interpersonal/Competencies) versus "knows how social, organizational, and technological systems work and operates effectively within them" (under Systems/Competencies). In summary, SCANS proved to be too limited in level and scope of expectations to serve the diverse arenas demanded of the SFS, too discursive in content and organization to interface with all subjects in a high school environment, and too specific to avoid the need for constant revalidation (e.g., Nash & Korte, chapter 4, this volume). The latter design problem can be resolved only by prescience or moving up the level of specificity to gain generalizability, which was the strategy adopted in the SFS. The level of specificity necessary for instruction and assessment comes from an interface with the content subjects and the indicators, which are subject to elaboration as needed.

INSTRUCTION AND ASSESSMENT

What role will the SFS play in instruction and assessment in Maryland's high school reform? From their inception, the SFS were never intended to be taught or assessed in isolation from the subject matters that gave them substance. No course in "higher order thinking" per se was planned, and no one envisioned tests in "learning to learn" or "communication skills" as such. Instead, the SFS were intended to be an integral part of instruction in academic and other disciplines in high school and tested, insofar as possible, in the normal course of assessing student competence in state-mandated learning goals in English, mathematics, science, and social studies and in the normal course of teaching and testing other subjects like fine arts and vocational–technical programs.

But that sounds suspiciously like the "infusion" strategy that led many "career-oriented" curricula to obscurity in the last several decades. Teachers often resist intrusions into their curricular domains, and with good reason. They can hardly do justice to their own subjects in the time available. The challenge is how to design a meaningful and enduring role for SFS in instruction and assessment in academic and other subjects without displacing or diluting the primary focus on competence in those disciplines. Said more positively, how to *contribute*, in the eyes of teachers, to efforts to develop student competence by drawing out of those disciplines traces of powerful, highly generalizable learning goals and giving them prominence as the SFS. Stretching the curricular domain in this sense is parallel with expanding the scope of application of the SFS from workplace readiness skills to wider, more diverse arenas.

Curriculum Transformation

How much of the SFS is already in the high school curricula, especially in new formulations of them? Mathematics, for example, as reconstructed by the National Council of the Teachers of Mathematics in the late 1980s, has prominent communications, problem-solving, and reasoning strands (National Council of Teachers of Mathematics, 1989). The National Council of the Teachers of English is tireless in promoting writing across the curriculum (Mayher, Lester, & Pradl, 1983)—Communication in the SFS. Cooperative learning (Slavin, 1987) and peer response groups (Graves, 1983) proliferate everywhere (Interpersonal and Communication Skills in the SFS). Laboratory and hands-on science promote Technology and Interpersonal Skills. Science grows ever more high tech and collaborative. Even learning to learn is changing teacher discourse; *learning* rather than *teaching* appears more in teacher talk, informed by a growing body of research on concepts like metacognition and self-regulation (Borkowski & Muthrukrishna, 1992; Brown, Bransford, Ferrara, & Campione, 1983). And what discipline will not claim to have made major advances in moving from rote learning toward higher order thinking during the last decade? The culture of teaching and learning is changing, and the SFS represent the leading edge of that change.

To test this hypothesis of the place of the SFS in current curricular frameworks, the subject specialists who had led the development of Maryland's state of the art core learning goals in each discipline developed matrices representing matches among the SFS and the learning goals in their disciplines. The notion was that students could not achieve the competence represented in the discipline without having the competence indicated in the SFS. Thinking, Communication, and Technology Skills were well represented in these matrices but not symmetrically, that is, not covered equally well in each learning goal within a discipline or across disciplines. Table 3.4 displays a summary of the perceived match between learning goals in the SFS and the four academic disciplines at the time of this writing. Some correspondences, or the lack thereof, are extreme—for example, no perceived representation of Learning Skills in social studies. However, some variation is to be expected. The SFS are intended to promote *cumulative opportunities to learn across learning goals and disciplines.* Additional correspondences within and across disciplines may be teased out over time.

Among the SFS, Learning and Interpersonal Skills were less well represented in the matrices. These results seem reasonable since these two learning goals represent the most recent wave of change coming from the research community (e.g., self-regulation, cooperative learning, conflict resolution, and so forth).

But several uncertainties remain. What does the claim of a relationship in a subject matrix mean? Someone thought the intent of the academic subject learning goal encompassed the intent of the learning goal from the SFS. To explore those kind of relationships further, academic and business communities need to develop learning and work-related activities embodying specific learning goals and expectations from the SFS. Comparisons of conceptions of the same goal from different applications will provide adjustments and a fuller sense of the learning goal across domains.

So far, so good. However, if the SFS are already represented or are in the process of being represented in curricular frameworks and consequently in school programs, why bother to develop the SFS as a separate

TABLE 3.4
Percent of Overlap Between Skills for Success and Academic Subjects

Academic Subjects	Skills for Success				
	Learning	Thinking	Communication	Technology	Interpersonal
English	35	46	57	17	31
Mathematics	63	79	64	67	19
Science	23	64	43	62	29
Social Studies	0	71	21	58	80

entity? The question, again, is what *role* do they play? From a curriculum and instruction perspective, the SFS are a template drawing learning opportunities toward a common, generic set of learning goals manifested in different ways in different disciplines and learning activities. They are *metaskills*. The SFS promote quality and coherence in learning for students. They demand application, transfer, and generalization—the vitals of learning. Rather than having the metaskills, which were validated in the business and other communities, occur willy-nilly in instruction, the SFS set a target and expectation for their occurrence in every discipline.

The SFS also give these "naturally" occurring changes in the goals of education a prominence they would not otherwise have if left unmarked across curricula. The SFS can be expected to occur, then, in the most recent advances in instruction in each discipline. At the same time, they will draw instruction toward those advances. From a curriculum perspective, however, the unanswered question is: Do they encompass the array of significant advances that will prevail in at least the first few decades of the 21st century or will they, like so many of their predecessors in the name of innovation, slip into obscurity before they are ever implemented?

Implications for Instruction

What would the impact of SFS be on instruction if they were fully implemented? Teaching at the high school level in Maryland will undergo significant positive transformation as the SFS become a part of the course of studies for all students. Educational reformers like Ernest Boyer (1983) and Theodore Sizer (1984) argue that an average 70% of classroom time in American high schools is devoted to teacher-dominated, sage-on-the-stage behavior—"teaching" at the expense of learning. This percentage should drop precipitously in Maryland as the SFS become a fundamental part of instruction in all public high schools. All the SFS draw students toward more active roles as learners.

High school classrooms will begin to reflect a commitment to explicit instruction (Dole, Duffy, Roehler, & Pearson, 1991) promoting student ability to plan, monitor, and evaluate their own learning experiences (Schuder, 1993). High school classrooms will become increasingly constructivist (Driver, Asoko, Leach, Mortimer, & Scott, 1994) in their design, and teaching practices will reflect greater emphasis on seminars, inquiry-based learning, open discussions, encouragement of divergent thinking, and acceptance of multiple forms of expression as students become more engaged in learning. Learning situations will include but not be limited to role playing, simulations, debates, teamwork, and student-centered teaching.

The emphasis in SFS on Communication and Interpersonal Skills will result in a much greater incidence of cooperative and experiential learning in Maryland high school classrooms. All students, not just those labeled gifted, will be actively engaged in problem-solving and decision-making activities in all subjects (e.g., Ford, 1990; Olshavsky, 1976–77; Schoenfeld, 1987). Group work, particularly within classrooms that are heterogeneously organized, will enable students to assume a variety of reality-based roles and responsibilities. Similarly, the emphasis on the learner will lead high school educators to pay much greater attention to the variety of learning styles, attention spans, memory, developmental pacing, and intelligences that characterize student populations. These kinds of changes in instruction may reinforce the movement toward four-period days in high school, smaller schools, schools within schools, and other structural innovations.

Technology will also play a much greater role in all facets of the Maryland high school instructional program. Beyond access to computer laboratories for traditional applications, students will be taught to understand, apply, and evaluate technologies as labor-enhancing, problem-solving, and sensory-extension tools (e.g., aviation simulators—O'Neil & Robertson, 1992). Students will be taught when to use their knowledge of technology, how to adapt technology, and how to manage their own learning by using it. More and more, high schools will provide students with real-world opportunities and simulations involving the application of SFS to authentic settings and situations. In addition, traditional classroom walls will become less restrictive, with students encouraged to explore and apply technology and other skills in field experience, including internships, mentoring programs, and other forms of real world applications. By avoiding the limitations of workplace readiness per se, the SFS will help dissolve the discontinuities between high school, the 21st century workplace, and lifelong learning.

None of this is possible, however, without genuine teacher commitment to the SFS. As previously described, the SFS Team ran focus groups with teachers in selected local school systems in Maryland during the development of the SFS. In the fall of 1995, however, the State Board of Education asked that *every* high school teacher in the State of Maryland have the opportunity to review and comment on the core learning goals, including the SFS, and the results be reported to the Board. Each high school, 200 of them, would provide consensus responses and individual comments on a series of questions regarding, among other issues, the clarity of the expression of the SFS and the importance of the SFS to the future success of students in postsecondary education, the workplace, and social/personal lives in the 21st century. More than three fourths of the high schools have responded at the time of this writing. One consensus response per school was counted. On the issue of clarity of expression, affirmative responses range from 100% (Thinking Skills) to 93% (Communication and Interpersonal Skills). On the issue of the importance of the goals to student success in the next century, affirmative responses range from 98% (Learning and Communication Skills) to 84% (Interpersonal Skills). These high percentages compare with affirmative responses to the academic subjects often in the range of 50%, 60%, and 70%.

But these overwhelmingly positive responses from teachers are accompanied by a great deal of concern in the commentary on a variety of issues. In other words, this is a qualified indication of support. The major areas of concern were instructional resources, staff development, instruction, and implementation in general; the need to begin instruction in the SFS in the elementary grades with support in the home and community; the effects of such high standards on low achieving, special needs, and non-native English speaking students; and above all else, concern about assessment and accountability. "How does the State of Maryland intend to assess student performance in the SFS?" was a constant refrain.

In the spring and summer of 1996, the SFS Team will move toward assessment by beginning the development of descriptions of different levels of performance for each core learning goal in the SFS. The first decision will be whether to write descriptions on core learning goals, expectations, or indicators. The second decision will be the number and titles of levels (e.g., basic, proficient, and advanced). These decisions will be made in a complex context: (a) the proficiency levels used in Maryland's performance assessments in Grades 3, 5, and 8; (b) the proficiency levels used in the design of Maryland's new high school assessments in English, mathematics, science, and social studies; (c) the performance levels used in assessments locally and nationally; and (d) the special purpose and content of the SFS. These descriptions will provide the next level of specificity and the staging for the development of authentic assessments (e.g., performance, extended project, and portfolio assessments) for the SFS.

The Context for Assessment

There is some truth to the twin axioms that "what gets tested gets taught" and "how it's tested determines how it's taught." The status of the SFS in Maryland's statewide accountability program will affect their status in the classroom.

Parts of Maryland's accountability program are already in place and demonstrate some of the influences referred to previously. For example, Maryland has been giving census administrations of performance assessments in reading, writing, language usage, mathematics, science, and social studies in Grades 3, 5, and 8 since 1990 (see Table 3.1). Performance standards are high. Progress has been slow but steady across the State. After 5 years, only one local school system (out of 24) has met the satisfactory level of performance in only one subject, science, at Grade 8 (Maryland State Department of Education, 1995). Schools are also held accountable for results. Thirty-seven schools are eligible for reconstitution in 1996.

Teachers recognize the learning outcomes as high but worthy standards and recognize the performance tasks as models of good learning activities. They are in the process of incorporating these models into their instruction. The Rand Corporation (Koretz, Mitchell, Barron, & Keith, 1995) is currently studying the effects of these assessments on schools, and the Maryland State Department of Education, in collaboration with the University of Pittsburgh, has received a research grant to study the consequential validity of its performance assessments. Maryland's performance assessment program is an instance of the advantages of teaching to an exemplary test, and these kinds of assessments offer a fertile arena for the SFS.

Maryland also has minimal competency tests in reading, writing, mathematics, and citizenship, first administered in Grade 9 and thereafter until students pass. (These tests were moved to Grade 8 or lower in 1996.) The minimal competency tests offer limited possibilities as models for assessing the SFS because performance standards are low, and the tests are in multiple-choice format. The writing test, however, requires students to produce two extended writing samples in response to prompts, and the effects on writing instruction have been positive. For example, supervisors report that students are writing more whole discourse in classrooms, and they are writing more across the curriculum. Emphasis on the writing process and use of writing response groups have flourished in classrooms in reaction to the writing test in Grade 9 and performance assessments in Grades 3, 5, and 8. These assessment models are instructive for the SFS. Writing in particular will offer more opportunities to infuse the SFS into the academic subject exams.

The most immediate influence on assessing the SFS, however, will be the design of the high school assessment system for the core learning goals in English, mathematics, science, and social studies. To achieve a high school diploma in Maryland, individual students will be held accountable for meeting high performance standards in these academic subjects, beginning with the graduating class of 2004. Current recommendations are for state-mandated, end-of-course tests in a format that allows maximum content coverage and minimal turn-around-time for test results so teachers can use test scores as part of their course grades. Reliability of individual student scores is a necessity. These recommendations and constraints imply multiple-choice tests with, perhaps, some short-answer and extended-response options (e.g., essays). These are not, unfortunately, wholly auspicious formats for incorporating the assessment of the SFS in content tests.

Possibilities for Assessing the SFS

While no one ever envisioned separate tests of the SFS in the high school assessment program in Maryland, the possibility of producing SFS scores from tests of academic skills may be worth exploring, given the potential positive effects of assessment on teaching and learning described in the preceding section. The apparent match between the learning goals in the SFS and academic skills make it possible to write specifications for items/tasks that measure both the SFS and academic skills. For example, measures of problem solving in mathematics might well include attributes of problem solving emphasized in SFS, that is, both math-specific and generic attributes. In other words, the selected indicators for problem solving in SFS are infused into the indicators for problem solving in mathematics, and the combined set becomes the basis for writing item specifications. For example, take the following indicator in mathematics: *The student will apply formulas and use matrices (arrays of numbers) to solve real-world problems.* The expectation for SFS is *The student will solve problems systematically and rationally,* which is elaborated in the indicators as a traditional problem-solving sequence in mathematics (e.g., Polya, 1957): *understand the situation (context); define the problem; identify and evaluate alternative solutions; select and use appropriate strategies to solve the problem; and evaluate the solution and the use of the strategy.* The SFS indicators push the indicator in mathematics in the direction of explaining, justifying, and evaluating the use of the formula and matrices. In other words, the SFS expand the scope of application and depth of understanding required in the mathematics indicator. Maryland has a successful record of double scoring responses to performance tasks to produce scores in two learning outcomes, often from very different domains, for instance, writing and mathematics (see, for example, Maryland State Department of Education, 1993, p. 5).

The opportunities for double scoring responses in multiple-choice or short-answer formats, however, are not so clear. Thinking *can* be measured in multiple-choice formats (e.g., Norris, 1989), but the SFS are oriented toward the use of knowledge in authentic situations, while Maryland's end-of-course tests for academic subjects may emphasize the assessment of knowledge per se in multiple-choice, short-answer, and extended-response formats. Some responses could be double scored, for instance in higher order thinking, and scores for SFS might accrue across academic domains. (Maryland has a research grant to explore the use of scores from different item formats and learning domains to produce estimates of student competence. A consortium of states, coordinated by the Chief State School Officers, is studying the same problem.) But SFS goals like Learning, Technology, and Interpersonal Skills require extended, constructed responses in authentic situations. Measures of these skills are not likely to come from multiple-choice or short-answer item formats, double scoring withal.

SFS, then, are best measured in performance assessments, broadly construed. Simulations of authentic experiences (O'Neil & Robertson, 1992), such as those in use for decades in medicine, flight training, drivers' training, law school, plumbing, and so forth, offer excellent models. Commercial programs have been used for years by businesses, community colleges, and job training programs to teach and assess thinking and other workplace skills (Densford, 1994). Live performances, scored by panels of judges, as in sports, music, or dance, are equally appropriate models (Baker & O'Neil, 1994). Several commercial models of simulations, using video presentations interrupted periodically to allow students to respond to prompts or questions, are already operational measures of SCANS skills (Learning Resources, Inc., 1994), and these assessment models could be adapted to measure SFS. In this volume, the work of McLarty and Vansickle takes this approach.

But these are still "tests" and do not provide an adequate arena for assessing the extended, constructed responses demanded by the SFS. Learning Skills, for instance, are represented in the SFS by sequences of indicators that usually occur over longer periods of time than any on-demand test: setting goals, planning, performing, monitoring performance, problem-solving, and evaluating the learning experience. The best possibilities for assessing these kinds of critical learning goals are extended projects, performances, and portfolios. And these are the assessments that are most intimately a part of instruction and therefore most likely to promote qualitative improvements in teaching and learning (Murphy, 1994). To embed the goals and concepts of educational reform in the classroom, the assessment method should match or parallel the intended outcomes in teaching and learning.

Using authentic measures for statewide accountability programs, however, raises difficult economic, management, political, and psychometric issues as evidenced by recent experiences in California (Lawton, 1995), Kentucky, and Vermont. Insufficient reliability indices have been a pivotal problem in these states (Koretz, 1993; Koretz, Stecher, Klein, & McCaffrey, 1994). On the other hand, LeMahieu, Gitomer, and Eresh (1995) demonstrated the feasibility of reliable scores from large scale portfolio assessments in statewide assessment programs. Maryland has been using reliable performance assessments in Grades 3, 5, and 8, but this is a school program rather than an individual student accountability system. Maryland's high school assessment system is intended to hold individual students responsible for meeting performance standards in English, mathematics, science, and social studies.

The better context for assessing SFS in academic subjects might be curriculum-embedded (Nitko, 1995) rather than statewide, on-demand assessments. Building locally controlled, authentic assessments into continuous teaching-learning-testing cycles may be ideal for promoting systemic change, without which there will be no substantive, enduring reform in education. Maryland's 24 school systems have already begun planning the development of local assessments through their assessment consortium. Curriculum-embedded assessments, in tandem with statewide, on-demand tests, will provide multiple measures of performance on the same learning goals and, consequently, better estimates of student competence. Evidence of student competence from different instruments can be used to cross-validate each other, leading to continuous cycles of improvement in teaching, testing, and learning.

CONCLUSION

Significant challenges remain for Maryland's SFS. Final approval must be obtained from the State Board of Education. Given acceptance of the core learning goals, collaborative work on the high school assessment design will begin in earnest. Experience suggests (e.g., Norris, 1989) that some aspects of the Thinking and Communication Skills could be integrated into the test designs currently under consideration. However, the authentic assessment requirements for the Learning, Technology, and Interpersonal Skills, as well as more fulsome notions of Thinking and Communication Skills, have yet to be resolved. Several local school systems in the State of Maryland are already at work on alternative assessments.

The SFS, the product of an ambitious and far-reaching collaboration, constitute challenging new standards that will present opportunities that are simultaneously exciting and fraught with complexity. As an attempt to eliminate the discontinuities between work and school, to draw out similarities across the disciplines, and to better link secondary and postsecondary learning levels, the SFS are a powerful catalyst for secondary school reform throughout the State of Maryland. The collaborative inquiry, mutual support, and shared vision that guided and informed their genesis will become even more significant as they mature into phases of assessment design and development, program restructuring, and standards-based accountability that will be closely linked with new student graduation requirements. Just as a powerful new spirit of cross-role and cross-structural strategic planning pervaded the development of these new standards, their implementation process will foster a virtual revolution in how high schools and their curricula are organized, managed, delivered, and assessed.

REFERENCES

Baker, E. L., & O'Neil, H. F. Jr. (1994). Performance assessment and equity: A view from the USA. *Assessment in Education, 1,* 11–26.

Borkowski, J. C., & Muthrukrishna, M. (1992). Moving metacognition into the classroom: "working models" and effective strategy teaching. In M. Pressley, K. R. Harris, & J. T. Guthrie (Eds.), *Promoting academic competence and literacy in schools* (pp. 477–501). San Diego, CA: Academic Press.

Boyer, E. (1983). *High school*. New York: Harper & Row.

Brown, A. L., Bransford, J. D., Ferrara, R. A., & Campione, J. C. (1983). Learning, remembering, and understanding. In J. H. Flavell & E. M. Markman (Eds.), *Handbook of child psychology: Vol. 3. Cognitive development* (pp. 177 266). New York: Wiley.

Carnevale, A. P., Gainer, L. J., & Meltzer, A. S. (1990). *Workplace basics: The essential skills employers want*. San Francisco, CA: Jossey-Bass.

Densford, L. E. (1994). Program enhances learning ability. *Workforce Training News. 2*(1), 1.

Dole, J., Duffy,G., Roehler, L., & Pearson, P. D. (1991). Moving from the old to the new: Research in reading comprehension instruction. *Review of Educational Research, 61*(2), 239–264.

Driver, R., Asoko, H., Leach, J., Mortimer, E., & Scott, P. (1994). Constructing scientific knowledge in the classroom. *Educational Research, 23*(7), 5–12.

Ford, M. I. (1990, Nov.). The writing process: A strategy for problem solvers. *Arithmetic Teacher, 38*(3), 35–38.

Graves, D. (1983). *Writing: Teachers and students at work*. Exeter, NH: Heineman.

Koretz, D. (1993). New report of the Vermont portfolio project documents challenges. *National Council on Measurement in Education Quarterly Newsletter, 1*(4), 1–2.

Koretz, D., Mitchell, K., Barron, S., & Keith, S. (1995, November). *Perceived effects of the Maryland School Performance Assessment Program* (Draft Deliverable). Los Angeles, CA: National Center for Research on Education, Standards, and Student Testing.

Koretz, D., Stecher, B., Klein, S., & McCaffrey, D. (1994, Fall). The Vermont portfolio assessment program: Findings and implications. *Educational Measurement: Issues and Practice, 13*(3), 5–16.

Lawton, M. (1995). Revived California assessment bill waits as Governor Wilson weighs his options. *Education Week (Oct 4), XV*(5), 13.

Learning Resources, Inc. (1994). *Workplace success skills*. Stamford, CT: author.

LeMahieu, P. G., Gitomer, D. H., & Eresh, J. L. (1995). Portfolios in large-scale assessment: Difficult but not impossible. *Educational Measurement: Issues and Practices (Fall)*, 11–28.

Maryland State Department of Education. (1990). *Learning outcomes in mathematics, reading, writing/language usage, social studies, and science for Maryland school performance assessment program* (p. 3). Baltimore, MD: author.

Maryland State Department of Education. (1993, January). *Scoring MSPAP: A teacher's guide*. Baltimore, MD: author.

Maryland State Department of Education. (1995). *Maryland school performance report 1995: State and school systems*. Baltimore, MD: author.

Mayher, J. S., Lester, N., & Pradl, G. M. (1983). *Learning to write/writing to learn*. Portsmouth, NH: Heineman.

Murphy, S. (1994). Portfolios and curriculum reform: Patterns in practice. *Assessing Writing, 1*(2), 175–206.

National Academy of Science. (1984). *High schools and the changing workplace: The employers' view*. Washington, DC: author.

National Center on Education and the Economy. (1990). *America's choice high skills or low wages*. Rochester, NY: author.

National Council of Teachers of Mathematics. (1989). *Curriculum and evaluation standards for school mathematics*. Reston, VA: Author.

Nitko, A. J. (1995, Fall). Is the curriculum a reasonable basis for assessment reform? *Educational Measurement: Issues and Practice, 14*(3), 5–10, 35.

Norris, S. P. (1989). Can we test validly for critical thinking? *Educational Researcher, 18*(9), 21–26.

O'Neil, H. F., Jr., Allred, K., & Baker, E. L. (1992). *Measurement of workforce readiness: Review of theoretical frameworks* (CSE Tech. Rep. 343). Los Angeles, CA: National Center for Research on Evaluation, Standards, and Student Testing.

O'Neil, H. F., Jr., & Robertson, M. M. (1992). Simulations: occupationally oriented. In M. C. Alkin (Ed.), *Encyclopedia of Educational Research*. Sixth edition (Vol. 4, pp. 1216–1222). New York: Macmillan.

Olshavsky, J. E. (1976–1977). Reading as problem solving: An investigation of strategies. *Reading Research Quarterly, 4*(12), 654–674.

Polya, G. (1957). *How to solve it: A new aspect of mathematical method* (2nd ed.). Garden City, NY: Doubleday Anchor Books.

Schoenfeld, A. H. (1987). What's all the fuss about metacognition? In A. H. Schoenfeld (Ed.), *Cognitive science and mathematics education* (pp. 189–215). Hillsdale, NJ: Lawrence Erlbaum Associates.

Schuder, R. T. (1993). The genesis of transactional strategies instruction in a reading program for at-risk students. *The Elementary School Journal, 94*(2), 183–200.

Sizer, T. (1984). *Horace's compromise: The dilemma of the American high school.* Boston, MA: Houghton-Mifflin.

Slavin, R. E. (1987). Cooperative learning: Where behavioral and humanistic approaches to classroom motivation meet. *The Elementary School Journal, 88*(1), 29–37.

State of Maryland. (1989). *The report of the governor's commission on school performance.* Annapolis, MD: Governor's Commission on School Performance.

State of Maryland. (1992). Code of Maryland Regulations (COMAR 13A.03.02.01), 19:19 *Maryland Register* 1790. Annapolis, MD: Division of State Documents.

U.S. Department of Labor. (1991, June). *What work requires of schools: A SCANS report for America 2000.* Washington, DC: U.S. Department of Labor, Secretary's Commission on Achieving Necessary Skills.

Wingspread Group on Higher Education. (1993). *An American imperative: Higher expectations for higher education.* Racine, WI: The Johnson Foundation, Inc.

Chapter 4

Validation of SCANS Competencies by a National Job Analysis Study

Beverly E. Nash
Robert C. Korte
American College Testing

CHANGING WORKPLACE DEMANDS: THE SKILLS GAP

Just as the industrial revolution dramatically changed the nature of work in the 19th century, technological growth and global competition are transforming organizations and the role of workers today (Offermann & Gowing, 1990). The advent of mass production in the late 19th century eventually created large, centralized organizations that produced increasing numbers of goods for the domestic market. Jobs were defined narrowly, and worker tasks became routine. Work skills were primarily learned on the job, with only the basic reading, writing, and math skills being taught in the schools (National Center on Education and the Economy [NCEE], 1990a).

The workplace of the early to middle 1900s was based on the Frederick Taylor approach to management, which emphasized piece-meal work in a highly controlled environment (Taylor, 1911). Performance was measured in terms of observable outcomes such as rate, time, and errors (Gerhart & Milkovich, 1992). The Taylor model served to standardize work and observable performance, and its effects on workers were significant, yet restrictive. Employees received little recognition and few rewards, such as bonuses or flexible work schedules, for their efforts and service to organizations. The organizational structure was hierarchical, therefore workers assumed little responsibility for their assignments and had virtually no autonomy in their jobs (Fisher, 1994). Higher order thinking skills, such as problem solving, critical thinking, and creativity, were not deemed essential for performing tasks (Stasz, Ramsey, Eden, Melamid, & Kaganoff, 1995). Technology existed in simple form, yet most workers had no reason to understand how the equipment around them worked, depending on maintenance and repair personnel to anticipate and solve problems (NCEE, 1990a).

The Taylor model was, however, successful because it fit industry's fledgling goal of producing more products through larger production (Fisher, 1994). The larger the industry, the more products could be manufactured and sold to an expanding domestic market. This philosophy made the nation rich and profitable because it fostered domestic trade and capital investment in large scale, mass manufacturing production (NCEE, 1990b). For half a century, the United States was the undisputed leader in setting standards of products, production, and performance.

Sixty years later, the globalization of the marketplace has directly impacted the economic position of the United States of America, which no longer is alone in setting the workplace standards. Businesses and industries in the United States must now compete with corporations in other countries to capture shares of the domestic and global markets (Cascio, 1986). To do this, American industries are changing their focus from production itself to producing quality goods and services in a timely manner that meet the tailored needs of customers (Appelbaum & Rosemary, 1994). This has produced a shift from mostly domestic trading to an integrated global marketplace (Offermann & Gowing, 1990). According to economist Anthony Carnevale (1991), the movement toward economic globalization is rooted in several factors: (a) the equalization of worldwide media, marketing, and travel; (b) higher incomes that are creating a rise in international markets for national products; (c) advances in communication and transportation technology that allow multinational companies to enter international markets without leaving their country; and (d) decentralized worldwide production and sales efforts (pp. 17–18). Carnevale (1991) also notes that information-based technologies are critical to increasing competitiveness in the global market through production and service quality.

This philosophy, which replaces the Taylor model (Taylor, 1911), is based on a new approach called "high performance" (Galagan, 1992; Gephart, 1995). American companies that follow a high-performance model are thriving, while those still working with the large scale, mass production model are struggling to compete (U.S. Department of Labor, 1993). The new models of high-performance companies share a number of characteristics, according to sources such as the National Center on Education and the Economy (1990a, 1990b) and the Department of Labor, Office of Work-Based Learning (1992a). Some of those characteristics, in abbreviated form, include: strong goal orientation, focus on continuous improvement, emphasis on quality and performance, decentralization of authority and responsibility, free flow of communication along the chain of command, integration of work into jobs rather than separation into discrete tasks, more direct worker control over job outcomes, more worker responsibility for quality, more worker flexibility due to job rotation and training, more coaching activities by supervisors and managers, greater worker opportunities for product development, more comprehensive formal and informal education programs, and compensation based on performance.

By the turn of the 20th century, national and business leaders envision that the United States of America will have even more high-performance workplaces (Gephart, 1995). The change to high performance is forcing organizations across the country to decentralize their chains of command and redefine the responsibilities of workers (King, 1995). To develop and maintain a productive and competitive edge, companies are orienting and grouping employees at all levels of the organizations to emphasize teamwork, problem solving, critical thinking, the ability to respond rapidly to change, and the ability to work effectively with little direct supervision (Stasz et al., 1995). This has led to the realization that American workers must now be capable of learning new skills and adapting their abilities as jobs are redefined and typically expanded by the economic and organizational models of the times (Carnevale, Gainer, & Meltzer, 1988).

Unfortunately, most of these new skills and abilities that are now being demanded were not emphasized in the past, learned on the job, or taught in the classroom, which has left workers lacking in the full range of skills necessary to operate effectively in high-performance jobs (NCEE, 1990b; Office of Technology Assessment [OTA], 1995). In many cases, workers must seek learning opportunities on their own in order to upgrade their skills and advance within their respective companies (Ferguson, 1995). In response, a number of businesses and industries are upgrading their training programs to meet these new demands, but these programs are few and slow in keeping up with industry's rapid pace.

Therefore, the problem lies in the urgent need to ensure that entry-level employees have the skills needed to assume job responsibilities in a time of change and that incumbent workers quickly receive the necessary training as soon as the need becomes clear (Ferguson, 1995; NCEE, 1990a). These needs place a large responsibility on the educational system to identify and teach employability/workplace skills to students, who must enter a variety of occupations, and on business organizations, who must develop and revise training curricula to incorporate essential workplace skills (Reynolds, 1994).

IDENTIFICATION OF SCANS COMPETENCIES

The identification of skills necessary for entry-level workers to succeed in today's most competitive workplaces is currently being investigated (Nash & Korte, 1994). A number of reports and studies have emerged during the past decade. One of the most provocative reports is the Secretary's Commission on Achieving Necessary Skills (SCANS, 1991) report, which defined competencies and foundation skills necessary for worker success and made recommendations as to how those skills could be infused in the nation's schools and workplaces. The SCANS initiative is now focused on creating valid assessments of these competencies in order to gauge the progress of workers and students (Nash & Korte, 1994).

One of the most important goals of SCANS, which convened in 1990 at the request of the Secretary of Labor, was to develop a taxonomy of higher order workplace competencies and foundation skills that promoted a high-performance economy. The SCANS work, in essence, served as a wake-up call for the nation to take a much closer look at its educational and training systems and approaches to assisting students and learners in making effective transitions into the world of work. Its message, which recognizes the need for, and value of, a highly skilled workforce to compete in a global marketplace, was also targeted to employers. From the outset, in reports such as *What Work Requires of Schools: A SCANS Report for America 2000* (1991) and *Learning a Living: A Blueprint for High Performance* (1992), SCANS has emphasized the need for students to learn not only the basic academic skills (i.e., foundation skills) but also the competencies required in a high-performance workplace (i.e., workplace know-how).

To this end, SCANS was successful in facilitating communication between educators and employers in an effort to bridge the skills gap. SCANS has provided a blueprint for guiding educators as they develop curricula that includes employability skills, and it has informed employers about the competencies they should expect entry level workers to possess upon employment. The SCANS' work has been one of the first initiatives taken to assist learners in making successful transitions from school to the workplace. The development process included interviews with employers, educators, labor leaders, and policymakers about the skills entry level workers should possess for success in high-performance workplaces. Existing skill lists were also reviewed, and workers were interviewed (approximately 4 in each of 50 occupations) to determine the application of the taxonomy elements to their jobs. The SCANS competency framework in Table 4.1 illustrates the five workplace competencies as they are needed by workers in high-performance workplaces. The foundation skills are ones that are supportive of the competencies.

According to SCANS, if this system is nationally recognized and provides voluntary assessment, educators will be encouraged to include the SCANS competencies and foundation skills in their curricula; students will be eager to learn and demonstrate the competencies if they are connected with successful work outcomes; and employers will acknowledge and recognize students and learners who are certified by SCANS in workplace know-how

TABLE 4.1
The SCANS Competency and Foundation Skill Framework

Workplace Know-How

The know-how identified by SCANS is made up of five workplace competencies and a three-part foundation of skills and personal qualities that are needed for solid job performance. These are:

WORKPLACE COMPETENCIES—Effective workers can productively use:

- Resources—They know how to allocate time, money, materials, space, and staff.
- Interpersonal Skills—They can work on teams, teach others, serve customers, lead, negotiate, and work well with people from culturally diverse backgrounds.
- Information—They can acquire and evaluate data, organize and maintain files, interpret and communicate, and use computers to process information.
- Systems—They understand social, organizational, and technological systems; they can monitor and correct performance; and they can design or improve systems.
- Technology—They can select equipment and tools, apply technology to specific tasks, and maintain and troubleshoot equipment.

FOUNDATION SKILLS—Competent workers in the high-performance workplace need

- Basic Skills—reading, writing, arithmetic and mathematics, speaking, and listening.
- Thinking Skills—the ability to learn, to reason, to think creatively, to make decisions, and to solve problems,
- Personal Qualities—individual responsibility, self-esteem and self-management, sociability, and integrity.

Source: Learning a Living: A Blueprint for High Performance. The Secretary's Commission on Achieving Necessary Skills, U.S. Department of Labor, April 1992, p. 6.

(SCANS, 1992). SCANS' goal was to create an assessment system based on performance levels for the competencies students or learners need in order to function effectively upon entering the workplace. Instead of determining a unitary standard for attaining the five competencies, SCANS proposed the following five levels of proficiency: (a) preparatory, (b) work-ready, (c) intermediate, (d) advanced, and (e) specialist. These proficiency levels could be reported on a student's or learner's resume for future employment opportunities.

DEVELOPMENT OF THE NATIONAL JOB ANALYSIS STUDY

In July 1992, American College Testing (ACT) was awarded a contract by the U.S. Departments of Labor and Education to develop assessments for the five SCANS competencies.

Evidence of the SCANS Skills

Early in the ACT assessment development process of the SCANS competencies, it became apparent to both the project's Steering Committee and
its Planning Committee that there was insufficient information about the
SCANS competencies to develop valid and reliable assessments. The steering and planning committees were convened to advise ACT and project
sponsors on how to best develop reliable and valid assessments that would
be consistent with the goals of SCANS. The planning committee was
charged with establishing the assessment framework, and the steering committee guided consensus building with the planning committee, reviewed
project outcomes, and made recommendations. It was recommended in
January 1993 that the ACT research team turn its attention from the
assessment development process and focus on developing a methodology
for collecting further empirical evidence for the SCANS framework. The
result was the National Job Analysis Study, designed to identify the content
requirements for the SCANS assessments (Nash & Korte, 1994).

The primary goal remained: to develop criterion-referenced assessments
based on an empirically established taxonomy of workplace competencies.
As the basis for a criterion-referenced assessment, the taxonomy must serve
as a precise, accurate, and predictive framework on which to measure what
an individual can and cannot do irrespective of the performance of others.
Criterion-referenced assessments require that the content of the SCANS
framework be valid so that test items yield an accurate interpretation of
what an examinee's performance means. If the specifications are imprecise,
incomplete, or inaccurate, or if the items are not congruent with the specifications, it is not possible to make credible interpretations of the results
(Popham, 1984). The costs of these deficiencies can be high. If the test
framework is not reflective of what is actually required in a work setting,
people may either be denied an opportunity for which they are qualified,
or placed in jobs for which they are not sufficiently prepared. The resulting
losses to individuals and organizations can be significant.

The National Job Analysis Study (Nash & Korte, 1994) is designed to
ensure that the assessment framework represents actual job behaviors that
are essential in high-performance workplaces across a wide range of occupations and geographic areas. The study helps to establish the relationship
between workplace competencies and basic foundation skills; and according to convention, the associated knowledge required to perform a competency. Observable and demonstrated physical and mental skills and abilities needed to acquire and utilize skill and knowledge will be identified
(Uniform Guidelines, 1978).

Using standard survey methodology (Gael, 1983), the study is designed
to identify a common core of job behaviors of current workers in a variety

of occupations and workplaces across the country. In addition, the study seeks specific information from organizations to determine the extent to which they meet the criteria for "high performance." The study will yield a taxonomy of core job behaviors at various job tenure levels (e.g., at entry, after 6 months, after 1 year) in numerous occupations in workplaces that will later be linked to high-performance practices. The finalized framework will establish a definitive foundation on which to base assessments, work training programs, educational curricula, and comprehensive descriptions of job requirements.

The methodology for the National Job Analysis Study is based on acceptable standards used for job analysis (Gael, 1983, 1988). However, the outcome will be different. Traditional job analysis (Gael, 1983, 1988) gathers information on all of the behaviors inherent in a particular occupation and determines what it is about the occupation that separates it from all others. The traditional outcome highlights the *differences* among occupations. The National Job Analysis Study uses the same methodology to gather information on what job behaviors are *similar* across numerous occupations and does not seek to identify job-specific tasks within individual occupations. The generic focus of the study illustrates the dynamic nature of today's jobs and reflects how critical it is for workers to have generalizable skills they can apply as changes occur within and across organizations and occupations.

THE NATIONAL JOB ANALYSIS STUDY

The National Job Analysis Study focuses on job, or work-oriented, behaviors that describe observable physical activities workers perform. Once behaviors are identified, it is relatively easy to determine the requisite knowledge, skills, and abilities. But in contrast to the more traditional task-based job analysis approach, the National Job Analysis Study seeks to identify cross-occupational, rather than job-specific, work behaviors. It focuses on specific work-related activities that are performed across jobs. This approach does not deny the cognitive basis of behavior, but from a measurement perspective, it is concerned foremost with behaviors that can be directly observed. Cognition/mental processes will be ascertained through a behavioral analysis to identify the knowledge, skills, and abilities required to perform the behaviors (Lesgold, Lajoie, Logan, & Eggan, 1990). As such, the study includes the development of behavioral statements that are generalizable across occupations. These common, or generic, behaviors will be derived from task statements contained in job analysis information databases and from the SCANS competencies and foundation skills. The study will further describe these behaviors in terms of levels of proficiency and the knowledge, skills, and abilities (KSAs) that are necessary for each level. A complementary approach to the same goal has been provided by Stasz et al. (1995).

The National Job Analysis Study is conducted in two phases. Phase 1 is designed to identify a broad set of common core job behaviors. Phase 2 is designed to verify and analyze those core behaviors relative to high-performance workplaces, relative weighting, proficiency levels with their associated knowledge, skills, and abilities, and the relationship of the behaviors to one another. Each phase will include the use of surveys to collect job analysis information from thousands of incumbent workers.

PHASE 1

Phase 1 involves two surveys. Survey 1, which also serves as a pilot test for the instrument, uses the initial empirical attempt to determine cross-occupational generalized behaviors and to identify behaviors that may have been omitted from the survey. To this end, job incumbents have the opportunity to add and rate generalized behaviors they perform that do not appear on the survey. Survey 2 cross-validates the core set of behaviors identified by Survey 1, including behaviors added by incumbents. The research team at ACT believes the two-survey approach will provide the most sound basis on which to develop assessments of workplace competencies.

The first phase is designed to produce the following outcomes:

1. An empirically derived set of core job behaviors common across occupations and related to high-performance workplaces. This set will consist of common core job behavior statements organized in a taxonomy. The set of core behaviors will form the basis for further analysis in Phase 2, at which time the behaviors will be described in terms of levels of proficiency. Necessary knowledge, skills, and abilities will be identified for each level. At that point, the content for making decisions about people and resources will have been identified, and assessments can be developed. The taxonomy will also be a vehicle for describing occupations and relating occupational standards in a language common across industries.

2. A matrix of core behaviors by job tenure levels (e.g., at entry, after 6 months, after 1 year). This matrix will provide empirical evidence for identifying the core behaviors needed at various levels of job tenure.

3. Models of the relationships between all core behaviors. This outcome is an exploratory model for relating all core behaviors from the study. Knowledge about these relationships will be of significant value to teachers and trainers who are preparing individuals for the workplace.

Specific Procedures for The National Job Analysis Study

A panel of experts was formed at the outset to serve throughout the course of the project. The panel is made up of members who have job-relevant experience in performing and managing jobs; members with expertise and experience in industrial–organizational psychology education; experts in education and training; and members with expertise and experience in testing and in developing causal models. One duty of the panel is to review all materials and assist with classification of the behaviors into a meaningful taxonomic system.

Identifying Common Job Statements

The research team conducted a search of ACT's Realistic Assessment of Vocational Experiences (RAVE) database (American College Testing [ACT], 1992), which contains information from the more than 12,000 jobs listed in the Dictionary of Occupational Titles (DOT) updated through 1991 (U.S. Department of Labor, 1991). The information in the database was electronically queried on the basis of key words to locate task statements containing those words. The search began with the use of 157 action verbs identified by Gael (1983, p. 60) as ones frequently included in task statements (e.g., adjusts). Other verbs were added as appropriate (e.g., communicates). The task statements were clustered by verb, and the corresponding DOT code numbers were preserved so that the task statements could be traced to the job or jobs from which they came.

The research team conducted a parallel search process in the database of job analysis information developed in 1992 for the Fort Worth, Texas, Independent School District and Chamber of Commerce (Fort Worth Independent Schools, 1992). This database contains task statements from approximately 240 jobs in the Fort Worth area. These statements, which are coded by DOT codes, were compared to those contained in the RAVE database to ensure that the job analysis information for Survey 1 is as current and as comprehensive as possible. Other databases, such as the one developed by the Office of Personnel Management (OPM, 1993), containing generalized work activities in the federal sector, were also consulted to ensure comprehensiveness.

Compiling, Translating, and Reviewing Lists of Behaviors

After common job task statements were identified, the research team reviewed the task statement clusters for meaning and level of specificity. Team members translated a sample of the statements into common, or generic, behaviors written in a common language (Nash & Korte, 1994) and at an appropriate level of specificity. The iterative process developed by ACT for writing the behavior statements includes the following steps:

- Group task statements by similarity of verbs.
- Group task statements by similarity of verb objects.
- Group task statements by similarity of outcomes.
- Group task statements by similarity of knowledge or physical skills required.
- Write behavior statements for task groups.

A common language was developed to ensure that the behaviors were as free of occupation-specific jargon as possible. It was imperative to use terminology that would be consistent across occupations. For example, a piece of equipment needed to be described so that it could be recognized by all occupations that use it. This process also included the assessment of specificity. The behavioral information could not be so general that it provided no new information (e.g., uses tools), nor could it be so specific that it would only be performed by select occupations. The goal throughout was to achieve a level of specialty for the behaviors that would be informative without excluding occupations that might perform it because of its unique wording and selectivity for particular occupations.

An example of the process is explained for the verb "signal." By querying the DOT on the verb signal, the following task statements from the DOT were found. For the position Construction Worker 1, the use of the verb was in terms of signaling operators of construction equipment to facilitate alignment, movement, and adjustment of machinery. For the position of Machinery Operator, signal was used in terms of notifying an Overhead Crane Operator to lower basic assembly unit, to bedplate and align unit. For the position of Tanker, signal was similarly used to notify an Overhead Crane Operator to lift load of material over tank and guide load into tank.

In keeping with the criteria used to combine the tasks into generalized statements, it was noted that in all cases the object of the jobs noted required signaling another person, and the outcome was to align equipment. The knowledge required to perform the tasks is the ability to align machinery by specified criteria, and the skill required is physical/short vocal signals. Taken together, a behavioral statement covering the set of tasks for the verb signal might be the following: Signals other persons regarding the alignment of equipment or machinery.

After the sample task statement groups were translated, ACT's research team reviewed the behavior statements for overlap, combining or eliminating statements as necessary. The statements were then edited to simplify the language as much as possible.

Using a modified Delphi technique, ACT's research team met with the panel of experts to reach a consensus regarding the sample behavior statements. This procedure was repeated until a consensus was reached by all members of the panel as to the appropriate level of specificity in which to write all behavior statements (Gael, 1983, 1988).

Once consensus was obtained, ACT's research team developed and refined behavior statements for all task groups, consulting experts in the corresponding occupations as needed. SCANS and other sources of generalized behaviors were compared to the content of the task-group behaviors. The panel of experts reviewed the full list of behavior statements to ensure that they reflected commonality of language and the level of specificity agreed upon through the consensus process. The panel also evaluated the behavior statements for content and comprehensiveness, expression in behavioral terms, and reading level, which must be easily understandable for a vast majority of job incumbents. Behavior statements were revised or added as necessary before being compiled for Survey 1. This resulted in 214 behavioral statements.

Developing Rating Scales

Survey 1 includes two scales (importance and frequency) for rating the common job behaviors: the SCANS illustrative behaviors, and the foundation skill behaviors. Figure 4.1 shows the rating scales for Survey 1. Both scales were 7-point scales containing absolute, rather than relative, anchors. The importance scale anchor points were from 1 (this behavior is of no importance to doing my job), to 7 (this behavior is essential to doing my job). The frequency scale anchor points were from 1 (I do this behavior on my job approximately once each year), to 7 (I do this behavior on my job at least once each hour). Because the survey was national and ratings were to be provided by incumbents in different jobs, the rating scale anchors were absolute (Harvey, 1991). It would have been inappropriate to use a scale that asked how much time is spent on a task relative to the time spent on other tasks. It could have produced ambiguous results when interpreted across jobs. The questions might have required a great deal of interpretation on the part of job incumbents and may have related only

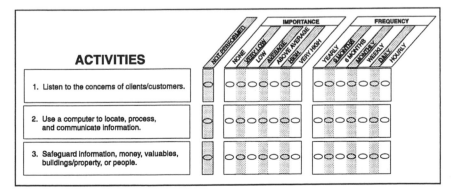

FIG. 4.1. The rating scales for the National Job Analysis Study Survey 1.

to a specific context. Each survey included specific instructions on how to interpret the rating scales.

Constructing Survey 1

Survey 1 was constructed from the list of common job behaviors, SCANS illustrative behaviors, and foundation skill behaviors finalized by the panel of experts.

The finalized behavior statements reflect the level of specificity determined by the panel of experts. The survey items were intermingled rather than grouped by organizing dimensions or duty categories. The rating scales of importance and frequency corresponded to each survey statement, and a blank section was provided so that job incumbents could list other general tasks they perform that are not included in the survey. Survey 1 also contained demographic statements on factors such as time on the job, job title, wages, gender, ethnicity, and age. The specific list was developed by the ACT research team in consultation with the project consultants and panel of experts. The demographic variables were treated as covariates for the data analysis. Because of the number of items, two forms of the survey were constructed, each containing half of the items.

Developing the Sampling Frame

To develop the survey sample, ACT's research team identified a representative sample of occupations, and then selected a sample of organizations in which workers in those occupations are found.

Sample of Occupations. Because the intent of this study is to identify behaviors that are common across jobs, the selection of occupations to be sampled is critical. It would be very costly and unnecessary to include all of the more than 12,000 occupations covered by the DOT. However, the sample must be representative of the total set of occupations so that the results of the study will be valid.

ACT's research team has based the selection of a representative set of occupations on three criteria that, together, are most likely to yield an adequate sample from which to generalize to most occupations in the country. The three criteria are:

• *Commonality.* ACT's research team selected occupations that as a group employ 80% of workers as counted by the Occupational Employment Statistics (OES) system (U.S. Department of Labor, 1992b). This will ensure that the behaviors identified in the study are performed by a substantial majority of the workforce. The sample was large enough to provide stable estimates regarding the behaviors in question. Based on the list of occupations described in the OES classification system data, the ACT research team determined that 140 occupations are required to meet this criterion.

• *Representativeness Across Occupations.* The sample for this study must cover the full range of occupations regardless of the relative size of occupational families. The ACT research team chose representativeness as a criterion to ensure that the sample is not restricted to a narrow set of occupations that employ large numbers of people. ACT used its "World-of-Work Map" (Prediger, 1976) to analyze the set of occupations selected by the first criterion and to add occupations in areas that are proportionally underrepresented.

• *Growth.* ACT's research team selected occupations that are expected to grow most rapidly in the next decade (U.S. Bureau of the Census, 1992). The sample included the 25 occupations with more than 100,000 workers that are expected to experience the most real growth (e.g., medical assistants, travel agents). If these occupations were not part of the sample derived from the first two criteria, they were added.

The research team's selection process for the occupational sample was iterative. The first criterion generated 140 occupations. This set was compared to the job families on the "World-of-Work Map" (American College Testing, 1988; Prediger, 1976) and to the list of the 25 fastest-growing occupations. The resulting sample contained 164 occupations, all of the 25 fastest-growing occupations, and at least one occupation in each of the 23 job families on the "World-of-Work Map." This representativeness indicates that the set of occupations is well distributed and, therefore, taps a breadth of skills that would represent all aspects of the labor market.

Sample of Organizations and Job Incumbents. The ACT research team selected a sample size of 12,000 job incumbents from 6,000 organizations to respond to Survey 1, and will select 6,000 job incumbents from 3,000 organizations to respond to Survey 2. (Because of the large number of items, Survey 1 was split into two forms and 6,000 job incumbents were asked to respond to each form.) The sample size was considered sufficient to make generalizations about cross-occupational work behaviors. However, the sample size for this study would not be sufficient to make inferences about specific occupations. The number of organizations in the sample was large enough to include a variety of types of organizations in diverse geographic areas. The large number of workers sampled increased the chances that the overall responses from incumbents represented an honest and accurate portrayal of the behaviors performed in their occupations. The sample included organizations from the private sector and the non-profit/public sector. The sample also included workplaces reported to be high performance throughout the full spectrum of industries. Given the lack of a comprehensive and mutually agreed upon definition of high performance, organizations or industries that reported themselves to be high-performance workplaces were included in the study. Many of these organizations were Malcolm Baldridge recipients, ISO 9000 registrants, or other quality/service award holders.

The ACT research team further chose organization size as an explicit stratification variable, which means that a certain number of large, medium-sized, and small organizations were included in the sample. In addition, implicit stratification of geographic region was accomplished through ordering the stratum by ZIP code and conducting a systematic sampling within the stratum.

To conduct the National Job Analysis Study, job incumbents were identified within the 164 occupations to be sampled. ACT is not aware of any databases that contain lists of job incumbents by occupation. However, there are databases that contain the names of organizations and contact persons indexed by Standard Industrial Classification (SIC) codes, which classify organizations by type of activity (U.S. Executive Office of the President, 1987).

The number of job incumbents chosen for each occupation was proportional to the total number of incumbents in each profession sampled. Contact persons in each of the selected organizations chose employees from the target occupations to complete the survey. Each contact person was informed of the occupational title to be selected and received a set of explicit instructions for selecting two survey respondents in that occupation. The selection of incumbents was random rather than based on specific criteria so that the responses were as representative as possible of the whole population of job incumbents. The contact persons were asked to select only job incumbents, not supervisors, in the target occupations. This helped to ensure that the surveys were completed by employees who actually perform the jobs and were therefore the best judges of the importance and frequency of the relevant behaviors. Contact persons were also responsible for distributing, collecting, and returning the surveys to ACT.

Administration and Analysis of Survey 1

The ACT research team conducted four separate mailings and follow-up telephone calls during the administration of Survey 1. All mailings and telephone calls were directed to the contact person within each organization in the sample.

Descriptive statistics were computed on both the importance and frequency ratings for items in Survey 1. A multiplicative composite index (importance × frequency) was also calculated on the raw importance and raw frequency ratings. Typically, job activities (i.e., tasks/behaviors) are rated on either a single rating scale, or on more than one scale (e.g., importance and frequency). When more than one scale is rated, a composite index is usually computed (Cascio & Ramos, 1986; Harvey, 1991;

Sanchez & Fraser, 1992). The means and standard deviations associated with the composite index were used to help determine the core set of behaviors common to the jobs surveyed. Following analyses, all behaviors from Survey 1 were rank ordered on the basis of their criticality. The criticality index reflected the composite found when importance and frequency ratings were multiplied together for each work activity or behavior in the survey. A cut score was then established to identify the core set of behaviors common to the occupations.

Initial cut scores clearly indicated which behaviors should be kept and which should be dropped. Each member of the panel of experts then established cut scores for the behaviors that were not determined as ones to keep or drop. Panel members then met to discuss their decisions and reach a consensus. To be inclusive rather than exclusive, the experts accepted the lowest cut score possible for including behaviors above that point. The panel further elected to keep some behaviors that would have been dropped if only the cut point had been used.

Constructing Survey 2

The research team at ACT will construct and administer a second survey that contains the core job behaviors identified from Survey 1. Survey 2 will be administered to a new sample of 3,000 organizations and 6,000 job incumbents in the same occupations as those sampled in Survey 1. In addition to the importance and frequency scales used in Survey 1, Survey 2 will include a need scale for incumbents to assess *when* each behavior is needed.

This scale will provide valuable information concerning when the behaviors are important at the entry level, within the first year on the job, and after one or more years on the job. The behaviors will then be sequenced on the basis of need, providing the ACT research team with a logical approach to writing and developing assessments that target specific behaviors necessary at different points in time on a job.

Analyzing Data and Assigning Behaviors to Dimensions

Data analysis of Survey 2 will help determine the final core behaviors to include in the workplace competency taxonomy. The panel of experts will again determine the cut score for the final behaviors, using the same method they used for the core behaviors from Survey 1. Additional data analyses of Survey 2 will form the basis for assigning behaviors to the taxonomy and developing models of the relationships between the behaviors. The utilization of cluster and factor analyses will inform panel decisions about establishing category dimensions and relating the behavioral information in meaningful ways.

*Data Analysis for Dimension Assignment
and Model Development*

Cluster analysis (Aldenderfer & Blashfield, 1984) will empirically group similar core behaviors from Survey 2. The results will be used to inform the panel of experts, who will make the final decisions on the assignment of behaviors to dimensions. A dimension is a category heading that describes the context of the behavioral information. An example of a possible dimension from the study could be "managing resources," and for that dimension there would be behaviors from the survey that correspond to that activity. In essence, dimensions provide a logical occupational category for the behaviors in the survey. The panel members will split randomly into two equal groups, and each group will assign clusters of behaviors to dimensions that best describe the clusters. After the two groups make their initial assignments, all panel members will reconvene to discuss the dimension assignments and reach a consensus on the final taxonomy.

Factor analysis (Gorsuch, 1974) of the Survey 2 information will provide a starting point for developing *a priori* hypothesized models of the relationship of the behaviors to the determined dimensions. To this end, confirmatory factor analytic models (Mulaik, 1992) will be developed and analyzed to provide preliminary information for further behavioral modeling efforts. These efforts will be conducted after assessment items are developed to measure the dimensions arrived at through the study.

PHASE 2

After the taxonomy of behaviors is identified, Phase 2 begins with the establishment of proficiency levels for the behaviors in each taxonomic dimension.

Establishing Proficiency Levels

Data analysis from Survey 2 will provide background for the initial step in rank ordering the core set of behaviors according to when they are needed on the jobs sampled (e.g., at entry, after 6 months, after 1 year). The core cross-occupational behaviors, ranked on a continuum from entry to one year and beyond, will be scaled along relevant anchor points using a format such as the Behaviorally-Anchored Rating Scale (BARS) developed by Smith and Kendall (1963).

BARS will be developed from core behaviors in each dimension, the relationships between all of the core behaviors, job tenure information, and input from the panel of experts. There are six steps in the development of BARS for specific dimensions:

Step 1: Organize the data into dimensions based on the NJAS outcomes and the Panel of Experts' judgements.

Step 2: Scale the behaviors along a continuum on the basis of criteria such as complexity, level of details, and skill requirements.

Step 3: Develop full level descriptions based on behaviors defining a level.

Step 4: Identify the knowledge, skills, and abilities (KSAs) for each level description.

Step 5: Pilot the level and KSAs descriptions.

Step 6: Validate the BARS.

Once the behaviors are scaled hierarchically, the panel of experts will establish elaborated descriptions for the proficiency levels and identify the knowledge, skills, and abilities (KSAs) associated with the behaviors for each proficiency level. Instrumentation will be developed and administered to ensure that the level assignments and KSAs are reliable and valid.

For example, the dimension "Informs Clients and Customers" represents behaviors at the lowest level of complexity. Directing customers to the location of a product or service can be as simple as telling a customer that the product is in aisle 3. Comparing and contrasting various products and services represents more conceptually complex behaviors. Proficiency levels are based on the assumption that successful attainment of the behaviors and KSAs is hierarchical; that attainment of level 3 assumes successful attainment of the elements involved in levels 1 and 2. If level 3 cannot be performed, it is assumed that level 4 cannot be performed.

At the conclusion of Phase 2, the project will have defined workplace behavioral domains that are common across occupations, identified levels of proficiency within each domain, and identified the knowledge, skills, and abilities for each level of a domain. These domains will represent, once validated, an invariant set of scales. The assessments will then be designed to measure levels of performance on each scale.

CHARACTERISTICS OF HIGH-PERFORMANCE WORKPLACES

It was also imperative to the National Job Analysis Study that the competencies be related to high-performance workplaces, because these organizations value the competencies and have workers who possess them. It also became clear to the ACT research team that it would be necessary to first define the concept of a high-performance workplace in order to relate the SCANS competencies to this type of workplace and others.

It was also recognized that there are many definitions of a high-performance workplace. Most of the definitions are based on the work of W. Edward Deming (Joiner & Gaudard, 1990), quality and performance standards such as the Malcolm Baldridge Quality award (Heaphy, 1992), and ISO-9000 standards registration (Aguayo, 1990; Peach, 1994; Spechler, 1993). It remains difficult to reach a consensus around a single definition for a high-performance workplace. There appear to be various characteristics or factors that seem to be operating in different contexts or workplaces (i.e., manufacturing versus service). In all likelihood, there are numerous approaches to defining the concept of a high-performance workplace, and given a finite set of characteristics, there may be a subset functioning within and across high-performance workplaces. The research challenge is to identify a comprehensive set of characteristics or factors that can stimulate the gathering of factual information about workplaces, so that a profile of characteristics can be created to establish an organization's status as a high-performance workplace.

To address this issue, the U.S. Department of Labor, Employment and Training Administration (ETA), and the U.S. Department of Education, National Center on Education Statistics (NCES), made funds available for a collaborative project to identify the key characteristics of high-performance work environments (Nash & Hausner, 1994). The project was conducted by American College Testing (ACT), the National Alliance of Business (NAB), and the Department of Labor, Office of the American Workplace (OAW). A group of representatives from business, industry, and government identified a list of 36 characteristics they considered essential in their own definitions of high performance. This list was incorporated into an interview protocol that was administered to executives of 44 organizations (approximately half manufacturing and half service) that were considered to be high performance. The protocol asked executives to rate the 36 characteristics in terms of importance, and also to provide their own definitions of what high performance entailed in their own organizations.

From analysis of the importance ratings and key elements of the executives' own definitions, eight characteristics emerged that were considered most essential. These characteristics, which appeared in the top tiers of rankings but were not assigned a priority order, are:

1. Focus on quality
2. Focus on customer satisfaction/input
3. Flexible culture/openness to change
4. Leadership with clear vision/commitment
5. Training/continuous learning/education
6. Information sharing

7. Profitability
8. Productivity

Several of these characteristics also were mentioned most frequently in the definitions provided by the group of business, industry, and government representatives. Manufacturing and service organizations differed somewhat on the relative importance of particular high-performance characteristics. Further study will help determine the significance of these differences. The results of this collaborative study will be used to inform the development of the High-Performance Workplace Survey for measuring high performance to be administered in conjunction with the Behaviorally-Anchored Rating Scales (BARS) developed from the outcomes of Phase 1 of the National Job Analysis Study.

OUTCOMES OF THE NATIONAL JOB ANALYSIS STUDY

The National Job Analysis Study (NJAS) and associated Behaviorally-Anchored Rating Scales (BARS) were initiated for the purpose of developing assessments. However, the NJAS/BARS information will be valuable to a number of other public and private initiatives whose goal is to equip workers with the skills they and their organizations need to compete successfully. The common language of the NJAS/BARS makes it possible to compare occupations, cluster them into broader areas, and develop profiles of the requisite behaviors, knowledge, skills, and abilities. These profiles can be used to facilitate worker transitions from one job to another, and to infuse workplace context into the K–12 educational system so that students may successfully make the transition from school to work. These elements are the building blocks for an integrated system that prepares workers for challenging jobs in high-performance environments and for the lifelong learning those jobs require.

OCCUPATIONAL CLUSTERS AND SKILL PROFILES

One problem with occupational information to date has been the idiosyncratic nature of the language used to describe the job content (Nash & Korte, 1994). In many cases, this information is only meaningful to the occupations described within a particular industry context. The lack of a standardized language not only makes it difficult to compare occupations across industry contexts, but it impedes efforts to train people in more general occupational areas and to help people make the sometimes necessary transitions from one occupation to another.

The NJAS/BARS, with its common language across occupations and its emphasis on behaviors, knowledge, skills, and abilities, provides the mechanism for identifying similarities and differences among occupations and across industries. This capacity makes it possible to develop profiles of occupations or occupational clusters that include generalized work behaviors, specific job skills, and standards of performance.

Linking Occupational Information

Industry- and occupation-specific information can embellish the skill profiles in terms of similarities and differences within and across occupational areas and industries. For example, a profile for a research technician in one industry can be compared to the profile for a research technician in another. One would expect these jobs to have similar profiles of behaviors and KSAs. But differences can also be noted, and further determinations can ascertain whether the differences are due to the industry in which the occupation is performed (context) or to true occupational differences.

The profiles can incorporate industry- and/or occupation-specific information because the NJAS/BARS generalized behaviors are closely linked to such information. As described earlier, most of the NJAS behaviors were systematically developed from specific task lists in the DOT. Each behavior can therefore be traced to the specific job tasks from which it was derived, allowing movement back and forth from job-specific tasks to generalized behaviors. As such, the common language of the NJAS provides the link between cross-occupational and occupation-specific information.

One can analyze generalized work behaviors identified by the NJAS to job-specific tasks, or statistically relate specific tasks to it. An example might be the behavior of comparing and contrasting products and services for a customer. A waiter in a restaurant might be asked by a patron to discuss several wine choices, while a concierge in a hotel might be asked by a guest about the relative merits of nearby restaurants. Both workers are performing job-specific tasks with associated KSAs. But they are also performing at a particular proficiency level of generalized behavior that requires its own KSAs. It is anticipated that the KSAs associated with the proficiency levels in the NJAS will be highly correlated with those for the job-specific tasks.

Profiles Common Across Industries

The NJAS/BARS process for relating job-specific to cross-occupational information forms the basis for developing skill cluster profiles with common elements across industries. The profiles would include the generalized work behaviors from the NJAS and other job-specific task information that is highly correlated with the NJAS/BARS behaviors and KSAs. The profiles would also include standards for specific occupations as well as broader occupational areas. These multilevel standards can be directly compared and communicated because they have been linked through the common language of the NJAS/BARS.

EDUCATION AND TRAINING

The skill profile clustering system from the NJAS/BARS will have direct relevance to school-to-work and work-to-work transitions. Through its use of a common language and its implications for skill profile development and occupational classification, the NJAS/BARS can in fact link these initiatives into one system.

School-to-Work Opportunities

The successful transition of people from school to high-paying jobs in globally competitive businesses and industries is essential both to the nation's status as a world leader and to all the economic and social benefits that status affords individuals as well as the nation. This need affirms that a primary focus of the curricula development system must be to identify behaviors and KSAs that can be taught and that can guide the progress of young people and adults into higher-paying, productive work.

Curriculum development has typically been a bottom-up process. Educators teach very discrete types of skills that, although critical to learning and future job success, do not always transfer directly to the work context. Information from the skill profiles identified through the NJAS/BARS can be infused into training and education curricula to make that skill transfer more direct.

The NJAS/BARS system will describe profiles of behaviors, tasks, and KSAs at increasing levels of specificity. This enables one to speak about the workplace skills that every student needs upon leaving school, the skills students need as they enter into training within general occupational clusters, and the skills they need as they enter into occupations within specific industries. The process will be informed by the type and levels of information contained in the NJAS/BARS and by emerging occupational standards.

There are sets of behaviors and KSAs that every student needs upon leaving school and that they must demonstrate at various levels of proficiency. A standard can be set for all students. As students enter training in more general occupational areas, they will typically develop higher levels of proficiency in these sets of behaviors and KSAs. They may also learn additional, yet generalizable, behaviors and KSAs to meet the standards set for the occupational cluster. As students move into specific industry settings, they learn the required job-specific tasks and KSAs that in most cases are more specific applications of the behaviors they learned in school and/or in training for an occupational area. By maintaining the link between the general and specific behaviors, tasks, and KSAs, students can begin to make transitions in efficient and predictable ways. They have learned the required elements not only at the specific job level, but also at levels more readily transferable to new learning and job contexts.

In summary, the NJAS/BARS system can support school-to-work transitions by:

- providing information about behaviors and KSAs that is written in a common language, can be organized by occupations/occupational areas, and is associated with standards for the occupational areas.

- linking information so that clear paths of learning from the more general school context to the more specific job context are defined. Working backward, curricula can be developed to bring the citizens through different levels of learning efficiently and successfully.

- providing identification of necessary or enabling KSAs for specific behavioral outcomes.

This information will be readily available for individual generalizable behaviors and for job tasks within and across occupations. The linkage can guide curricula and training plans, and can be used to assist individuals who are having difficulty with a behavior by identifying and remediating the KSAs necessary for performing that behavior.

Job-to-Job Transition

Because jobs change and people change jobs, the NJAS/BARS system affords the identification of generalizable behaviors and KSAs that help with workplace transitions. Built-in linkages between general behaviors and KSAs and their more specific counterparts are critical to allowing these transitions to happen. The reemployment of workers and the transition to better jobs is greatly aided by the ability to match a worker's current repertoire of behaviors and KSAs to those required for available or desired jobs. Workers who meet the standards for other occupations can be placed in new jobs, and those who do not can be directed to appropriate training programs. This matching process requires that occupations be described in such a way that their required behaviors and KSAs can be compared, and that workers can be matched to multiple occupations.

The common language and organizational structure of the NJAS/BARS system offers the vehicle needed to match workers to jobs and to required job training. This information can form the basis for a labor market information system that matches workers to available jobs and to training programs. The result would be a matrix of occupational profiles linked to a current labor market database.

Knowledge that a person has been a successful carpenter provides an estimate of his or her minimum set of behaviors and KSAs. This set can then be matched to occupations with similar profiles because the occupational information in the system is expressed in a common language. The system identifies the extent to which the previous job matches the requirements of new occupations and the residual training the person may need.

Matching can be done for occupations that may be specific to a given context, such as the military. The job of tank operator appears to have few direct matches with civilian occupations. However, a close assessment of the required behaviors and KSAs reveals that tank operators must use lasers and complex automated equipment. These skills have applications in a number of civilian occupations. Skill profiles for broad occupational areas will highlight such similarities to better match workers to new jobs.

A COMPREHENSIVE SYSTEM

The common language provided by the NJAS/BARS can combine the initiatives described in this chapter into a single system. The result would be a comprehensive workforce learning system.

The strategy unites the educational system (K–12) with the work context by infusing the clusters of behaviors and KSAs identified by the NJAS/BARS into school curricula. The common language of the NJAS/BARS connects occupations and occupational areas within and across industries. Workers who meet the performance standards for an occupational area will be able to move readily from one industry to another because the behaviors and KSAs that are common across the clusters are known. A customer service representative, for example, is likely to have a constellation of behaviors and KSAs that transfer across numerous contexts (e.g. retail and hotel/restaurant). Attainment of these transferable skills will increase the probability that the job of choice will be in a context best suited to the person's skill profile, thereby increasing one's income security as well as providing employers with a more productive employee.

This comprehensive strategy empowers learners with a profile of behaviors and KSAs necessary for lifelong learning in the educational and work contexts, and for entry into the workplace—specifically, high-performance work environments.

ACKNOWLEDGMENTS

The authors wish to acknowledge the following employees of ACT for their contributions to the book chapter. Kevin Johannsen, Test Specialist for the National Job Analysis Projects, was instrumental in the final editing and formatting of the manuscript. Dr. Darryl Stark, Industrial/Organizational Psychologist for the National Job Analysis Projects, assisted with expertise in the area of high-performance workplaces. The authors also wish to thank Barb Ciha, Michele Mitchell, and Anne Price for their assistance with the chapter.

REFERENCES

Aguayo, R. (1990). *Dr. Deming: The American who taught the Japanese about quality.* Secaucus, NJ: Carol Publishing Group.

Aldenderfer, M. S., & Blashfield, R. K. (1984). *Cluster analysis.* Beverly Hills, CA: Sage.

American College Testing. (1988). World-of-work structure. *Interim psychometric handbook* (3rd ed., pp. 2–9). ACT Career Planning Program. Iowa City, IA: Author.

American College Testing. (1992). *Realistic assessment of vocational experiences (RAVE)* (Vols. 1 & 2). Hunt Valley, MD: American College Testing, Educational Technology Center.

Appelbaum, E., & Rosemary, B. (1994). *High-performance work systems: American models of workplace transformation.* Washington, DC: Economic Policy Institute.

Carnevale, A. P. (1991). *America and the new economy.* Washington, DC: U.S. Government Printing Office.

Carnevale, A. P., Gainer, L. J., & Meltzer, A. S. (1988). *Workplace basics and the skills employers want.* Washington, DC: U.S. Government Printing Office.

Cascio, W. F. (1986). *Managing human resources.* New York: McGraw-Hill.

Cascio, W. F., & Ramos, R. A. (1986). Development and application of a new method for assessing job performance in behavioral/economic terms. *Journal of Applied Psychology, 71,* 20–28.

Ferguson, R. L. (1995). Workforce improvements and lifelong learning: A new paradigm for education and training in the USA. *Industry and Higher Education, 9*(8), 241–247.

Fisher, K. (1994). Diagnostic issues for work teams. In A. Howard (Ed.), *Diagnosis for organizational change* (pp. 239–264). New York: Guilford Press.

Fort Worth Independent Schools. (1992). *Project C: Identification of workplace skills.* Fort Worth, TX: Author.

Gael, S. (1983). *Job analysis: A guide to assessing work activities.* San Francisco: Jossey-Bass.

Gael, S. (1988). Subject matter expert conferences. In S. Gael (Ed.), *The job analysis handbook for business, industry, and government* (Vol. 1, pp. 432–445). New York: Wiley.

Galagan, P. A. (1992). Beyond hierarchy: The search for high performance. *Training and Development, 46*(8), 21–25.

Gephart, M. A. (1995). The road to high performance. *Training and Development, 49*(6), 29–38.

Gerhart, B., & Milkovich, G. (1992). Employee compensation. Research and practice. In M. D. Dunnette & L. M. Hough (Eds.), *Handbook of industrial and organizational psychology* (2nd ed., Vol. 3, pp. 481–569). Palo Alto, CA: Consulting Psychologists Press.

Gorsuch, R. L. (1974). *Factor analysis.* Philadelphia: W. B. Saunders.

Harvey, R. J. (1991). Job analysis. In M. D. Dunnette & L. M. Hough (Eds.), *Handbook of industrial & organizational psychology* (Vol. 2, pp. 71–163). Palo Alto, CA: Consulting Psychologists Press.

Heaphy, M. S. (1992). Inside the Baldridge Award guidelines. *Quality Progress, 25*(10), 74–91.

Joiner, B. L., & Gaudard, M. A. (1990). Variation, management, and W. Edwards Deming. *Quality Progress, 23*(12), 29–37.

King, J. (1995). High performance work systems and firm performance. *Monthly Labor Review, 118*(5), 29–36.

Lesgold, A., Lajoie, S., Logan, D., & Eggan, G. (1990). Applying cognitive task analysis and research methods to assessment. In N. Frederiksen, R. Glaser, A. Lesgold, & M. Shafton (Eds.), *Diagnostic monitoring of skill and knowledge acquisition.* Hillsdale, NJ: Lawrence Erlbaum Associates.

Mulaik, S. A. (1992). *The foundations of factor analysis.* New York: McGraw-Hill.

Nash, B., & Hausner, L. (1995). *Identifying characteristics of high-performance workplaces.* Iowa City, IA: American College Testing. Unpublished report presented to the U.S. Department of Labor, Employment and Training Administration, and the U.S. Department of Education, National Center on Education Statistics.

Nash, B., & Korte, R. (1994). *Performing a national job analysis study: Overview of methodology and procedures.* Iowa City, IA: American College Testing.

National Center on Education and the Economy, Commission on the Skills of the American Workforce. (1990a). *America's choice: High skills or low wages!* Rochester, NY: Author.

National Center on Education and the Economy, Commission on the Skills of the American Workforce. (1990b). *What is high performance work?* Rochester, NY: Author.

Offermann, L. R., & Gowing, M. K. (1990). Organizations of the future: Changes and challenges. *American Psychologist, 45*(2), 95–108.

Office of Technology Assessment, Congress of the United States. (1995). *Learning to work: Making the transition from school to work.* Washington, DC: U.S. Government Printing Office.

Peach, R. W. (1994). *The ISO 9000 handbook.* Fairfax, VA: CEEM Information Services.

Popham, W. J. (1984). Specifying the domain of content or behaviors. In R. A. Berk (Ed.), *A guide to criterion-referenced test construction* (pp. 29–48). Baltimore: Johns Hopkins University Press.

Prediger, D. J. (1976). A world-of-work map for career exploration. *Vocational Guidance Quarterly, 24,* 198–208.

Reynolds, L. (1994). School-to-work program is ready to roll. *HR focus, 71*(5), 7–8.

Sanchez, J. I., & Fraser, S. L. (1992). On the choice of scales for task analysis. *Journal of Applied Psychology, 77,* 545–553.

Secretary's Commission on Achieving Necessary Skills. (1991). *What work requires of schools: SCANS report for America 2000.* U.S. Department of Labor. Washington, DC: U.S. Government Printing Office.

Secretary's Commission on Achieving Necessary Skills. (1992). *Learning a living: A blueprint for high performance: SCANS report for America 2000.* U.S. Department of Labor. Washington, DC: U.S. Government Printing Office.

Smith, P. C., & Kendall, L. M. (1963). Retranslation of expectations: An approach to the construction of ambiguous anchors for rating scales. *Journal of Applied Psychology, 47,* 149–155.

Spechler, J. W. (1993). *Managing quality in America's most admired companies.* San Francisco: Berrett-Koehler.

Stasz, C., Ramsey, K., Eden, R., Melamid, E., & Kaganoff, T. (1995). *Generic skills and abilities in the workplace: Case studies of technical work.* Santa Monica, CA: Rand.

Taylor, F. W. (1911). *Principles of scientific management.* New York: Harper.

Uniform guidelines on employee selection procedures. (1978). *Federal Register, 43,* 38290–38350.

U.S. Bureau of the Census, Statistical Abstract of the United States. (1992). *Occupational Employment Projections* (112th ed., p. 395). Washington, DC: U.S. Government Printing Office.

U.S. Department of Labor. (1991). *Dictionary of occupational titles* (5th ed.). Washington, DC: U.S. Government Printing Office.

U.S. Department of Labor, Office of Work-Based Learning. (1992a). *Reorganizing work for high performance.* Washington, DC: U.S. Government Printing Office.

U.S. Department of Labor. (1992b). *Occupational Outlook Quarterly, 36*(1). Washington, DC: U.S. Government Printing Office.

U.S. Department of Labor, Office of the American Work Place. (1993). *High performance work practices and firm performance.* Washington, DC: U.S. Government Printing Office.

U.S. Executive Office of the President, Office of Management and Budget. (1987). *Standard industrial classification manual.* Washington, DC: U.S. Government Printing Office.

U.S. Office of Personnel Management, Office of Personnel Research and Development. (1993). *Crosswalk of the DOT content model with the Department of Labor's (DOL) illustrative behaviors and OPM's generalized work behaviors* (ACWA). Washington, DC: U.S. Government Printing Office.

Chapter 5

Workforce Competencies of College Graduates

Kenneth E. Sinclair
The University of Sydney

In recent years considerable attention has focused on the education/business interface and, particularly, on transitions from school and postschool education to the workplace. In Australia there have been a number of influential reports addressing the question of competencies needed for effective workplace participation (Finn Report, 1991; Mayer Report, 1992). In the Mayer Report, seven key competencies were identified—collecting, analyzing and organizing information; communicating ideas and information; planning and organizing activities; working with others and in teams; using mathematical ideas and techniques; solving problems; and using technology. The reports have fueled a lively debate, but have also been influential in finding application at the school and postschool levels of education. As in the United Kingdom, as Norris (1991) has noted, it is as if "Everybody is talking about competence. It is an El Dorado of a word with a wealth of meanings and the appropriate connotations for utilitarian times" (p. 331).

In the debate that has ensued, a generally agreed on definition of competencies which can be used to guide curriculum development in school and postschool education has been particularly difficult to find. Two main conceptions of competence have been used (Hager, Gonczi, & Athanasou, 1994). In one, competencies refer to discrete behaviors identified as being needed to perform the tasks and functions of a job with proficiency. In the other, competencies refer to generic skills, abilities, knowledge, attitudes, and other attributes which together are needed for successful job performance. The second of the two conceptions appears to be gaining increasing support and is the sense in which the term *competencies* is used in this chapter.

A number of studies and reports in Australia have focused on the transition of graduates from university to the workplace. The question of objectives for university education and the competencies and outcomes desired of graduates when they enter the workplace has been of particular interest to the Business/Higher Education Round Table, a group of university vice-chancellors and chief executive officers from prominent business enterprises in Australia. Four investigations have been conducted by the writer for the Round Table (Business/Higher Education Round Table, 1991, 1992, 1993, 1995). In the initial study the views of Round Table members themselves were sampled about objectives of secondary and university education, characteristics desired of university graduates on entry to the workforce, and partnership possibilities between business and universities. This study was then replicated with a sample of lecturers from universities and managers from business who had responsibility for supervising newly recruited university graduates. The third study involved a national sample of business students and focused on their views about the objectives and outcomes of their university education. The final study examined factors influencing the career success and progression of a sample of university graduates in the first ten years of their work experience. In that study, graduates discussed the outcomes of their university education in terms of its relevance for career success and the way in which competencies need to change as more responsible management positions are achieved.

OBJECTIVES OF UNIVERSITY EDUCATION

A question common to all the studies conducted for the Round Table asked respondents to rank a list of objectives for university education in order of importance. The list included professional and general knowledge objectives, the development of generic skills, and objectives relating to business knowledge and work experiences. The results are remarkably consistent across the quite different samples of Round Table members, university lecturers, business managers, business students, and university graduates working in business. The results are summarized in Table 5.1.

With surprising consistency the general skills of thinking/decision-making, and communicating are accorded highest importance for objectives of university education, while the remaining generic skill objective—learning skills of cooperation and collaboration—is ranked near the middle of the list. Next in importance after the two highest ranked generic skills are professional studies at university, and after them (although still regarded as very important objectives) are objectives about work experience and

TABLE 5.1
Ranking of Objectives for University Education in Order of Importance: Responses Across
Samples From Four Studies

	Sample					
	1a	1b	2a	2b	3	4
Learning thinking/decision-making skills	1	1	1	1	1	1
Learning communication skills (e.g., writing and speaking)	2	2	2	2	2	2
Learning professional skills--practical studies	5	5	3	5	3	3
Learning professional skills--theoretical studies	3	3	4	3	5	4
Learning skills of cooperation and teamwork	6	6	*	*	4	5
Developing standards of personal and business conduct	4	7	5	6	7	6
Learning about work and career choice	8	8	7	7	6	7
Learning a broad range of general academic subjects	7	4	8	4	9	8
Receiving on-the-job work experience	9	9	6	8	8	9

Note. 1a: Study 1 (1991) CEOs ($N = 56$); Study 1 (1991) Vice Chancellors ($N = 17$); 2a: Study 2 (1992) Business Managers ($N = 147$); 2b: Study 2 (1992) University Lecturers ($N = 122$); Study 3 (1993) University Business Students ($N = 535$); 4; Study 4 (1995) Graduates in Business ($N = 261$); * Response not available.

knowledge and skills for the workplace. A notable difference between the business and university respondents, however, is evident with respect to the importance given to learning a broad range of general academic subjects as part of a professional education. This objective was rated as much more important by university vice-chancellors and lecturers (ranked 4th) than by business CEOs or managers, who ranked it near the bottom of their list. It will also be observed that the business students and graduates in the workforce also ranked it near the bottom of their lists.

The result reinforces the view that, first and foremost, professionals need to be educated, at least in their first degree, in general skills and professional knowledge and skills. Those outcomes are consistently regarded as more important than outcomes focusing on business knowledge and work experience. Such a finding is consistent with the view that the workforce of the future needs to be adaptable and flexible to meet emerging challenges and changes. It is a finding that has been supported in a number of other reports as well (Higher Education Council, 1992; National Board of Employment, Education, and Training, 1992). For instance in the influential *Higher Education: Achieving Quality* report of the Higher Education Council in Australia (1992), and after considerable consultation in the business and education communities, it is concluded that:

Discipline-specific skills in many areas have only a short life, and what will be needed in even the medium term cannot be predicted with any great precision. The groups consulted were as one on this issue—while discipline skills and technical proficiency were seen as important . . . the so-called higher level generic skills were seen as critically important, and sometimes lacking. While it would not be claimed that these characteristics are found only in graduates, . . . if universities are to add value, they must take responsibility for the specific development and refinement of these skills. (p. 20)

DESIRED CHARACTERISTICS OF GRADUATES

In the second of the studies (Business/Higher Education Round Table, 1992), university lecturers were asked to indicate what emphasis was being given in their programs to developing particular graduate characteristics. Business respondents were asked to indicate what emphasis they gave each of those characteristics in choosing graduate staff. Responses were on a five-point scale: 1 (no emphasis), 2 (little emphasis), 3 (moderate emphasis), 4 (strong emphasis), 5 (very strong emphasis). These results are summarized in Table 5.2.

The importance of the capacity to make decisions and solve problems, the capacity to learn new skills and procedures, the ability to apply knowl-

TABLE 5.2

Ranking of Emphasis Given to Suggested Characteristics of University Graduates: Business and University Respondents

	Business		University	
	Rank	Mean	Rank	Mean
Communication skills (e.g., writing, speaking)	1	4.29	7	3.26
Capacity to learn new skills and procedures	2	4.21	5	3.36
Capacity for cooperation and teamwork	3	4.18	8	3.24
Capacity to make decisions and solve problems	4	4.17	3	3.54
Ability to apply knowledge to the workplace	5	4.05	4	3.51
Capacity to work with minimum supervision	6	3.65	6	3.33
Theoretical knowledge in professional field	7	3.52	1	4.02
Capacity to use computer technology	8	3.43	2	3.65
Understanding of business ethics	9	3.21	12	2.24
General business knowledge	10	3.03	11	2.33
Specific work skills	=11	3.02	9	3.13
A broad background of general knowledge	=11	3.02	10	2.93

Note. University respondents were asked: "In educating undergraduates in your faculty . . . what emphasis is given to developing each of the following characteristics?" Business respondents were asked: "In selecting newly graduated professionals to work in your company . . . what emphasis do you consider should be given to each of the following characteristics of applicants?"

TABLE 5.3
Emphasis given to Developing communication Skills in University Courses and to the
Importance of Communication Skills in Selecting Company Recruits

	Little or No Emphasis (%)	Moderate Emphasis (%)	Strong or Very Strong Emphasis (%)
Emphasis in university courses (university respondents)	21.5	38.0	40.5
Emphasis in business recruitment (business respondents)	0.9	6.3	92.8

edge to the workplace, and the capacity to work with minimum supervision are all rated highly both in course emphases and company recruitment. Some significant differences also emerged, however, in the responses of the two groups. It is interesting that, while recognizing the importance of developing communication skills as an objective of university education, the lecturers indicate that only a moderate amount of emphasis is given to it in professional courses. For business, in contrast, communication skills are clearly the number one criterion of those listed characteristics in selecting staff. The frequencies for this response are provided in Table 5.3. Similarly, capacity for cooperation and teamwork is ranked third by business but only eighth by university respondents. On the other hand, for universities the greatest emphasis is given to providing theoretical knowledge in the professional field followed by capacity to use computers, yet these are ranked only seventh and eighth by the business respondents for purposes of staff recruitment. There are, then, some sharp differences to be observed between university teaching priorities and the priorities business gives to the outcomes of that teaching in selecting staff. Universities have traditionally focused on knowledge outcomes in teaching. Furthermore, large lecture classes in many undergraduate courses do not lend themselves particularly well to skill development, and it is often assumed that important skills such as communication skills will be well developed on entry to university.

For the business respondents there was a further write-in question associated with recruiting graduates from university. The write-in responses generally confirmed the rankings just noted and sometimes went beyond them. The most frequently nominated characteristics were:

• strong academic background;
• appropriate communication skills;

- high motivation to succeed;
- ability to work as part of a team;
- initiative and decision-making skills;
- interpersonal skills;
- appearance and manner.

Typical examples of responses to the open-ended question were these:

- "Good, articulate team player with a strong desire to win. Good understanding of how to apply theoretical knowledge in the working environment."
- "A self-starter who has displayed his or her ability to achieve. Very strong people skills and one who is motivated by achievement."
- "High academic qualifications and ability to learn new skills. Capacity to make decisions and solve problems. Good presentation and communication skills. Capacity to work in a team."

In the study, therefore, the graduate characteristics desired by business give particular emphasis to communication skills, capacity to learn new skills and procedures, capacity for cooperation and teamwork, capacity to make decisions and solve problems, ability to apply knowledge to the workplace, and motivation to succeed. We have also seen that the emphasis given to the development of such characteristics at university does not always reflect the importance given by business to those characteristics. Other studies confirm the responses provided by the business respondents. For instance, in one recent survey of 80 Australian employers (National Board of Employment, Education, and Training, 1992), the skills and personal qualities employers sought in graduates fell into the three broad areas of communication skills, social skills (including interpersonal and teamwork skills), and the ability to apply academic learning to a work environment. In the report of that survey, mention is also made of a study conducted by the University of Sheffield (Green, 1990) in which employers indicted that they wanted graduates who:

1. had excellent communication skills (both oral and written);
2. were able to work effectively in teams;
3. were quick to take sound initiatives;
4. were adept at managing and organizing themselves, their workload and others;
5. were enthusiastic and highly self-motivated and able to motivate and inspire others.

As well as asking business and university respondents in the second study (Business/Higher Education Round Table, 1992) about the desired characteristics of graduates, a second element of the question asked them for an assessment of the current standards reached by university graduates in those characteristics (see Table 5.4). Participants rated the standards achieved on a five point scale: 1 (very poor), 2 (poor), 3 (adequate), 4 (good), 5 (very good). A mean score of 3 indicates that the responses cluster around the "adequate" score, while a mean of greater than 3 or less than 3 indicates that responses tend towards a judgment of "good" or "poor" respectively. The lecturers and business supervisors in the survey judged most characteristics to be in the adequate range.

While no characteristics were assessed as being clearly good or very good (mean of 4 or greater), those judged adequate to good by both groups of respondents were:

- theoretical knowledge in the professional field;
- capacity to learn new skills and procedures;
- capacity to use computer technology.

Characteristics judged as being of poorest standard include:

- general business knowledge;
- understanding business ethics;
- communication skills (business but not university respondents).

TABLE 5.4
Current Standards Reached by Graduates From Universities: Mean Scores for Business and University Respondents

	Business Mean	University Mean
Communication skills (e.g., writing, speaking)	2.65	3.04
Capacity to learn new skills and procedures	3.70	3.45
Capacity for cooperation and teamwork	2.95	3.22
Capacity to make decisions and solve problems	3.22	3.43
Ability to apply knowledge to the workplace	2.83	3.44
Capacity to work with minimum supervision	3.16	3.24
Theoretical knowledge in professional field	3.60	3.59
Capacity to use computer technology	3.69	3.55
Understanding of business ethics	2.68	2.32
General business knowledge	2.88	3.18
Specific work skills	2.92	2.83
A broad background of general knowledge	2.59	2.49

The question about standards achieved in developing a range of desirable characteristics in graduates was asked of the business students who participated in the third study (Business/Higher Education Round Table, 1993) as well. The results revealed that a third or more of respondents considered that standards achieved by graduates from their faculties were poor or very poor with respect to:

- developing specific work skills (46.6%);
- understanding business ethics (46.0%);
- ability to apply knowledge to the workplace (43.9%);
- general business knowledge (34.7%).

It is significant that these characteristics are predominantly ones directly relevant to workplace career goals. It is also significant that as many as one fourth of the respondents considered the standards achieved by graduates in their faculties to be poor in relation to the highly regarded general skills areas of communicating, capacity to make decisions and solve problems, and capacity for cooperation and teamwork. The result confirms that views about lower than desirable standards of achievement among university graduates in areas of key importance are not confined to people from the university and business communities but are acknowledged by the students themselves.

COMPETENCIES AND ATTRIBUTES ASSOCIATED WITH GRADUATE CAREER SUCCESS

In the last of the studies (Business/Higher Education Round Table, 1995), graduates with up to ten years of business experience were surveyed as to the qualities and attributes they considered to be important in their own career achievements and in attaining high level positions in their companies. Several questions asked respondents to assess the importance of a set of competencies and other attributes for present and future career success. The attributes included communication skills, thinking and decision-making skills, professional expertise, motivation, gaining a breadth of job experience, management and leadership abilities, corporate vision, and coping with change. The attributes are listed in order of importance as assessed by the graduate respondents in Table 5.5.

The results reveal a strong level of agreement between the attributes selected as most important for the respondents' careers so far and those chosen as likely to be important in their future careers. In particular, skills such as are involved in communication, thinking and decision-making, motivational qualities such as motivation to succeed and setting high per-

TABLE 5.5
Attributes Ordered in Terms of Importance for Current and Future Career Success:
Graduate Responses

Rank	Current Success	Rank	Mean
1	setting high standards for achievement	1	communication skills
2	communication skills	2	capacity to cope with change
3	thinking and decision-making skills	3	thinking and decision-making skills
4	capacity to cope with change	5	capacity to see the big picture
5	ability to work independently	5	leadership abilities
6	motivation to succeed	6	motivation to succeed
7	expertise using professional knowledge	7	setting high achievement standards
8	skills in using information technology	8	skills in managing staff
9	professional knowledge	=9	ability to develop new ideas, products
10	social skills	=9	gaining attention of senior managers
11	variety of job experience and roles	11	ability to work independently
12	ability to develop new ideas, products	=12	social skills
13	capacity to see the big picture	=12	expertise using professional knowledge
14	leadership abilities	14	skills using information technology
15	gaining attention of senior managers	15	variety of job experience and roles
16	skills in managing staff	16	professional knowledge

Note. = denotes equal ranks.

sonal standards, and capacity to cope with change feature prominently. Expertise in using professional knowledge is ranked seventh among the attributes selected as being important in the respondents' careers so far but is well down the list in terms of importance for their respective future careers. On the other hand, as might be expected, leadership abilities and capacity to see the big picture in terms of company goals and strategies feature prominently for the respondents' future careers but are not selected as being of particular importance in the respondents' careers so far.

In addition to the graduates, 16 personnel managers from companies employing the graduates responded to a parallel question about characteristics associated with career success. They ranked the same set of characteristics in terms of their importance for (a) the first 10 years of a graduate's career, and (b) beyond the first 10 years of a graduate's career. The results are summarized in Table 5.6.

TABLE 5.6
Attributes Ordered in Terms of Importance in the First 10 and Subsequent Years of a
Graduate's Career: Manager Responses

Rank	First 10 Years	Rank	Beyond First 10 Years
1	communication skills	=1	communication skills
2	capacity to cope with change	=1	thinking and decision-making skills
3	motivation to succeed	3	capacity to cope with change
4	setting high self-achievement standards	4	setting high self-achievement standards
5	professional knowledge	=5	motivation to succeed
6	thinking and decision-making skills	=5	capacity to see the big picture
7	skills using information technology	=7	leadership abilities
8	variety of job experiences and roles	=7	skills in managing staff
9	expertise using professional knowledge	9	expertise using professional knowledge
10	social skills	10	professional knowledge
11	ability to work independently	11	ability to develop new ideas, products
12	gaining attention of senior managers	12	ability to work independently
=13	ability to develop new ideas, product	13	social skills
=13	capacity to see the big picture	14	gaining attention of senior managers
15	leadership abilities	15	skills using information technology
16	skills in managing staff	16	variety of job experiences and roles

There is considerable agreement between the rankings of the graduates and the managers. Again the importance of self-motivating forces found in setting high standards for personal achievement and motivation for success are highlighted, together with communication skills and skills in thinking and decision-making, and the ability to cope with change. Similarly, a capacity to see the big picture in terms of company goals and strategies, leadership abilities and skills in managing staff are ranked towards the bottom of the list for graduates in the first 10 years of their careers but are considered to be of very high importance for their careers beyond that point. It will be noted, however, that professional knowledge is rated by the managers (ranked 5) as being of greater importance in the first 10 years of a graduate's career than the graduates acknowledge for their career so far (ranked 9), while ability to work independently is ranked by the graduates (ranked 5) as being of greater importance than considered by the managers (ranked 11).

ACHIEVING THE MOST SENIOR POSITIONS
IN A COMPANY

A follow-up question asked the graduates to select the seven most important of the attributes listed and rank them in order of their importance for achieving the "most senior positions" in their company (see Table 5.7).

The attributes which were most commonly ranked first by the graduate respondents were leadership abilities (ranked first by 19.9% of respondents) and capacity to see the big picture (15.7%). These were followed in importance by gaining the attention of senior staff, motivation to succeed, communication skills, and skills in thinking and decision-making. If the importance of the attributes is assessed in terms of whether they are nominated as one of the respondents' first seven choices, the same attributes emerge although the ranking changes somewhat. This time communication skills are the most commonly selected attribute (selected as one of the seven most important attributes by 76.5% of the respondents), followed by capacity to see the big picture (74.1%), leadership skills (73.6%), thinking and decision-making skills (68.1%), motivation to succeed (57.4%), and gaining the attention of senior management (55.5%). Respondents were able to add attributes not listed in the questionnaire. A range of attributes was suggested with multiple responses being made for assertiveness/competitiveness/risk taking, empathy/approachability/compassion, personal integrity, "political" skills, networking skills, self-management skills, and luck/being in the right place at the right time. The managers' rankings of attributes important for achieving the most senior positions in a company were very similar to those of the graduates. Communication skills, capacity to see the big picture, thinking and decision-making skills, and leadership skills received the highest rankings.

TABLE 5.7
Attributes Ranked in Order of Percent Response (in Parentheses) for Achieving the Most Senior Position in a Company

Selected First		Selected in First Seven Choices	
leadership abilities	(19.9%)	communication skills	(76.5%)
capacity to see the big picture	(15.7%)	thinking/decision-making skills	(68.1%)
gaining attention of senior managers	(12.6%)	capacity to see the big picture	(74.1%)
motivation to succeed	(11.9%)	leadership abilities	(73.6%)
communication skills	(10.3%)	motivation to succeed	(57.4%)
thinking/decision-making skills	(8.8%)	gaining attention of senior managers	(55.5%)

The managers, however, gave relatively more importance to the attributes of setting high standards for personal achievement and capacity to cope with change than did the graduates.

These findings suggest that as staff progress in their careers and assume positions of greater authority, so do the attributes needed change in their relative importance. In particular, leadership abilities and the capacity to see the big picture in terms of company goals and strategies rise in importance. High levels of motivation and of communication and thinking skills remain important, but these latter skills perhaps will be often directed in rather different ways and towards rather different ends than before. For instance, they are likely to be more directed towards defining and gaining agreement for statements of company vision and mission and in thinking through strategies and plans to realise company goals.

The finding also raises the question of how attributes such as leadership abilities and capacity to see the big picture may be identified and/or developed and nurtured through professional development programs and other selected experience. Should companies recruit graduates with such attributes in mind, or should they identify staff with such attributes after some years with the company, or should they select high performing staff and seek to develop those attributes in preparation for later appointment to executive management positions? In the recently published Enterprising Nation report (Industry Taskforce on Leadership and Management Skills, 1995), a serious weakness identified is the way companies handle the transition of staff from positions as technical or functional specialists (for which they are typically hired) to positions of responsibility in managing others. Too often, it appears the transition is one of "sink or swim," with staff promoted on the basis of high level current performance and only slight regard for other needed qualities and skills. On the basis of a review of literature, Callen (1995) argues, "The transition from specialist to manager is clearly one of the most significant challenges and transitions that most employees will experience in the workplace. New managers over time reframe their understanding of what being a manager means. The adjustment is a transformation from an individual actor, to a generalist, agenda-setter and network builder. New managers have to learn how to think, feel, and act as managers instead of being individual contributors" (p. 120).

Interviews with company personnel managers (Business/Higher Education Round Table, 1995) indicate that, in some companies, potential for leadership is an attribute sought when graduates are first recruited, while in others it is identified early in a staff member's career and then nurtured through particular work assignments and experience, including overseas experience, and leadership education programs. It was argued that such a planned approach was needed to prepare graduates for senior management positions by age 45–50. Increasingly, therefore, leadership attributes are likely to be sought in graduate recruitment, perhaps encouraging university programs to provide specific experiences designed to nurture such attributes.

INFLUENCE OF HOME, SCHOOL, UNIVERSITY, AND WORK ON WORKPLACE ATTRIBUTES

In the most recent study (Business/Higher Education Round Table, 1995) an attempt was also made to gauge the graduates' perceptions of the relative influence of the home, school, university, and workplace on the development of selected attributes chosen for their occupational importance (see Table 5.8). The results are very interesting and reveal that, at this point in their careers, the workplace was most commonly considered to be of greater influence than the home, school, and university in relation to most attributes. This applied with respect to the development of communication skills, thinking and decision-making skills, professional knowledge, ability to apply professional knowledge, and skills of cooperation and teamwork. For instance, in the case of development of communication skills, 28.52% of total influence was attributed to the workplace, 24.78% to home, 21.59% to school, 18.53% to university and 6.58% to other influences. The home was considered to be of greatest influence in developing motivation to succeed (34.91% relative influence) and ethical values (49.59% relative influence), while the university was considered to be of similar influence to the workplace in the development of ability to work independently (29.39% relative influence).

The results stress the importance of learning and development that takes place in the workplace after schooling and university have been completed. For the most part such learning will occur through the expe-

TABLE 5.8
Influences Attributable for Present Capacities: Mean Percent

	Home %	School %	University %	Work %	Other %
Ability to apply professional knowledge	8.37	7.85	26.94	52.96	3.88
Communication skills	24.78	21.59	18.53	28.52	6.58
Thinking and decision-making skills	20.54	17.68	26.96	30.73	4.09
Professional knowledge	6.74	9.98	32.93	46.65	3.70
Motivation to succeed	34.91	15.40	16.77	26.26	6.66
Skills of cooperation and teamwork	22.05	17.35	18.97	32.93	8.70
Ability to work independently	19.35	16.66	29.39	30.02	4.58
Ethical values	49.59	13.54	12.22	18.33	6.32

rience of work, supplemented only to a small extent by workplace training. It is of particular interest that the respondents attributed, on average, 46.65% of their professional knowledge to the workplace and only 32.93% to university. This reaffirms the importance given by the business students in an earlier survey to the importance of on-the-job experience as part of their university program (Business/Higher Education Round Table, 1993).

THE CONTRIBUTION OF UNIVERSITY AND BUSINESS EDUCATION AND EXPERIENCE TOWARDS CAREER SUCCESS

Open-ended questions were also asked about the workplace value and standards achieved in relation to aspects of the graduates' university education and work-related training courses not provided as part of a university degree program (Business/Higher Education Round Table, 1995).

University Education

In one such question respondents were invited to list the three most valuable outcomes of their university education. Particularly valued were the information processing/decision-making skills developed (nominated as one of the three values by 76.6% of respondents), the personal development and development of self-confidence that took place (34.9%), the business relevance of the program and the technical expertise developed (32.5%), the qualifications gained and the career opportunities opened up (32.1%), and the broad background of knowledge and experience developed (29.0%). A second open-ended question asked them to list the three most serious deficiencies of their university education. By far the most common deficiency related to the lack of business relevance and technical expertise developed (94%). Other commonly nominated deficiencies included quality of teaching (32.4%), and the lack of attention given to the development of communication skills (31.4%) and to career guidance (21.1%).

It is of interest that business relevance and technical expertise was perceived both as a value (by 32.5% of respondents) and as a deficiency (by a very large 94%). It seems that only a relatively small proportion see this important aspect of their university education as being of value. The very large majority sees it as a serious deficiency, and it appears that even some of those considering it to be a value believe that it needs much more emphasis. The evidence is clear that substantial numbers (20%–30%) of university graduates find serious deficiencies in the quality of teaching at university and in the development of communication skills and provision of career guidance. On the other hand, it is also clear that, in addition to the value gained in the development of general information processing skills and the business expertise developed, considerable value from a university education comes from the opportunities it provides for personal development and goal achievement and for broadening one's knowledge and experience.

Work-Related Training Courses

Another question asked those surveyed to rate the overall importance of, and standards achieved in, various objectives of work-related training courses not provided as part of a university degree program. The objectives listed referred to the intellectual challenge of courses, learning new professional knowledge and its application, developing skills in communicating, thinking, and managing others, increasing workplace motivation and so on. The predominant response for all the objectives listed was in the moderate to high importance range. Greatest importance, however, was given to learning new applications and skills for the workplace, with 76% of respondents considering that objective to be of high or very high importance, 17% as being of moderate importance, and 7% as low or very low importance. Other objectives considered to be of particular importance were the improvement of skills in communication, in thinking/decision-making, in cooperation and teamwork, learning new professional knowledge, and improving skills needed for the management of others. The objective considered to be of least importance for work-related training courses was "increasing your understanding of workplace ethics" with 23% regarding that objective as being of only low or very low importance, 40% considering it to be of moderate importance and 37% regarding it as being of high or very high importance.

GRADUATE COMPETENCIES FOR WORKPLACE SUCCESS

Together the studies assist our understanding of the competencies needed for graduate career success. Furthermore, the view that emerges receives wide support from such different perspectives as business managers, university lecturers, business students, and university graduates in the workforce. Generic skills such as thinking and decision making skills, communication skills, and skills in cooperation and teamwork are accorded highest priority. However, professional knowledge and its application are also considered to be of high overall importance, and of greater relative importance than business knowledge and work experience. Such a finding is consistent with the view that the workforce of the future needs to be adaptable and flexible to meet emerging challenges and a rapidly changing workplace. Given the increasing rate with which specific jobs in the workplace disappear, are modified, or are added to, staff will be expected to be particularly adaptable with well developed general skills in learning how to learn and think and a well developed and up-to-date mastery of core professional knowledge. This finding is at odds with the more behavioristic approaches used to define competencies as discrete behavioral skills, but is consistent with the view that competencies should refer to generic skills, abilities, knowledge, attitudes and other attributes which *together* are needed for successful job performance. The finding also reinforces the importance of a career-long approach to learning. As we have seen, graduates in business responded that the workplace is responsible for a relatively greater influence than the school or university on such matters as the development of communication and thinking skills, skills of cooperation and teamwork, and professional knowledge and its application. Planned experiences encouraging such development in the workplace and through postgraduate study at university will greatly assist this process.

The findings also reveal that the emphasis given to competency development at university does not always coincide with the emphasis given to particular competencies in recruiting graduates for the workplace. Despite the importance given by both business and university respondents to the objective of developing communication skills and the importance of communication skills in securing employment, the amount of emphasis given to its development in university courses is considered to be only moderate to low. University lecturers give greatest importance in courses to providing theoretical knowledge in the professional field. The emphasis on knowledge outcomes has been a traditional focus of universities and one which they have performed with distinction. A sharper focus on being able to use that knowledge, and to use communication skills, collaboration and teamwork skills, and thinking and decision making skills in both creating and using that knowledge is needed in preparing graduates for professional lives in the decades ahead. Recent implementation of problem-based learning approaches in such university programs as medicine, business, nursing, and education demonstrate that universities are gradually moving in the needed direction.

The findings of the studies discussed also reveal that the competencies needed for successful career advancement change as positions of greater management responsibility are reached. While high levels of motivation and communication and thinking skills remain important throughout a career, as staff progress to positions of greater authority, other attributes such as leadership skills, skills in managing other staff, and a capacity to see the big picture in terms of company goals and strategies become very important. Increasingly, companies are seeking to identify staff with potential for such leadership positions at the time of recruitment or very early in the graduate's career so that the necessary experience and professional development of the needed attributes may be provided before staff assume such positions. Management courses have typically been regarded as part of postgraduate programs. The trends identified suggest that the development of leadership and management skills and attributes might also become part of an undergraduate professional program.

ACKNOWLEDGMENT

The research reported in this chapter was commissioned by the Business/Higher Education Round Table whose support is gratefully acknowledged.

REFERENCES

Business/Higher Education Round Table. (1991). *Aiming higher: The concerns and attitudes of leading business executives and university heads to education priorities in Australia in the 1990s* (Commissioned Report No. 1). Melbourne: Author.

Business/Higher Education Round Table. (1992*). Educating for excellence. Part 2: Achieving excellence in university professional education* (Commissioned Report No. 2). Melbourne: Author.

Business/Higher Education Round Table. (1993). *Graduating to the Workplace: Business students' views about their education* (Commissioned Report No. 3). Melbourne: Author.

Business/Higher Education Round Table. (1995). *Identifying future leaders: A study of career progression and development* (Commissioned Report No. 4). Melbourne: Author.

Callan, V. (1995). The transition from specialist to manager. In Industry Task Force on Leadership and Management Skills *Enterprising nation: Research report. Vol. 1* (pp. 117–154). Canberra, Australian Capital Territory: Australian Government Publishing Service.

Finn Report (Australian Education Council Review Committee). (1991). *Young people's participation in post-compulsory education and training.* Canberra, Australian Capital Territory: Australian Government Publishing Service.

Green, S. (1990). *Analysis of transferable personal skills requested by employers in graduate recruitment advertisements in June 1989.* Sheffield University.

Hager, P., Gonczi, A., & Athanasou, J. (1994). General issues about assessment of competence. *Assessment and Evaluation in Higher Education, 19*(1), 3–16.

Higher Education Council. (1992). *The quality of higher education: Discussion papers* (National Board of Employment, Education and Training Publication). Canberra, Australian Capital Territory: Australian Government Publishing Service.

Industry Task Force on Leadership and Management Skills. (1995). *Enterprising nation: Report of the industry task force on leadership and management skills.* Canberra, Australian Capital Territory: Australian Government Publishing Service.

Mayer Report. (1992). *Employment related key competencies: A proposal for consultation.* Melbourne: Mayer Committee.

National Board of Employment, Education, and Training. (1992). *Skills sought by employers of graduates* (Commissioned Report No. 20). Canberra, Australian Capital Territory: Australian Government Publishing Service.

Norris, N. (1991). The trouble with competence. *Cambridge Journal of Education, 21*(3), 331–341.

Chapter 6

Workforce Competencies of Teachers

Ken J. Eltis
The University of Sydney, Australia

The emergence of competencies as a "new" force in education and training is described in this chapter, including the particular interest of government in using competencies as a way of attempting to increase the skills and productivity levels of the workforce. Addressing the case of the teaching profession, a contrast is drawn between earlier notions of competencies as pursued through Competency-Based Teacher Education in the 1960s and 1970s and more complex definitions and understandings now being put forward. Data from Australian case studies are used to describe how competencies can be developed to portray dimensions of teachers' work needing to be addressed in teacher education programs, both preservice and inservice. Also considered is the significant contribution competencies might make to ensuring that all teachers entering the profession have met appropriate criteria for initial and ongoing employment as a teacher. Finally, emphasis is given to the value of using competencies to promote critical discussion of teaching and standards of teaching at all levels of the profession.

A NEW INTEREST IN COMPETENCIES

Competencies and Economic Reform

During the last decade and a half governments around the world have been introducing significant reforms to achieve stronger links between education, training, and employment in an attempt to enhance their country's economic competitiveness. A significant component of these reforms has been a focus on defining competencies seen as necessary to enable individual workers to perform their daily tasks more efficiently and thereby achieve greater productivity (see as an example developments in the United Kingdom, as described in Jessup, 1991).

The new competency agenda has been pursued, not necessarily as a result of soundly based educational theory being translated into practice, but more as a result of sets of initiatives taken by governments in an attempt to make their countries much more competitive and successful in world markets. The assumption is that by specifying for those engaged in various occupations and professions generic competencies and essential learnings at various levels which can be focused on in training and development programs, countries can build for themselves a bridge to the future. There is not a universal acceptance of the wisdom of the approach being adopted, with arguments being raised that efforts to upgrade workers' skills and their salaries will not prevent business from chasing cheaper labor elsewhere. Further, as work practices and processes are changing so rapidly, some argue it is a false hope to rely on a "standing still" model of competencies (Knight, McWilliam, & Bartlett, 1993, p. 149).

According to the model being followed, workers are seen as "human capital" able to be developed and to whom value can be added through appropriate training with a focus on competencies.

> The assumption is that people's skills and knowledge can be invested and that returns to the individual and the economy can be expected. What is needed to resurrect the nation from its economic decline is a 'multi-skilled and flexible workforce' for greater international competitiveness. Hence the implementation of the 'structural efficiency principle' which balances increased wages against new competencies for greater productivity. Education and training become the means to achieve this and the central tools for micro-economic reform. (Bartlett, 1993, p. 72)

Australia as a Case Study

Australia provides a very interesting case study for understanding how decisions by government have provided the structures for the development of work-related competencies at all levels and in all industries, including what has come to be called the "education industry." The following is an accurate portrayal of what has occurred in Australia since early 1983 (and the thrust is continuing):

> The federal Labor government has insisted that radical measures are needed in order to make the Australian economy more export-oriented and internationally competitive . . . Having deregulated the economy, Labor has

turned its attention to micro-economic reform, carried through in the context of a nationally created economic infrastructure . . . **The national push towards competencies is linked to the implementation of award restructuring proposals throughout various industries** (my emphasis) that have been negotiated between the government, the Australian Council of Trade Unions and various employers. (Porter, Rizvi, Knight, & Lingard, 1992, p. 51)

The award restructuring proposals referred to have been designed to make it easier for individual workers to move across different levels of education and training, and to ensure that training is consistent with the various labor requirements of a particular industry. So, great importance is placed on comparability and portability of qualifications and on the need to recognize prior learning as workers take up training opportunities to enhance their skills. Competencies are seen as the key to achieving these goals.

A significant step in establishing the industry reform agenda was the establishment of a National Training Board in 1989 as a joint venture between the Australian Government and the various State Governments, with the brief to establish for industry an Australian Standards Framework.

The National Training Board developed the Australian Standards Framework as a means of relating groups of units of competencies in industry standards to qualifications. This enables enterprises and industries to have a common understanding of what a course is delivering and providers to have confidence that the course will be accepted and valued by industry. (National Training Board, 1993, quoted in *Australian Training Reform*, Keating, 1995, p. 39)

The Australian Standards Framework has eight levels of competency which serve as a reference point for the development and recognition of competency standards. These are shown in Table 6.1.

The National Training Board has had responsibility for elaborating competencies for occupations encompassed by Levels 1–6 while a separate organization, the National Office of Overseas Skills Recognition, established to coordinate the implementation of the migrant skills reform strategy, was given a key role in the development of competencies for the professions (encompassing Levels 7 and 8), including teaching.

The term "competency standards" is now in common use and describes the standards to be developed by industry and occupations and to be pursued in education and training provision. "Competency standards reflect the specification of the knowledge and skill and the application of that knowledge and skill to the standard of performance required in employment" (National Training Board, 1991, p. 7). It will be noted that this definition has a very narrow focus. That developed by the National Office

TABLE 6.1
Australian Standards Framework

Australian Standards Framework		Approximate Levels of Current Courses
Level 8	Competent senior professional manager	Degrees
Level 7	Competent professional or manager	
Level 6	Competent senior administrator, specialist, technologist, or paraprofessional	
Level 5	Competent administrator, specialist, technologist, or paraprofessional	Diplomas
Level 4	Advanced skilled autonomous worker	
Level 3	A competent skilled autonomous worker	
Level 2	Advanced operative or service sector worker	Certificates
Level 1	Competent operative or service sector worker	

Note. The original Australian Standards Framework descriptors from the National Training Board (NTB, 1991, p. 11), linked to Existing Course Levels. Source: From Keating (1995, p. 40), Australian training reform: Implications for schools. Permission has been given by the publisher. *Australian training reform: Implications for schools* is available from the publisher, Curriculum Corporation, 141 Rathdowne St, Carlton Vic 3053 Australia.

of Overseas Skills Recognition differs somewhat, emphasizing aspects of particular (but not exclusive) significance to the professions, for which it has responsibility:

> The competence of professionals derives from their possessing a set of relevant attributes such as knowledge, abilities and skills and attributes. These attributes which jointly underlie competence are often referred to as competencies. So a competency is a combination of attributes underlying some aspects of successful performance. (Gonczi, Hager, & Oliver, 1990, p. 9)

In summary, the development of competency standards and their general application to education and training, known as the training reform agenda, is seen as imperative for the country's future. This training reform agenda is a process of fundamental reform to make the education and training system more responsive to industry needs. It has extended into the school system, as well, with work being undertaken to develop eight key competencies to be incorporated in school curricula following the publication of two major reports on postcompulsory education provision (Finn, 1991; Mayer, 1992).

The Finn Report indicated that there should be developed a set of six "employment-related key competencies" (p. x) seen as "essential for all young people engaged in post-compulsory education and training" (p. 57). A broadly representative committee, known as the Mayer Committee, was subsequently established to develop these key competencies and report to the Australian Education Council and to the Ministers of Vocational Education, Employment, and Training. The outcomes are shown in Table 6.2. As the summary shows, thinking in Australia parallels what has been happening elsewhere. As an example, a comparison has been given with the National Council for Vocational Qualifications core skills developed in the United Kingdom (Jessup, 1991).

The professions are not being allowed to sit outside the reform agenda, and work has proceeded on the development of competency standards on a profession-by-profession basis. Entry level standards being developed are

TABLE 6.2
Proposed Key Competencies

Australia		United Kingdom
Areas of Competence (Finn)	Key Competencies (Mayer)	Core Skills (National Council for Vocational Qualifications)
• Language and communication	• Collecting, analyzing, and organizing information	• Communication
• Personal and interpersonal characteristics	• Communicating ideas and information	• Communication and Personal Skills (improving own learning and performance)
• Cultural understanding	• Planning and organizing activities	• Personal Skills (improving own learning and performance)
• Mathematics	• Working with others in teams	• Personal skills (working with others)
• Problem solving	• Using mathematical ideas and techniques	• Numeracy: Application of number
• Scientific and technological understanding	• Solving problems	• Problem solving
	• Using technology	• Information technology

Note. Finn, 1991; Mayer, 1992; Jessup, 1991.

designed to reflect the basic criteria on which individuals are recognized within their professions. Such recognition may be based on staged progression and professional experience required after graduation. There are expectations that university courses should establish that specific competencies must have been acquired by those wishing to enter specific professions, including teaching.

It is important to note that the extension of principles underpinning the training reform agenda to the professions has not been universally well received. The emphasis on economic rationalist and corporate managerialist approaches to resolving complex educational questions, including specification of competencies for the professions, has been subject to particular criticism (Bartlett, 1993). So, too, has been the potential misuse of power and system of control by the state as it becomes more involved in the specification of standards and of the underlying knowledge base of competencies (Magnusson & Osborne, 1990). For a profession like teaching, this is a particular concern, as has been pointed out in criticisms of reforms of initial teacher education in England during recent years (see Barton, Barrett, Whitty, Miles, & Furlong, 1994).

The Current Professional Response

Initiatives designed to enhance the quality of teaching, through the clearer specification of teachers' work using competencies, have not been greeted with enthusiasm by all who work in the profession. The fact that part of the thrust (if not a major impetus) is coming from government-driven economic reform agendas is a major problem. Equally, earlier experiences with competencies in the 1960s and 1970s have left their mark. As well, those who work in universities have often expressed strong opposition to competencies, as they fear the vocationalizing of education at the expense of a sound generalist education (see Hager, 1994; Hager & Beckett, 1995).

Over the last few years many involved in teacher education have experienced a sense of déjà vu as they have heard discussions about the desirability of defining teachers' work in terms of competencies. They are aware of the experiences in the 1960s and 1970s when competencies and performance-based training were being hailed as a way to provide a new platform for teacher education programs (Houston, 1974; Howsam, 1976). They are hearing again the high degree of skepticism about attempts to define the complex professional work in which teachers engage in terms of competencies (see, e.g., Evans, 1992; Knight & McWilliam, 1995), and have been reminded of earlier dismissals of the theoretical underpinnings of Competency-Based (or Performance-Based) Teacher Education (e.g., Broudy, 1984; Johnson, 1975; Kaminsky, 1975).

What needs to be recognized is that we have before us a new opportunity to consider the merits of what is now being proposed for development of the professions and the establishment of professional competence. It would be wrong, for example, to assume that, because the current resurrection of interest in competencies has come, in a large part, not from those with highly refined knowledge and experience in education and training, but from those who are perceived to have more immediate, pragmatic economic concerns, there is little to be gained from a critical reappraisal of potential opportunities which might accrue if we can successfully use the competency drive as a basis for revitalizing our current teacher education efforts. Here is a clear example of how educators should be keen to join the educational debate and bring to bear on the issues their informed and critical perspectives, rather than remain aloof.

In later sections of this chapter descriptions will be given of a case study undertaken to develop teacher competencies. While the competencies proposed are of special interest, an equally important outcome of the discussion is the way the proposed competencies can be used to foster further discussions about the work teachers do and how to assist teachers at all levels to carry out their various responsibilities as "extended professionals" (Hoyle, 1975). It is important to note, for example, the vital contribution beginning teachers make by identifying areas of competence seen to be inadequately addressed in current teacher education programs. Such gaps reflect, in part, new circumstances in schools where teachers' roles have changed dramatically in recent years following curriculum reform and restructuring, including moves to have schools more involved in decision making, and for policy decisions to be less centrally driven. In the case of secondary schools, there has been a major impact on teachers' work created by schools having increased numbers of students staying on to the end of secondary school. The gaps identified may also reveal a failure to some degree of teacher education institutions to keep pace with change. Whatever the reasons, it is important for teacher educators to see present moves to define and to focus more closely on the competencies needed to function successfully as a member of the teaching profession as a welcome opportunity to review and improve current practices. Using new insights as a basis for developing more holistic approaches to understanding teaching and teacher competencies, and to assist the ongoing professional development of teachers, will be very helpful in the long run as an educational response to the use of competencies in education and training programs.

COMPETENCIES AND TEACHER EDUCATION

Reference has already been made to the emergence in the 1960s of competency-based teacher education programs, originally developed mainly in North America. As Kennedy (1996) pointed out "conceptually these programs were of the one variety—they focused on discrete tasks and were underpinned by behaviouristic psychology" (p. 16). It is interesting to note, as Kennedy points out, that while competency-based teacher education did not exert any significant long-term impact, its influence lingered on. The residual impression coming from the movement was mainly a negative one, relating particularly to the emphasis on using performance-based and behaviorist notions of education as an appropriate way to train teachers.

Critics of competency approaches in education are quick to target the behaviorist underpinnings of earlier programs, emphasizing, as they did, observable behavior and explicit criteria to enable performance to be measured. The major complaint is that "(competency) models can be highly reductive, providing atomised lists of tasks and functions, or they can be highly generalised, offering descriptions of motivational dispositions or cognitive abilities such as problem solving" (Norris, 1991, p. 334).

It is a mistake to accept that our understanding of competency-based approaches in education remains at levels which existed in the 1960s. This is what Hager has called the *genetic fallacy*: "arguing that because something had certain characteristics at an earlier time, therefore it must have those characteristics now" (Hager, 1994, p. 4). There can be little doubt that, in formulating our views in the 1990s, we need to reject the kinds of technicist approaches apparent in some of the earlier American conceptions of competency-based teacher education (Tuxworth, 1982). The recent focus on competencies has led to much work being done in relation to defining competencies, and to analyzing their relevance and applicability in professional preparation programs. Considerable advances are being made.

Preston and Walker (1993) offered the following definition of "a competencies approach" before going on to contrast the earlier behaviorist approach to competencies with what they term a "holistic" approach:

> A competencies approach can provide a common framework for developing and linking many aspects of work and education—work organisation; deployment; career structures; development and improvement of individuals, groups and workplaces; the labour market; credentials; research and development of the knowledge base of occupation; curriculum development and assessment in initial and post-initial education; and individuals' and groups' access to education, training and work. (Preston & Walker, 1993, p. 117)

Having put forth the view that it is possible to range from comprehensive to partial or limited application of competency approaches to areas they describe, they go on to compare the characteristics of *behaviorist* and *holistic* approaches to competencies, shown in Table 6.3.

To support the case for adopting a holistic approach to competencies in education, Preston and Walker call on the following definition of competence developed by Walker (1993):

TABLE 6.3

Characteristics of Behaviorist and Holistic Approaches to Competencies: A Summary

	Behaviorist Approach	*Holistic Approach*
Nature of competencies and relations between competencies	Individual, specific, discrete defined in terms of behavior only.	Competencies are complex combinations of personal attributes, enabling the performance of a variety of tasks. They form coherent "structures of competence," and attributes have a distinct coherent structure.
Evidence of competencies	Direct observation of performance of relevant activities--assumed to give direct and clear indication of whether or not the competency is held.	A range of evidence may be sought, in general none can give certainty that the relevant competencies are held. What evidence to use and what to make of it would be indicated by relevant theories.
Relation between knowledge and competencies	Required knowledge is inferred directly from behaviorally defined competencies.	Knowledge exists and can be understood separately from the exercise of competencies. Knowledge and understanding can be understood as having a complex and coherent structure.
Relation between competency statements and the education or training program	Competency statements indicate directly the content, structure, and assessment criteria of education and training programs. There can be little diversity, local flexibility, experimentation, and development	There is broad coherence between structures of competence and education and training programs, and programs will generally have overall coherence. Programs can, however, be diverse in their structure and curriculum, be flexible, and involve experimentation and research.
Variation in specification of competency-based standards according to purpose	There can be little variation in the way standards are specified.	The way standards are specified can vary significantly according to purpose. In particular, for "summative" purposes standards can be explicit and public, and assessment procedures rigorous, valid, and reliable; for "formative" purposes a more flexible and open approach is possible.

Note. Preston and Walker (1993, p. 119).

Competence involves the combination of **attributes** (knowledge, capabilities, skills, attitudes) structured into competencies which enable an individual or group to **perform** a role or set of tasks to an appropriate level or grade of quality or achievement (that is, an appropriate **standard**) in a particular type of situation, and thus make the individual or group **competent** in that role. (p. 94)

The writers maintain that a holistic approach:

- incorporates an understanding of competencies as complex and structured, and constituted by personal attributes (such as knowledge) which have coherence and structure;
- allows a perspective wider than individual performance . . . and takes into account the impact of group processes and culture on the level of performance, and on the development and exercise of competencies;
- allows for the role of judgement in competent performance.

(Preston & Walker, 1993, p. 118)

The propositions being put forward here go well beyond the limited notions of competence found in competency-based teacher education programs in the 1960s. There are two aspects of particular importance. First is the notion that competence is much more than performance (as Noddings [1984] has pointed out). Second, the focus is not just on individual action but on group processes and their significance in terms of performance and assessments of performance.

Of relevance for this discussion is further work undertaken by Kennedy (1996). He suggested that the distinctions offered between behaviorist and holistic, or, for him, "integrated" approaches "take the notion of 'competency' far beyond behaviourism and are informed by more recent developments in cognitive psychology, learning theory and related understandings of purposeful action in complex real life situations—including the world of work where initiative, judgement and collaboration are important" (p. 17). He made the point that:

Competency is the relation between an individual's personal *attributes* (such as knowledge, physical and social skills, values and dispositions), the performance of *tasks* (that can be broadly defined and can involve professional judgement), in the *context* of practice (which can be complex and unpredictable). (Kennedy, 1996, p. 17)

By offering his more complex definition of competency, Kennedy has shown how a more elaborate holistic (or integrated) approach can respond to the following plea by Hager and Beckett:

Only by taking proper account of the essentially relational nature of the concept of competence can the holistic richness of work be captured in competency standards. (Hager & Beckett, 1995, p. 2)

The discussion to this point has focused on how it is possible to broaden the approach to competencies in education and go well beyond the more technicist and performance-oriented notions of the 1960s. There is the further question of how to devise appropriate programs to enable those undertaking professional preparation to become professionally competent in relevant areas.

Heywood (1994) has argued that the traditional approach to professional preparation which is "principally concerned with inputs from various established disciplines may not concern itself with the overall qualities and professional abilities of graduates beyond this knowledge base" (p. 3). Heywood is really asking the question: Can we define a set of generic competencies for the professions which can then be used as a framework for developing professional preparation programs? That is, can we start by looking at what is required to be a highly competent professional, then see the contribution that theory and practice can make to achieve this goal?

Hager (1994) has investigated the same question, and he has developed his "integrated or task *plus* attribute approach" to the use of competencies for incorporation in programs of professional preparation. As shown in Table 6.4, Hager suggests that such an integrated or "task *plus* attribute" approach is a powerful way of improving content delivery and assessment of the curriculum.

On the basis of his analysis, Hager (1994) concluded:

> . . . integrated competency standards offer a correction to the all too familiar mistake of education being centred on information absorption and recall. Instead, the structure of knowledge and its relationships and applications is highlighted, thereby promoting a more productive theory of learning. (p. 15)

This analysis of more recent work on competencies, and particularly of the potential significance of competencies for the professions, highlights what advances are being made in thinking about their relevance for professional preparation. Having accepted that there is a significant contribution competency-based approaches can make to professional preparation, the issues now are:

- what competencies to include in programs of professional preparation and further professional development;
- how to structure professional preparation programs to take into account more complex notions of competencies;
- how to enhance opportunities for integrating theory and practice, and enable assessments to be made of progress in mastering competencies.

TABLE 6.4
Three Different Conceptions of Competence and Their Implications

Behaviorist or Specific Tasks Approach	Attribute or Generic Skills Approach	Integrated or Task + Attribute Approach
1. Overt performance is competence	General attributes as predictors of future performance	Competence inferred from performance
2. Atomistic, reductive, trivial, mechanistic, standardized, routine, discrete tasks or skills	Abstract, remote from actual practice, problem of transfer--overall rationale often lacking	Holistic, richness of practice captured
3. Large number of specific competencies--list lengthens with complexity of work, e.g., professions	Small number of generic competencies	Manageable number of key competencies
4. Uniformity (1 right way)	Diversity (> 1 right way)	Diversity (> 1 right way)
5. "Doing" curriculum Practical modules. Jettison current curriculum	Conventional curriculum Fragmented into subjects	Powerful device for improving content delivery and assessment of current curriculum
6. Central control of curriculum	Provider autonomy in curriculum	Profession/provider joint control of curriculum
7. Checklist for ticking--invalid assessment	Traditional assessment—has its limitations	Competence demonstrated over time, portfolios, etc. Assessment needs careful planning
8. Minimum competence "Lowest common denominator discourages excellence" "Deskilling"	Encourages excellence that is remote from professional practice	Richness of quality professional performance is captured

Note. From Hager (1994). Is there a cogent philosophical argument against competency standards? *Australian Journal of Education*, Vol. 38, No. 1, p. 16. Reproduced by permission of the Australian Council for Educational Research Ltd.

DERIVING COMPETENCIES
FOR THE TEACHING PROFESSION

In this section Australia is again taken as a case study to demonstrate approaches adopted to develop generic competencies for beginning teachers. The work to be described was commissioned by a body called the National Project on the Quality of Teaching and Learning (NPQTL), established by the Australian Minister for Employment, Education, and Training and which existed from 1991–1993.

While the work undertaken to develop national competency standards for teachers took place against the broad background of award restructuring and microeconomic reform, the National Project on the Quality of Teaching and Learning recognized that, if properly and appropriately used, national competency standards could:

- assist teachers to improve their work organization and their workplace performance by encouraging them to reflect critically on their own practice, individually and collaboratively;
- inform professional development to support improvements to teaching;
- boost teachers' self-esteem and their commitment to teaching by enhancing their awareness of the nature of their teaching competence;
- underpin a national approach to improving teacher education programs, including curriculum and pedagogy;
- underpin a national approach to improving induction programs in schools and systems;
- possibly form the basis for a nationally consistent approach to registration for employment and probation;
- provide a good basis for communication about the nature of teachers' work and the quality of teaching and learning within the education community and among education interest groups. (Peacock, 1993, p. 8)

Additionally, it was believed that "the development of national competency standards, of explicit and public statements of professional competence should provide a considerable fillip to teachers' professional self-confidence and stimulate restored public confidence in teaching" (Peacock, 1993, p. 8).

Rather than commission a single study to develop teacher competencies, the National Project on the Quality of Teaching and Learning set up three research projects, and each adopted its own approach. The studies are described in Abbott-Chapman, Radford, and Hughes (1993), Eltis and Turney (1992, 1993), and Louden (1993). The discussion that follows is derived from the particular study in which the present author was involved.

The Study

In our study, we (Eltis & Turney, 1992) approached our task conscious of the fact that not all teacher educators (or educators more generally) are comfortable with efforts to define teachers' work in terms of competencies.

The research undertaken was seen as opening up new horizons, rather than limiting interpretations of what teachers do in a kind of *reductio ad absurdum* exercise. The view we had of our work was that our analysis would and should inform the development of appropriate programs to prepare teachers for the profession and to support their further development.

The fact that teaching is complex should not prevent us from attempting to understand and define as precisely as we can what effective teachers do, so that those wishing to become members of the profession can be helped to master, along the lines suggested by Heywood (1994), the skills and knowledge required for effective performance, while reflecting on how they can develop their own professional response to being a teacher. The ultimate significance of any efforts to define teaching competencies will lie not so much in how successful we might be in isolating and defining the various components of what teachers do, but rather in how we use the outcomes to develop effective programs and enhance the quality of the profession.

Principles and Assumptions Underpinning the Study

Underpinning the study were several important principles, and these are restated here as they have significance for any attempt to develop competencies for the profession. It is the absence of such underpinning principles in work undertaken in other countries which has been strongly criticized. Barton and his colleagues (1994), for example, point out that, though they might not strongly object to areas of competence being proposed as a basis for initial teacher education in England, they see no obvious theoretical rationale for the selection of particular competencies.

Validity. A combination of methods should be used to ensure the validity of competencies, their relevance to the workplace, and their contribution to the mission and goals of the profession.

Wide Acceptance. The methodology followed to develop competencies should incorporate wide consultation within the profession so that all stakeholders appreciate and widely endorse the competency standards developed.

Comprehensiveness. The focus should not be exclusively on classroom teaching competencies, but should relate to teacher responsibilities in the whole school and to those in connection with parents and the local community.

Present and Future Orientation. The competencies developed for the profession should reflect not only present work expectations of teachers, but should display an awareness of future or emergent expectations.

Coherence. The competencies proposed should be mindful of the direct relationships between the competencies and the goals and mission of the profession, and of the interlocking and overlapping relationships between the competencies themselves. They should contribute to a coherent view of the profession's work.

Teacher Individuality. The competencies proposed should recognize the individuality and creativeness of teachers as they will make decisions about applying the competencies in their particular school contexts.

Concern for Quality. The identification of competencies should take into account the wide variety of contextual factors which shape the teacher's work.

In addition, particular assumptions were made about *effective teaching*, to which any specification of competencies should be related. The following assumptions were derived from the extensive literature on effective teaching and underpin earlier work on teaching skills undertaken by the authors (see Turney et al., 1983). These assumptions highlight significant aspects needing to be taken into account in programs of professional preparation if the introduction of competencies is to contribute to the development of effective teachers (see earlier discussion and references to Heywood, 1994, and Hager, 1994).

Effective Teaching is Purposeful. Good teaching is not routine. It occurs to achieve particular learning objectives—objectives consistent with the goals and mission of the profession.

Effective Teaching is Flexible and Dynamic. Good teaching is adjusted to the teacher's knowledge of individual students, the curriculum, and other aspects of the teaching context.

Effective Teaching is Responsive to Individuals. Good teaching takes into account the needs and differences of individual students and maximizes the chances of success for all.

Effective Teaching is Reflective and Rational. Good teaching involves planning and evaluation, including critical reflection on performance. It invariably involves teachers in making decisions and judgments about its implementation in particular contexts.

Effective Teaching is Highly Complicated. Careful analysis can be used to simplify the activity of teaching for preparing professionals, but must not trivialize it. Atomistic analyses are totally inappropriate.

Effective Teaching is Integrative. The activity of teaching integrates the use of various teacher competencies.

Effective Teaching is Based on General Principles. Teacher decisions about teaching should be guided by soundly based principles of teaching and learning and by the ethics and values of the profession.

Effective Teaching is Knowledge-Based. Teachers' performance must be informed by knowledge of the subject matter and of the learners, as well as by knowledge of the curriculum and pedagogical practices.

Effective Teaching Goes Beyond Classrooms. The teacher's work in promoting students' learning and their general welfare can be strengthened by the understanding and support of parents and by the cooperation of colleagues.

Effective Teaching Can Be Developed. Systematic teacher education programs, in which pedagogical theory and practice are suitably integrated and sound grounding in subject matter knowledge is achieved, can produce effective beginning teachers, provided they deal with a comprehensive range of competencies covering the responsibilities of teachers in today's schools. An effective teacher will also have both the inclination and capacity to continue to develop professionally.

Conducting the Project

An extensive literature search was conducted covering general teaching competencies and competencies seen as desirable for beginning teachers. The search embraced all aspects of the teacher's work, in school and out, and covered the range from early childhood through elementary and secondary school. The detailed literature search was designed to update earlier work (Turney & Eltis, 1986). In this earlier work, teachers' roles and tasks had been extensively analyzed in three domains: the classroom, the school, and the community.

By integrating the outcomes of the literature search with the earlier analysis, it was possible to arrive at a set of initial competencies in each domain. Work being carried out in other States of Australia on teacher competencies was also reviewed, as well as work being undertaken for other professions (law, nursing, architecture, and veterinary science). Reference was also made to the *Guide to Development of Competency Standards* prepared for the National Office of Overseas Skills Recognition (Heywood, Gonczi, & Hager, 1992), which provided a format for presenting the key competencies, their elements, and performance criteria.

Note was also taken of the definition of the terms *competency* and *competent* in the National Office of Overseas Skills Recognition Guidelines, and the following definitions (which reflect more broadly based notions of competency and competence, described previously) were agreed to as the basis of the study.

Competency focuses on the performance of a significant role expected of teachers. A competency contributes to the achievement of worthwhile educational objectives and integrates its objectives and principles of performance with a combination of knowledge, skills, and attitudes. A competency is not fixed. It may change as the expectations of teachers change and knowledge of teaching changes. Each competency will need to be adapted to the context in which it is implemented. Competencies are not discrete. They overlap and interlock and merge as the activity of teaching proceeds.

Becoming competent involves the acquisition of the knowledge, skills, and attitudes of a particular competency so that an individual can perform it at an appropriate standard.

Consultations were held in the States of New South Wales and South Australia and in the Australian Capital Territory. Involved in consultations were beginning and experienced teachers, including principals from government and nongovernment schools, employer representatives, union representatives, representatives of community groups and parent organizations, and teacher educators. The vast majority of participants contributed to the validation process positively and with considerable interest. This was particularly the case with the beginning teachers, who regarded the process as an opportunity to communicate their professional needs and to comment directly on the quality and adequacy of their professional training. Many saw great promise in competency statements as a means of ensuring that teacher educators become more closely attuned to the realities of the workplace and to emerging developments in schools.

Outcomes

In the final report (Eltis & Turney, 1992), prepared for the National Project on the Quality of Teaching and Learning, the following were included:

- a statement describing the teaching profession and its purpose;
- a set of guiding principles seen as providing a necessary underpinning of teachers' work;
- a set of goals of teaching;
- key competencies set out in four domains.

It will be recalled that initial work to describe teachers' professional activities had included only three domains: the classroom, the school, and the community domain. Feedback given during the study, particularly from beginning teachers, led to the inclusion of one whole new domain alongside the original three, the professional domain. All groups of beginning teachers felt this was an important omission from our initial analysis and one in which they had been inadequately prepared. The professional domain was seen generally to embrace teachers' understanding of the social significance of their work. More particularly, through preparation in this domain, teachers should realize the need for professional self-improvement, for considering professional ethics, for recognizing their legal obligations, and for promoting the status of the profession.

An overview of the proposed competencies in each domain is set out in Table 6.5. For each competency a detailed description is provided along with elements and suggested performance criteria. As an example, a key competency in the classroom domain is entitled "Managing the Learning Environment." In Table 6.6, details are given to show how this competency is described in the final analysis.

Summing Up the Study

As will be evident from the principles and assumptions developed to underpin this developmental work, it is important to remember that preparation of teachers for their roles needs to go well beyond the provision of opportunities for those in training to demonstrate fundamental levels of performance. Personal attributes, values, and ethical considerations still need to be brought into context. That explains why, as part of the study, a statement of purpose for the profession and a set of goals for the teaching profession (not included here) were also developed.

What will really matter, in the long run, will be the way we continue to encourage teachers to be critically reflective upon their practice so that they refuse to be satisfied with minimalist performance on specific performance indicators. At least three levels of reflection have been suggested as the basis for comprehensive analyses: technical, situational or practical, and critical (Zeichner & Liston, 1987). At the technical level, the concern is with the efficient and effective application of educational knowledge to achieve goals not seen as being problematic. At the practical or situational level, reflection requires the examination of underlying assumptions with a view to determining the educational consequences of competing possibilities. Critical reflection assumes that both teaching and the context in which it occurs are viewed as problematic. The work of Zeichner and Liston has as its motivation the need to break away from a "conventional" teacher education approach which "inhibits the self-directed growth of student teachers and thereby fails to promote their full professional development. "Through their work they have promoted a move toward" the goals of reflective teaching, greater autonomy, and increasing democratic

TABLE 6.5
Overview of Competencies

The competencies are listed below without their elements and related performance criteria. The symbol * indicated those competencies which should be developed both in initial teacher education and in further programs of professional development. The competencies are set out in four domains, which encompass the various aspects of professional work undertaken by teachers.

The Classroom Domain

*1	Demonstrating and enhancing curriculum expertise
*2	Planning for learning
*3	Developing and integrating theoretical understandings
*4	Initiating and guiding learning
*5	Facilitating independent learning
*6	Fostering interpersonal relationships and student welfare
*7	Responding to special needs of students
*8	Managing the learning environment
*9	Assessing student learning and evaluating programs

The School Domain

*1	Identifying and mobilizing resources
*2	Cooperating in curriculum development
*3	Organizing and pursuing cocurricular activities
4	Knowing, administering, and improving school policy and procedures
5	Participating in school-wide development programs

The Community Domain

*1	Promoting information exchange
2	Gaining cross-cultural knowledge and understanding
3	Opening the class and school to the community

The Professional Domain

1	Acquiring knowledge of the professional context
*2	Pursuing professional self-improvement
*3	Understanding ethical considerations
*4	Meeting legal obligations
*5	Enhancing the status of the profession

Note. From Eltis and Turney, 1993, p. 33.

participation in systems of educational governance" (Zeichner & Liston, 1987, p. 23).

Teaching continues to be the sum total of performance on particular indicators or competencies. Reflecting on practice, as Knight and McWilliam (1995) point out, is more than "looking back"—its meaning should be expanded to include the tools of critical analysis so that reflection is oriented to social as well as personal professional change. Accepting this

TABLE 6.6
Example of the Presentation of Competencies
Key Competency 8: Classroom Domain
Managing the Learning Environment

Managing the learning environment, most commonly a classroom, so that learning can effectively occur is a fundamental teacher competency. This competency seeks to establish and sustain conditions under which students feel secure and unthreatened and in which there are high levels of student involvement in classroom activities and teaching time is efficiently use.

In pursuing such ends, managing the learning environment minimizes student disruption and off-task student behavior and encourages students to assume responsibility for their own behavior as individuals and as group members. Successful classroom management then, not only involves responding effectively to problems when they occur but preventing problems from occurring.

The six main elements of this competency are as follows:

8.1   *Creating a safe, secure, and supportive environment*

Performance criteria include the ability to

- organize the classroom to meet the safety and security needs of students and to reduce stress and opportunities for inappropriate behavior.
- create an attractive and stimulating learning environment.
- promote a climate in the classroom in which all students respect the contributions of each other and accept the need to be mutually supportive.
- be responsive to students who are anxious or distressed and offer them support

8.2   *Using effective management strategies*

Performance criteria include the ability to

- structure and supervise the learning of individuals and groups so that the work proceeds with smoothness, momentum and strong student interest and involvement.
- plan daily learning programs which will accommodate students' learning needs and are suited to the time of day.
- handle disruptions in an appropriate way so that attention to learning is not distracted.
- recognize and commend appropriate student behavior.

8.3   *Establishing routines and standards*

Performance criteria include the ability to

- establish classroom routines so as to use time efficiently and establish cooperation with students.
- negotiate with students positive and clear expectations of behavior

8.4   *Observing and recognizing patterns of behavior*

Performance criteria include the ability to

(Continued)

TABLE 6.6
(Continued)

- observe individual students to identify their unique characteristics.
- identify trends or patterns of student behavior over time.
- notice discrepancies or departures from typical behavior patterns.
- identify causes of atypical behavior and help the student overcome or cope with these causes and know where to obtain support to help seriously troubled students.

8.5   *Encouraging student responsibility*

Performance criteria include the ability to

- encourage students to work independently as individuals or in groups and to accept responsibility for their own behavior and learning.
- enlist the cooperation of students and parents, caregivers, and other interested groups if necessary in dealing with issues and problems being expressed by students.

8.6   *Managing resources and facilities*

Performance criteria include the ability to

- maximize the time students spend actively engaged in learning.
- locate and prepare in advance appropriate and sufficient resources needed by students.
- arrange classroom furniture and equipment to facilitate learning activities.
- ensure all students have equitable access to equipment and resources.

Note. From Eltis and Turney, 1993, p. 34-35.

point of view implies that, as we use our knowledge derived from analyzing teachers' work to arrive at possible competencies, we need to go well beyond a limited focus on establishing, performing, and testing particular behaviors in isolated contexts in order to demonstrate competence.

THE FUTURE

Australian experience has been used in this chapter to highlight a number of critical issues relating to the development and use of competencies in educational settings. It remains to pull together various threads and suggest desirable directions for future thinking about uses of competencies within the teaching profession.

Competencies as They Relate to Teachers' Work

The case study just presented highlights the need for close and careful consultation to occur when attempts are made to define in detail the work teachers undertake. As Ozga and Lawn (1988) argued, for a study of teachers' work to be adequate, it needs to encompass teacher "practices, struggles, lived experience and contradictions" (p. 334). To paint such a comprehensive picture, discussions need to occur at all levels of the profession—with beginning and experienced teachers, with those in executive positions including principals and administrators, and most importantly, with members of the community. And we should not forget the students. That is, we need to talk to all of those with whom teachers work or who have responsibility for the circumstances in which they carry out their professional responsibilities.

It is important, too, to consider how best to provide clear specifications of the competencies teachers need for successful performance of their work. It is one thing to arrive at a set of specifications; it is another to present these in a way that makes them useful for the teachers themselves and also for course developers. Whereas Eltis and Turney (1992) analyzed teaching competencies in terms of four broad domains in which teachers can be seen to operate, Louden, in his study (1993) conducted for the National Project on the Quality of Teaching and Learning, suggested five broad *units of competence* for teachers:

- Teaching practice
- Student needs
- Relationships
- Evaluating and Planning
- Professional responsibilities.

He then broke each of these units of competence into a series of *elements of competence*; for example, in the case of "student needs" he suggested the following five elements:

- Understands how students develop and how they learn;
- Recognizes and responds to individual differences among students;
- Fosters independent learning;
- Believes that all students have the right to learn;
- Takes action to eliminate harassment among students.

To the elements are then added performance criteria and brief *case summaries* to illustrate the competencies. The use of the stories was "adopted in order to add the flavour of the complex and indeterminate problems teachers face in their work with students" (Louden, 1993, p. 17). Most importantly "the chief value of the story is not in *defining* a standard of competence, but in *exemplifying* the quality alluded to in the element of the competence" (p. 19). Louden makes an important point in presenting his findings and one which needs to be taken into account as further work is undertaken:

> In place of the search for precise and scientifically warranted effective be-
> haviours, many researchers have focused on teachers' practical knowledge
> in particular disciplines. The difference between these research programs
> and the teacher effectiveness program emanates from contrasting assump-
> tions about the nature of the knowledge teachers have and use. Rather than
> assuming that teachers' knowledge may be reduced to a set of scientifically
> validated propositions about effective teaching, these researchers have em-
> phasised the importance of context, subject content and personal experience
> in determining what counts as effective teaching for individual teachers
> working with particular groups of students. From this perspective, lists of
> generic skills, and standards expressed in context-free behavioural terms,
> do not make much sense. (Louden, 1993, p. 19)

It will be obvious from this comment that there is much still to be debated in the area of specifying teacher competencies. It is incontestable that lists of generic skills, on their own, will not greatly assist teachers. It remains to be seen, however, whether the power of case examples will be an adequate way of overcoming the somewhat sterile nature of long lists of competencies. While cases may help by stimulating debate, it seems more probable that the use of competency statements to provoke discussions among teachers about what constitutes effective teaching will be the key. If such discussions are to be meaningful and have impact, then the competency statements themselves need to be provided in a language accessible to teachers. Similarly, perform-ance criteria need to be capable of easy interpretation and written so that they invite consideration of minimal and optimal levels of performance and encourage teachers, working collaboratively, to enhance the quality of their efforts. In workplace discussions the "case studies" teachers generate for themselves as they consider aspects of competent teaching will be the key to improving the quality of the profession.

One of the underlying principles put forward as a basis for the case study described in detail earlier was that definitions of teacher competen-cies should have **a present and future orientation**. Conditions under which teachers now work have changed dramatically since the mid-1980s, par-ticularly as a result of the increased powers given to schools for decision-making and policy establishment through the process known as devolution. In a perceptive analysis of changing professional behaviors of teachers, Hargreaves (1992) suggested the following dimensions of what he calls "The New Professionalism":

- shifts from individualism in schools to collaborative endeavors, at all levels;

- the sharing of power "to get the job done";
- the "flattening" of school hierarchies and the promotion of a more collegial culture in schools;
- changing principles of professional development, with a closer reconciliation of individual professional needs and school needs;
- schools "seizing the opportunity" and making full use of the power and authority now being devolved to them.

To Hargreaves' list, we could add rapid changes affecting teachers' work as advances in technology play an ever-increasing role in our lives and bring new challenges to the curriculum and to the way schooling is managed.

Competencies developed for the teaching profession should take into account, as best we can, not only changing circumstances; they should encourage speculation about how the profession can manage its future by being more anticipatory. There is, then, a further implication that definitions of competencies need to be regularly reviewed for their relevance.

Finally, in discussing competencies as they relate to teachers' work, we need to consider how to move forward with analyses which go beyond the general. Further work needs to be explored with teachers in various sectors and at various levels to see how clear definitions of roles and tasks might ease transition into particular roles (e.g., that of school principal) and to determine how teachers might be better prepared for and supported in these positions.

Integration of Competencies into Initial and Further Teacher Education

Earlier in this chapter, reference was made to how programs of professional preparation often concentrate more on knowledge transmission and less on desirable qualities and professional competencies of graduates beyond knowledge bases.

New ways need to be found to enable those collaborating in teacher education programs (lecturers in universities and colleges, teachers in schools, and the students themselves) to assist students to identify needed competencies and how to master them. That is, careful consideration needs to be given to stages through which student teachers might progress, how to ensure there is critical reflection (as described earlier) on what occurs at each stage, and how to give support to enable student teachers to reach their potential. Attention needs to be given not only to the nature of the experiences students might undergo, but also to the changing roles lecturers and teachers in schools might play in supporting the development of young teachers.

As appropriate learning experiences are developed, consideration of competency statements will be helpful in ensuring that all appropriate areas are addressed. More importantly, by using as a basis for program planning competencies derived from teachers' professional roles and responsibilities, it will be possible to arrive at a more integrated set of professional experiences, bringing together components from areas which often exist as separate course components. As Tomlinson (1995) has argued: "Having a clear conception of the intelligent practical capabilities we wish to engender in student teachers in no way commits us to a behavourist approach to training. If anything, it gives us a better starting point for seeking relevant and well-grounded pedagogy, since explicitly or implicitly we always need a pedagogy in addition to learning outcome intentions if we are actually to design a curriculum, whether for student teachers or for school students" (p. 184). Moreover, if integrated course components built around competencies are offered in both on-campus and school settings, much will have been achieved to address the perennial difficulty in teacher education—how to integrate theory and practice.

An analysis of the roles and tasks teachers perform to arrive at competencies associated with professional performance will also provide an opportunity to determine what aspects of teacher development are best addressed in initial teacher education and how far a student teacher might progress in these areas, and what dimensions are best left for induction and further professional development programs (see Turney, Eltis, Towler, & Wright, 1985). In the case study presented earlier, it was suggested that certain competencies (though not ignored in initial teacher education) will be better addressed in detail when the new teacher enters a school (for example: knowing, administering, and improving school policy and procedures; gaining cross-cultural knowledge and understanding). This point is well taken up by Nance and Fawns (1993), who propose that it is important "to develop arrangements within schools that increase the sophistication of our teaching for, and appraisal of, teacher competence" (p. 161). Having argued that there are two "faces" of teaching competence—skills and knowledge, and contextual understandings—they go on to suggest that it is important to establish programs which foster growth at various levels.

If school systems become more aware of the extent to which student teachers have addressed certain issues and developed certain competencies in their initial preparation, with an explicit expectation that areas addressed will be further pursued while others will need much closer attention, then we shall have made an important step forward in offering an appropriate set of graduated experiences designed to support teacher development. Much is said about the desirability of establishing a continuum in teacher education, extending from preservice to induction and further professional development. Creative partnerships involving teacher educators, teachers, and school authorities are needed to make such a continuum a reality. Those working in such creative partnerships would do well to look at how competencies linked to the professional responsibilities of teachers can help them to develop an integrated approach to teacher education delivery.

Competencies and Assessment

From earlier discussion it will be obvious that much store is placed on the need for teachers to engage in discussions of performance and to share experiences, work samples, and teaching programs to determine how well criteria established in relation to particular competencies are being met. An essential component of such discussions is the way teachers can be encouraged to reflect critically on their achievements in order to determine how they can enhance their performance. Agreed understandings, arrived at through consideration of definitions of competencies and related performance criteria, will ensure that those working with young teachers will be in a sound position to be able to offer constructive advice and assistance, and that young teachers themselves should be able to make a positive contribution to their own growth. If student teachers in their initial training are accustomed to approaches which encourage thoughtful and critical reflection on practice, then they will be able to respond positively to further support when they begin teaching and feel confident as they approach assessment for appointment to a school on a permanent basis.

Competencies have been seen as a way of achieving appropriate standards for the teaching profession. It was pointed out earlier that in Australia, when the National Project on the Quality of Teaching and Learning decided to carry out work on the development of national competency standards, one of the perceived outcomes was that such standards could possibly form the basis for a nationally consistent approach to the way teachers are initially assessed for employment and then confirmed in their positions at the end of a defined probationary period. In some places, and especially Australia where the States and Territories have constitutional responsibility for education and schools, this is a vexed issue. If the community generally is to have greater confidence in and respect for the teaching profession, then we must be certain to ensure that standards for entry into the profession are suitably demanding, understood by the public, and applied with rigor. The use of competencies can assist in this process. All teachers should be required to meet appropriately determined standards both on entry and for continuing employment. It is also worth considering how initial standards for entry and probation purposes may be expanded and a set of performance criteria established for teachers to meet when they seek to continue their professional practice and/or move into other areas of responsibility.

It is to be regretted that in many places inadequate measures are taken to ensure that all teachers who work in schools are assessed against appropriate criteria and registered before they begin teaching. It is to be hoped that the current emphasis on being competent and meeting appropriate performance criteria will result in renewed pressure to ensure that systems are established which lead to efficient performance appraisal procedures which involve not only school authorities, but also those who have responsibility for teacher education programs. A strengthening of ties between employers and teacher education institutions on this issue should also result in stronger efforts to enhance the quality of teacher education programs themselves.

REFERENCES

Abbott-Chapman, J., Radford, R., & Hughes, P. (1993). Teacher competencies—a developmental model. *Unicorn, 19*(3), 37–48.

Bartlett, L. (1993). One nation: The regulation of teacher quality and the deregulation of the teaching profession. In J. Knight, L. Bartlett, & E. McWilliam (Eds.), *Unfinished business: Reshaping the teacher education industry for the 1990s* (pp. 66–92). Rockhampton, Australia: University of Central Queensland Press.

Barton, L., Barrett, E., Whitty, G., Miles, S., & Furlong, J. (1994). Teacher education and teacher professionalism in England: some emerging issues. *British Journal of Sociology of Education, 15*(4), 529–543.

Broudy, H. S. (1984). The university and the preparation of teachers. In L. Katz & J. Rath (Eds.), *Advances in teacher education* (Vol. 1). Norwood, NJ: Ablex.

Eltis, K. J., & Turney, C. (1992). *Generic competencies for beginning teachers.* Report to the National Project on the Quality of Teaching and Learning. Sydney: University of Sydney.

Eltis, K. J., & Turney, C. (1993). Defining generic competencies for beginning teachers. *Unicorn, 19*(3), 24–36.

Evans, G. (1992). Competence, competencies and the knowledge-base of beginning teachers. *The Journal of Teaching Practice, 12*(1), 81–98.

Finn, B. (1991). *Young people's participation in post-compulsory education and training.* Report of the Australian Education Council Review Committee. Canberra: Australian Government Publishing Service.

Gonczi, A., Hager, P., & Oliver, L. (1990). *Establishing competency standards in the professions.* National Office of Overseas Skills Recognition (Research Paper No. 1). Canberra: Australian Government Publishing Service.

Hager, P. (1994). Is there a cogent philosophical argument against competency standards? *Australian Journal of Education, 38*(1), 3–18.

Hager, P., & Beckett, D. (1995). Philosophical underpinnings of the integrated conception of competence. *Educational Philosophy and Theory, 27*(1), 1–24.

Hargreaves, D. H. (1992, July). *The new professionalism.* Paper presented to Fourth International Symposium on Teachers' Learning and School Development, University of New England, New South Wales.

Heywood, L., Gonczi, A., & Hager, P. (1992). *A guide to the development of competency standards for professions.* National Office of Overseas Skills Recognition (Research Paper No. 7). Canberra: Australian Government Publishing Service.

Heywood, L. (1994, September). *How competency standards could guide curriculum development.* Paper presented at the Professional Education and Development Conference, Sydney, Australia.

Houston, W. R. (1974). *Exploring competency-based education.* Berkeley, CA: McCutchan.

Howsam, R. (1976). Education for the teaching profession: Up-start or start-up? *South Pacific Journal of Teacher Education, 4*(3), 232–243.

Hoyle, E. (1975). Professionality professionalism and control. In V. Houghton, R. McHugh, & C. Morgan (Eds.), *Management in education reader I.* London: Ward Lock.

Jessup, G. (1991). *Outcomes: NVQs and the emerging model of education and training.* London: Falmer Press.

Johnson, H. C. (1975). Not one "unnecessary wriggle": some questions about the presuppositions of C/PBTE. *Educational Theory, 25,* 156–167.

Kaminsky, J. S. (1975). C/PBTE. An investigation in the philosophy of social science and competency/performance based teacher education. *Educational Theory, 25,* 303–313.

Keating, J. (1995). *Australian training reform: Implications for schools.* Carlton, Victoria: Curriculum Corporation.

Kennedy, K. J. (1996, July). *Professional education and the teaching profession in Australia: Towards the establishment of professional standards.* Invited paper, Annual Conference of the Japan–US Teacher Education Consortium, Naruto City, Tokushima Prefecture in Shikoko Island, Japan.

Knight, J., & McWilliam, E. (1995). New carpets for old brooms? Training teachers for the education industry. *South Pacific Journal of Teacher Education, 23*(1), 83–96.

Knight, J., McWilliam, E., & Bartlett, L. (1993). The road ahead: Refashioning Australian teacher education for the twenty-first century. In J. Knight, L. Bartlett, & E. McWilliam (Eds.), *Unfinished business: Reshaping the teacher education industry for the 1990s* (pp. 139–153). Rockhampton, Australia: University of Central Queensland Press.

Louden, W. (1993). Portraying competent teaching: Can competency-based standards help? *Unicorn, 19*(3), 13–23.

Magnusson, K., & Osborne, J. (1990). The rise of competency-based education: a deconstructionist analysis. *Journal of Educational Thought, 24*(1), 5–13.

Mayer, E. (1992). *Putting general education to work: The key competencies report.* Melbourne: The Australian Education Council and the Ministers for Vocational Education, Employment, and Training.

Nance, D., & Fawns, R. (1993). Teachers' working knowledge and training: The Australian agenda for reform of teacher education. *Journal of Education for Teaching, 19*(2), 159–173.

National Training Board. (1991). *Policy and guidelines.* Canberra: Australian Government Publishing Service.

Noddings, N. (1984). Competence in teaching: A linguistic analysis. In E. C. Short (Ed.), *Competence: Inquiries into its meaning and acquisition in educational settings.* Lanham, MD: University Press of America.

Norris, N. (1991). The trouble with competence. *Cambridge Journal of Education, 21*(3), 331–341.

Ozga, J., & Lawn, M. (1988). Schoolwork: Interpreting the labour process of teaching. *British Journal of Sociology of Education, 9,* 323–336.

Peacock, D. (1993). The development of national competency standards for teaching. *Unicorn, 19*(3), 7–12.

Porter, P., Rizvi, F., Knight, J., & Lingard, R. (1992). Competencies for a clever country: Building a house of cards. *Unicorn, 18*(3), 50–58.

Preston, B., & Walker, J. (1993). Competency-based standards in the professions and higher education: A holistic approach. In C. Collins (Ed.), *Competencies: The competencies debate in Australian education and training* (pp. 116–130). Deakin: The Australian College of Education.

Tomlinson, P. (1995). Can competence profiling work for effective teacher preparation? Part I: General issues. *Oxford Review of Education, 21*(2), 179–194.

Turney, C., & Eltis, K. J. (1986). *The teacher's world of work.* Sydney: Sydmac Academic Press.

Turney, C., Eltis, K. J., Hatton, N., Owens, L. C., Towler, J., & Wright, R. (1983) *Sydney micro-skills redeveloped: Series 1 and 2.* Sydney: Sydney University Press.

Turney, C., Eltis, K. J., Towler, J., & Wright, R. (1985). *A new basis for teacher education: The practicum curriculum.* Sydney: Sydmac Academic Press.

Tuxworth, E. (1982). *Competency in teaching: A review of competency and performance-based staff development.* London: Further Education Curriculum Review and Development Unit.

Walker, J. (1993). A general rationale and conceptual approach to the application of competency-based standards in teaching. In Schools Council, *Agenda papers: Issues arising from 'Australia's Teachers: An agenda for the next decade'.* Canberra: Australian Government Publicity Service.

Zeichner, K., & Liston, D. (1987). Teaching student teachers to reflect. *Harvard Educational Review, 57*(1), 23–48.

Chapter 7

Teamwork Competencies: The Interaction of Team Member Knowledge, Skills, and Attitudes

Janis A. Cannon-Bowers
Eduardo Salas
Naval Air Warfare Center
Training Systems Division

The importance of teams in modern organizations has increased in recent years (Guzzo & Salas, 1995). This trend has several sources. First, many organizations are electing to move to team-based organization as a means to eliminate layers of management and as a strategy to increase quality and productivity (Tannenbaum, Salas, & Cannon-Bowers, 1996). The reasoning here is that teams can be responsible for their own products and portion of work; as such they are "empowered" to maximize their own outcomes. Second, several costly incidents involving teams occurred in the late 1980s and early 1990s. These included the downing of an Iranian airliner by a U.S. warship (Rouse, Cannon-Bowers, & Salas, 1992) and a number of aviation mishaps and accidents (Lauber, 1987). Taken together, the analyses of these incidents have all suggested that faulty teamwork was at least a contributing factor in the tragedy. Third, the advent of advanced technology—particularly telecommunications—allows for individuals at remote locations to work together in "virtual" teams (Urban, Bowers, Cannon-Bowers, & Salas, 1995). Such situations are becoming more common, a trend that is likely to continue for many years.

Increased interest in, and use of, teams to perform crucial, dangerous, and stressful tasks in organizations requires that team-centered human resource systems be developed. In particular, mechanisms to select, train, reward, and maintain teams must be developed and implemented. At the core of these systems is the concept of team competencies—that is, the knowledge, skills, and attitudes required to be an effective team member.

In fact, it is essential to understand the nature of competencies required to function in a team as a means to define selection criteria, design and conduct training, and assess team performance.

The purpose of this chapter is to review recent work pertaining to team competencies, then focus and expand on a subset of these; namely those competencies that are *shared or compatible* among team members. To accomplish this goal, we first briefly review the literature regarding team competencies, with emphasis on recent work by Cannon-Bowers, Tannenbaum, Salas, and Volpe (1995). Next, we delineate our notion of shared team competencies, and define a set of these that we believe are essential to effective teamwork. Based on these competencies, we define a series of propositions for teamwork and present directions for future research that we believe move the field forward, from both a researcher's and a practitioner's perspective.

REVIEW OF TEAM COMPETENCIES LITERATURE

As noted, Cannon-Bowers et al. (1995) recently presented a detailed treatment of the team competencies area. Due to the extent and timeliness of this work, we review it in some detail here. To begin with, Cannon-Bowers et al. (1995) defined teamwork competencies as consisting of the knowledge, skills, and attitudes (KSAs) required for effective teamwork. In this context, *knowledge* refers to the necessary understanding of facts, concepts, relations, and underlying foundation of information a trainee needs to perform a task (Goldstein, 1993); e.g., understanding the rules of the road when learning to drive. *Skills* are those behavioral and cognitive sequences and procedures necessary for task performance (Cannon-Bowers et al., 1995); e.g., being able to shift gears while driving. *Attitudes* refer to the necessary affective components of the task (Cannon-Bowers et al., 1995); e.g., a willingness to obey traffic signals while driving.

Cannon-Bowers et al. (1995) went on to suggest that in teams, competencies could be delineated on two bases: (a) whether they are *task specific or generic*, and (b) whether they are *team specific or generic*. According to the authors, these delineations are important because they determine how best to train and maintain team performance. Therefore, what Cannon-Bowers et al. are suggesting is that the specific types of knowledge, skill, and attitudes required to be an effective team member vary depending on whether the KSAs are task-specific or task-generic, and team-specific or team-generic. This issue is addressed further after we describe the categories set out by the Cannon-Bowers et al. framework.

Turning first to *task-specific competencies*, Cannon-Bowers et al. (1995) defined these as competencies (or KSAs) that are executed in a manner that is particular to a task. An example here would be task interaction procedures, where the nature of the interaction among members is a function of specific task requirements. *Task-generic competencies*, on the other hand, are more generalized. These are KSAs that can be transported across a variety of tasks. Examples here include communication skills and interpersonal skills, both of which may be applicable across a host of tasks.

Team-specific competencies are defined as those that are specific to a particular set of team members; that is, they only have meaning for a single configuration of members. An example is shared knowledge. In this case, the level of the competency changes when team members change, so that shared knowledge may be high with one set of team members, but low if one of those members is replaced. In contrast, *team-generic competencies* can generalize to settings with different teammates (e.g., attitudes toward teamwork, assertiveness). Here the competency remains constant even if the team configuration changes. Cannon-Bowers et al. (1995) used these concepts to create four categories of teamwork competencies; these are shown in Fig. 7.1.

According to Fig. 7.1, KSAs that are both task-generic and team-generic are called *transportable* competencies; those that are task-specific and team-generic are called *task-contingent*; those that are task-generic and team-specific are called *team-contingent*; and those that are both task-specific and team-specific are called *context-driven*. The definition of each of these categories is also indicated in Fig. 7.1. Cannon-Bowers et al. (1995) went on to categorize a host of knowledge, skills, and attitudes for teamwork into the four categories shown in Fig. 7.1. Table 7.1 presents a modified version

	TEAM GENERIC	**TEAM SPECIFIC**
TASK GENERIC	*Transportable Competencies:* Generic competencies that generalize to many tasks and team situations	*Team-Contingent Competencies:* Competencies that are specific to a particular configuration of team members, but not to any particular task situation
TASK SPECIFIC	*Task-Contingent Competencies:* Competencies that are related to a specific task, but that hold across different team member configurations	*Context-Driven Competencies:* Competencies that are dependent both on a particular task and team configuration; these vary as either the task or team members change

FIG. 7.1. Definition of team competencies.

TABLE 7.1
Proposed Competencies for Teams

Nature of Team competency	Description of Team Competency	Knowledge	Skills	Attitudes
Context-Driven	Team-Specific & Task-Specific	Cue/strategy associations Task-specific teammate characteristics Shared task models Knowledge of team mission; objectives; norms; resources	Task organization Mutual performance monitoring Shared problem model development Flexibility Compensatory behavior Information exchange Dynamic reallocation of function Mission analysis Task structuring Task interaction Motivating others	Team orientation (morale) Collective efficacy Shared vision
Team-Contingent	Team-Specific & Task-Generic	Teammate characteristics Knowledge of team mission; objectives; norms; resources Relationship to larger organization	Conflict resolution Motivating others Information exchange Intrateam feedback Compensatory behavior Assertiveness	Team cohesion Interpersonal relations Mutual trust

		Knowledge	Skill	Attitude
Task-Contingent	*Team-Generic & Task-Specific*	task-specific role responsibilities Task sequencing Team role interaction patterns Procedures for task accomplishment Accurate task models Accurate problem models Accurate problem models Knowledge of boundary spanning role Cue/strategy associations	Task structuring Mission analysis Mutual performance monitoring Compensatory behavior Information exchange Intrateam feedback Assertiveness Flexibility Planning Task interaction Situational awareness	Planning Flexibility Morale building Cooperation Task-specific teamwork attitudes
Transportable	*Team-Generic & Task-Generic*	Understanding of teamwork skills	Morale building Conflict resolution Information exchange Task motivation Cooperation Consulting with others Assertiveness	Collective orientation Importance of teamwork

Note. Adapted from Canon-Bowers, Tannenbaum, Salas, and Volpe (1995). Reprinted with permission of Jossey-Bass.

of their results. Inspection of Table 7.1 reveals that each category contains a variety of competencies (i.e., KSAs) that are essential for effective team-work. Turning to column 3, this table suggests that several types of knowl-edge are required that can be considered context-driven (i.e., specific to both the task and team), while a slightly different set is required when the competencies are team-contingent (i.e., specific to the team, but not to the task). In column 4, the skills associated with each category are listed; and in column 5, attitudes are shown. These are explained more fully in the following sections.

Context-Driven Competencies

As noted, context-driven competencies are those that are specific to both the task and the team. According to Cannon-Bowers et al. (1995), these competencies are particularly important for teams who perform tasks that are highly demanding. Examples here include military teams, aircrews, emergency medical teams, and sports teams. Such tasks require that teams are flexible and quick to adapt their strategies; hence, they must have competencies that are specific to both the task and team. In addition, when team members have stable membership and perform a single task together, they are likely to benefit from context-driven competencies. This is because they can develop highly specific competencies over time since team membership and task demands remain fairly constant. An example of this type of team is an emergency medical team. In this case, team members must make quick, interdependent decisions, and rely on their knowledge of the task and teammates in order to be successful. If team members under such conditions had to make explicit all of their resource needs to teammates, precious seconds would be lost. Instead, effective emergency medical team members learn to share their conception of task demands and of their teammates in order to maximize effectiveness.

Moving to the third column of Table 7.1, Cannon-Bowers et al. (1995) contended that context-driven knowledge competencies in teams include the need for team members to hold accurate, detailed knowledge about the knowledge, skills, style, preferences, strength, and weaknesses of their teammates. Such knowledge is both team- and task-specific. It allows team members to anticipate the behavior and information needs of teammates and to perform in accordance with those expectations. Along the same lines, Cannon-Bowers et al. also argue that context-driven teams demand members who possess shared, accurate knowledge of task and environ-mental demands and of the role responsibilities of themselves and others. Once again, the reasoning here is that shared knowledge among team members allows them to anticipate appropriate behavior (based on what they expect teammates to do, and how they expect the task to unfold).

Moreover, in high workload or stressful situations, shared knowledge allows team members to coordinate implicitly, without the need to communicate overtly (Kleinman & Serfaty, 1989). Finally, team members must have (and share) cue-strategy associations so that they understand correct team strategies given particular task cues.

Skill competencies (column 4 in Table 7.1) in the context-driven category include task organization (i.e., sequencing and integrating task inputs according to team and task demands), which is specific to both the task and team. It is also crucial that team members have the ability to reallocate functions dynamically. The importance of this skill is that it allows team members to adapt to varying levels of task demands since the "load" can be shared—and optimized—among members. In highly demanding tasks, such adaptation must occur in "real-time" as the task is unfolding. Related to this, team members must be able to monitor each other's performance, so that when one member is becoming overloaded, others may rearrange the load so as to reduce the impact on that member. Context-driven team members must also be flexible, efficient in information exchange, and able to assess their task/mission requirements accurately. All of these skills contribute to the ability of team members to make adjustments necessary in highly demanding task contexts.

Attitude competencies (column 5 in Table 7.1) appropriate for context-driven teams are those that are specific to both the task and team. Such attitudes are necessary to ensure that team members can meet difficult environmental demands. For example, team members must develop collective efficacy—that is, the belief on the part of team members that the team has the resources to cope with the task. Such attitudes are specific to both the task and team, since team members' beliefs would be different if either of these changed. In addition, team orientation—or morale—is an attitude hypothesized to be necessary in context-driven teams.

Team-Contingent Competencies

This category refers to competencies that are specific to a particular set of team members, but not to a single task. According to Cannon-Bowers et al. (1995), task situations that require such competencies occur when a particular team must perform together across a variety of tasks. Several examples of such teams are currently popular in industry: self-managed work teams, management teams, quality circles, and functional department teams. The common characteristic of these teams is that members can build accurate knowledge about one another, but are less able to predict specific task demands since the task being performed is not consistent. This is because members work together consistently, but on several (or many) different tasks. Therefore, competencies in this category are specific to the team, but not to the task.

To begin with, knowledge competencies in this category include shared knowledge about the characteristics of teammates, including knowledge, skills, attitudes, preferences, styles, strengths, weaknesses, and the like. Knowledge of such variables in teammates allows team members to adjust their behavior accordingly. For example, a team member may not monitor closely a teammate who is perceived to be particularly competent, but may watch closely another who is not perceived to be as competent. Team members performing in team-contingent situations must also have a shared sense of the team norms, resources, missions, and objectives. Because these factors all affect how the team functions together, it is important that team members have a common conception of them.

Skill competencies in the team-contingent category are expected to affect team performance, but are not limited to a particular task. They include information exchange strategies, intra-team feedback (i.e., feedback that team members give to one another during or after task performance), compensatory behavior (i.e., backing each other up when necessary), interpersonal relations (since team members are called upon to perform together over extended periods of time), and leadership skills. In addition, team members in this category must be flexible and be able to plan for task contingencies.

Team-contingent attitude competencies are those associated particularly with how team members regard one another; they depend specifically on the team members involved. Examples of such competencies include: team cohesion, interpersonal relations, and mutual trust, all of which are likely to hold across task areas.

Task-Contingent Competencies

Task-contingent competencies are described by Cannon-Bowers et al. (1995) as those that are dependent on a particular task, but not limited to particular team members. Examples of when such competencies apply include any team situation where team membership changes frequently as a function of turnover or as an organizational policy. For example, many military cockpit crews rotate membership in order to accommodate duty schedules. In such cases, teammates do not have much opportunity to learn about one another, or to use this knowledge to their advantage in task performance. Therefore, team members in such situations must be trained in team work competencies that are not specific to a set of teammates, but rather are specific to the interaction demands of the *task*.

Teams in this category are not able to have rich knowledge about their teammates, but can hold detailed task and/or problem models. These models allow them to draw common expectations about the task, even though the players may change. They must also have an extensive understanding of task-specific role responsibilities, that is, how various roles interact and the relationship and interdependency among roles.

With respect to skills, task-contingent skills are those that are associated with a particular team task, regardless of who the team members are (i.e., they remain relatively constant across team members). These include leadership, performance monitoring and giving feedback, situational awareness, planning, and coordination skills. In some sense, team members in teams whose membership is unstable must depend on task cues in order to predict the behavior and information needs of their teammates since they have little knowledge of the other members.

Finally, attitude competencies in the task-contingent category are associated with how team members feel about working on teams in general, rather than on any specific team. Therefore, a member's attitudes toward teamwork are important because they indicate whether the team member values being part of a team, and whether he or she believes the work is best accomplished by a team.

Transportable Competencies

Transportable team competencies in the Cannon-Bowers et al. (1995) framework are truly generic—they are not dependent on any particular task or team. Rather, they can be transported between tasks and teams. Many teams in industry require transportable competencies: task forces, process-action teams, project teams, and the like. Common to these teams is the notion that they perform many different tasks, and that their membership is not stable.

Since team and task knowledge must be generic in this category, the only knowledge that team members can hold is a knowledge of teamwork skills—that is, an understanding of appropriate behavior and of the task or environmental cues that trigger it.

Moving to skills, several transportable team skills are necessary to optimize team performance. For example, team members can have generic interpersonal skills (i.e., a general understanding of how to manage relations, resolve conflict, build morale, etc.), communication skills, and task motivation. In all cases, these general skills must be honed and applied to the particular task and team at hand.

Finally, a couple of transportable attitudes exist that can be expected to affect team performance. The first is collective orientation, which is defined as an individual's propensity to be a team member. Collectively-oriented individuals are more likely to appreciate being on a team than egocentric individuals. Team members must also understand the importance of teamwork for task success, as their motivation to learn team knowledge and skill is associated with it.

Implications of the Framework

Cannon-Bowers et al. (1995) suggest that these categories of team competencies are differentially applicable depending on the nature of the task and environment in which the team performs. In fact, they provided a series of propositions that link the categories to selected task and environmental variables. First, they argue that tasks that are highly interdependent require context-driven competencies. This is because interdependency demands a greater reliance on teammates and a greater need for coordinated action. Therefore, team members must have extensive knowledge of each other and of the task to perform optimally. In fact, shared knowledge in such situations allows team members to react without needing to discuss strategy ahead of time. An example of where such functioning may be necessary would be in an emergency room, where team members must execute coordinated behavioral strategies with little or no time to discuss them in advance of being confronted with the patient.

Cannon-Bowers et al. (1995) also contended that highly stable tasks require task-contingent competencies, because for such tasks behavioral discretion on the part of team members is reduced. In other words, team members who operate in tasks that are not very dynamic have fairly predictable inputs. Therefore, it is not as crucial for team members to develop detailed knowledge of their teammates, as task performance will not change much regardless of who the teammates are. Highly proceduralized tasks would fall into this category.

In addition, Cannon-Bowers et al. (1995) argued that task-contingent competencies are required when turnover in the team is high, because it is impossible for team members to gain a detailed understanding of each other. Therefore, when team membership rotates often, team members are best prepared when they understand the association between task cues and appropriate team response. In this manner, team members are likely to behave in a coordinated manner, inasmuch as they are all reacting to similarly-perceived cues.

When team members are involved in multiple teams, Cannon-Bowers et al. (1995) hypothesize that transportable skills are required. This is because transportable team competencies are likely to benefit the team member regardless of which team he or she is a member. Task forces, committees, and tiger teams fall into this category.

According to Cannon-Bowers et al. (1995), when the team performs a variety of tasks, team-contingent competencies are required because team members have a chance to build knowledge bases about one another. Even when performing unfamiliar tasks, such knowledge can be useful because it helps team members to anticipate what their teammates will do. Functional and departmental teams fall into this category.

Summary

The framework and associated propositions put forth by Cannon-Bowers et al. (1995) represent a first attempt to decompose the notion of team competencies and to relate these to training and management issues. In order to be useful, several areas of research are needed to validate the competencies and framework, and also to test propositions stemming from it. The following sections take this initial work a step further conceptually, by focusing on the notion of "sharedness" among teamwork KSAs.

SHARED AND COMPATIBLE TEAM COMPETENCIES

Perhaps the most important point made by Cannon-Bowers et al. (1995) is that team competencies are not simply individual competencies applied to team tasks. Instead, they suggest that several team competencies only have meaning at the team level. For example, it is not possible to talk about knowledge being *shared* among members when considering only an individual working alone. In fact, it may be that true team competencies exist only at the *intersection of two or more team members.* The following sections examine more closely the nature of team competencies in light of the Cannon-Bowers et al. framework.

According to Cannon-Bowers, Salas, and Converse (1993), and others, an important aspect of team functioning is that team members share knowledge about the task and team. In fact, a major underlying theoretical framework used by Cannon-Bowers et al. (1995) in defining team competencies involves shared mental models (Cannon-Bowers et al., 1993; Klimoski & Mohammed, 1994; Orasanu & Salas, 1993). According to recent thinking, a crucial characteristic of successful teams is that they have members who share a common conception of the task and team. Specifically, shared mental models can be defined as shared or common knowledge about the task and/or team held by at least two team members. This common knowledge allows teams to coordinate implicitly (Kleinman & Serfaty, 1989), i.e., without the need to strategize overtly.

According to Cannon-Bowers et al. (1993), several categories of knowledge must be shared in support of task performance. These include knowledge of the equipment, task, team interactions, and teammates. Each of these categories contributes to task performance; for example, sharing knowledge about equipment allows team members to step in and back one another up; knowledge of teammates (e.g., style) allows team members to predict the information needs of teammates. Evidence to support shared mental model theory is both direct and indirect. Indirect support for the efficacy of shared mental models includes several studies that have suggested that team performance may be enhanced when team members hold common information about the task and team (see Cannon-Bowers et al., 1993, and Klimoski & Mohammed, 1994, for more detail). Further, some direct tests of the shared mental model hypothesis have also been conducted lately (Minionis, Zaccaro, & Perez, 1995). While these studies cannot be considered conclusive, they suggest that knowledge overlap may be important to effective team performance.

Recently, there has been much discussion regarding exactly what is meant by "shared mental models" (e.g., Klimoski & Mohammed, 1994) in teams. In fact, the term "shared" has been employed rather vaguely, and may not capture completely the nature of the relationship required among team members' mental models. For example, in saying that knowledge must be shared in team members, does this imply that the knowledge must be similar or identical among team members? Further, is it more accurate to say that knowledge among team members must be compatible, complementary, mutual, common, or similar? The point is that these terms suggest different relationships among team member knowledge.

According to Klimoski and Mohammed (1994), the term *shared* has been used to mean: (a) identical knowledge among members, (b) distributed knowledge among members, and (c) overlapping knowledge and expectations. In addition, the term *compatible* has been used in past formulations to describe the necessary relationship among team member knowledge (Cannon-Bowers et al., 1993). Each of these notions implies that a slightly different type of relationship must exist between team member knowledge.

It may be that the answer to the question of how intrateam knowledge must be related to ensure effective performance is not singular because several different purposes of shared knowledge might be required. According to Klimoski and Mohammed (1994), several common themes regarding shared mental model content have appeared in the literature. First, several authors suggest that team members employ mental models as a means to conceptualize things around them—as an interpretative mechanism. This implies that team member mental models contain information about situations and environmental events. Second, others emphasize the use of mental models as action scripts which contain information about task activities and action sequences. Klimoski and Mohammed also add that team mental models may also contain what have been traditionally referred to as *role expectations*. Given that team member mental models contain these various types of content, it follows that some types of team member knowledge may need to be shared, similar, or overlapping, while other types might need to be compatible or complementary.

Even more specifically, it may be that the nature of *expectations* required for task performance dictates the relationship required among team member knowledge. Team members act in accordance with their expectations for the task and based on their assessment of what their teammates will do at any given time. Therefore, it is the accuracy of expectations team members have for one another that determines the appropriateness (and effectiveness) of their behavior. Going back to what was said earlier, we contend that expectations are derived from at least two sources within team member mental models: (a) based on knowledge about situations and the interpretation of them, and (b) based on knowledge about task actions, behavioral sequences, and scripts.

Given what has been said thus far, we now contend that there are two general types of relationships among team member knowledge that are of interest to team performance because of the manner in which they affect team member expectations. First is the notion of *shared* knowledge. This term suggests that there is similarity among team members' knowledge (i.e., that there is a single pool of knowledge that is shared among members). In terms of expectations, shared knowledge allows team members to develop similar (and therefore accurate) expectations for the task and team. It is related to the first content of mental models noted above; namely, the situation and its interpretation. The terms *common* and *mutual* have similar connotations.

In contrast, the term *compatible knowledge* does not necessarily suggest that the knowledge is similar. Instead, it connotes that the knowledge must be such that it allows team members to execute coordinated behavioral sequences (and hence is related to the notion of action scripts constituting team mental model content). In such cases, it follows that similar cues in the environment would trigger separate, yet complementary behavioral strategies or sequences in order to accomplish the task. Hence, expectations for the performance of teammates are accurate, even though they are not necessarily based on shared knowledge.

This refinement in the conceptual meaning of shared knowledge has significant implications for the measurement of shared mental models. Currently, debate is taking place regarding how to measure shared mental models (see Klimoski & Mohammed, 1994). Some advocate the use of multidimensional scaling techniques such as Pathfinder (e.g., Kraiger, Salas, & Cannon-Bowers, 1995; Rentsch, Heffner, & Duffy, in press). In such cases, the "sharedness" of the mental models is assessed by C, a similarity measure. Essentially, shared mental models are defined in these cases as similar semantic nets among team members. Others employ concept graphs (e.g., Minionis et al., 1995; Duncan, Rouse, Johnston, Cannon-Bowers, Salas, & Burns, 1966). Typically, similarity is measured in such cases by counting the number and nature of like "nodes" in team members' graphs. Hence, it represents an assessment of how well team members agree regarding more procedural aspects of the task.

Our conceptual concern with these attempts to measure shared mental models is that, as noted, it is unclear whether the similarity of mental models is what is crucial among team members. Instead, it may be that in some cases what is important is what each team member's mental model produces by way of specific expectations for task performance. Under this formulation, we would argue that as long as environmental (and internal) cues trigger accurate predictions for teammate behavior and information needs, it may not be important what the form, structure, or exact contents of the underlying mental model is (except to the extent that understanding this may help to provide a target for training). The implications for this position are numerous—essentially, it takes the emphasis off of simply measuring hypothetical constructs (mental models), and instead seeks to understand the relationship between knowledge structures and behavior via expectations. In the simplest sense, this means that it may be possible in some cases to focus on the manner in which various cues trigger expectations and whether these expectations—for teammates' behavior—are correct in effective teams.

A second aspect of shared mental model theory that we believe requires further attention involves its focus on team member *knowledge*. In fact, it may be that team members must share skills and/or attitudes as well. Our contention here is that the notion of shared cognition in team members can be extended to include a number of KSAs besides shared problems models, equipment models, or teammate knowledge as suggested in past work. In fact, we believe that the notion of shared or compatible KSAs is pivotal to effective teamwork. As noted, this is due to the fact that in team performance, team members often act in accordance with what they expect will happen (a function of the task) and what they expect their teammates will do (or what their teammates are thinking). Therefore, we contend that teamwork is a function of both the team members' own KSAs and also of *their expectations for the behavior of teammates and for the task itself*. This implies that team members may need to share skills and attitudes as well. In the case of skills, this may mean that team members need to be able to execute their own skills in a manner that is consistent with their teammates' actions. Shared attitudes may be required to ensure that team members are motivated to work well together. In addition, many tasks require subtle, detailed, and/or complicated coordination in situations that do not allow team members to discuss their strategies. In such cases, shared KSAs help ensure that team members will make decisions for behavior that are compatible with one another.

Bringing together what we have said to this point, we argue that while team members must share some aspects of task and team relevant knowledge, skill, and attitude, other aspects of team knowledge, skill, and attitude must be compatible, but not necessarily shared (i.e., implying similar).

Given these notions, we reassessed the competencies presented by Cannon-Bowers et al. (1995) (scc Table 7.1) to determine which must be shared, which must be compatible, and which are more independent (i.e., that do not exist at the intersection of two team members). Table 7.2 displays the results of this effort, which included a reconsideration and streamlining of the initial set of competencies.

In constructing Table 7.2, our goal was to examine each of the original competencies with a slightly richer conceptual understanding of what is meant by the terms shared and compatible. It was a logical application of what we have proposed thus far, but clearly requires empirical investigation, as does the original framework. Inspection of Table 7.2 reveals that all of the competencies retained from the original formulation could be classified as either being shared among members (denoted with an S) or compatible among members (denoted with a C). In no cases were the surviving KSAs judged to be completely independent. In the following sections we describe further each of the KSAs in Table 7.2, emphasizing the requirements for them to be either shared or compatible among members.

Knowledge Competencies

As is evident from Table 7.2, knowledge competencies in teams can be either shared or compatible. We reasoned this to be the case based on what was stated earlier regarding expectations. Specifically, we have stated that the value of shared KSAs lies in their ability to create common, accurate expectations for task performance. Therefore, it is not always necessary that team members *share* team knowledge (implying that it is similar); it may be possible that compatible knowledge bases which lead to accurate expectations may be sufficient. For example, two team members may have slightly different conceptions about the task; however, this knowledge is compatible in the sense that it leads to accurate predictions for required behavior.

Given what has been said, we would hypothesize that team members must share basic knowledge about cue and/or strategy associations inasmuch as these define how environmental cues are interpreted. In order for team performance to be coordinated, team members must reach similar interpretations of what they perceive in the environment and what it means for task performance. In addition, basic task models, notions about role responsibilities, and knowledge of interaction patterns must be shared among team members so that they interpret situations in a like manner and have similar expectations for what is expected of them and their teammates. In a more dynamic sense, team members must have shared situational awareness, which generally refers to the team's shared interpretation of the situation at any given time. Team members must also share declarative knowledge about the team's mission, objectives, norms, and

TABLE 7.2
Shared and Compatible KSAs in Team Members

Nature of Team competency	Description of Team Competency	Knowledge	Skills	Attitudes
Context-Driven	*Team-Specific & Task-Specific*	Cue/strategy associations (S) Teammate characteristics (C) Role responsibilities (S) Shared task models (S) Interaction patterns (S) Knowledge of team mission; objectives; norms; resources (S) Situational awareness (S)	Mutual performance monitoring (C) Flexibility (C) Compensatory behavior (C) Information exchange (C) Dynamic reallocation of function (C) Mission analysis (C) Motivating others (C)	Team orientation (morale) (C) Collective efficacy (S) Shared vision (S)
Team-Contingent	*Team-Specific & Task-Generic*	Teammate characteristics (C) Knowledge of team mission; objectives; norms; resources (S) Relationship to larger organization (S)	Conflict resolution (C) Motivating others (C) Information exchange (C) Intrateam feedback (C) Compensatory behavior (C) Assertiveness (C) Planning (C)	Team cohesion (S) Interpersonal relations (C) Mutual trust (S)

Task-Contingent	*Team-Generic & Task-Specific*	Role responsibilities (S) Interaction patterns (S) Procedures for task accomplishment (C) Accurate task models (S) Cue/strategy associations (S)	Task structuring (C) Mission analysis (C) Mutual performance monitoring (C) Compensatory behavior (C) Information exchange (C) Intrateam feedback (C) Assertiveness (C) Flexibility (C) Planning (C)	Flexibility (C) Morale building (C) Cooperation (C) Task-specific teamwork attitudes (C)
Transportable	*Team-Generic & Task-Generic*	Understanding of teamwork skills (C)	Morale building (C) Conflict resolution (C) Information exchange (C) Task motivation (C) Cooperation (C) Consulting with others (C) Assertiveness (C)	Collective orientation (C) Importance of teamwork (C)

Note. C = Compatible; S = Shared.

resources. This assertion is based on the fact that these categories of knowledge provide a foundation for team member interpretation of situational cues, and in turn trigger action sequences; as such, they should be similar among members.

In contrast, team members need only hold compatible notions about their teammate's task relevant characteristics, so long as these lead to accurate expectations. Likewise, each team member's understanding of teamwork skills may differ so long as the notions are compatible and lead to common expectations. In both cases, the implication is that knowledge held by team members may be different so long as it leads to accurate expectations.

Skill Competencies

According to Table 2, team members must have compatible, but not necessarily shared skill competencies. This is due to the fact that behavioral actions (that result from skilled performance) do not necessarily need to be similar among members. Instead, team member behavior must be complimentary or compatible in order for coordinated action to occur. For example, in order to be effective team members may need to reallocate function dynamically. This implies that shared knowledge exists (i.e., a similarity in the interpretation of situational cues that trigger an awareness that reallocation is necessary), but not that similar behavior is exhibited. Rather, it requires compatible behavior (i.e., team member A shifts his/her responsibilities to team member B so as to reduce task load, whereas team member B assumes added tasks, thereby increasing load). Likewise, other team skills such as mutual performance monitoring, compensatory behavior, information exchange, assertiveness, and flexibility (as well as others shown in Table 7.2)—and the manner in which they are manifested in task performance—need to be compatible among team members.

Attitude Competencies

In terms of attitude competencies, some of these must be shared, while others must be compatible (see Table 7.2). For example, shared vision must, by definition, be similar among members. Team cohesion, mutual trust, and collective efficacy must also be shared in order for team effectiveness to be optimized. On the other hand, team members do not necessarily need to hold similar attitudes such as team orientation, teamwork attitudes, collective orientation, or beliefs about the importance of teamwork. These only need to be compatible so that they allow for effective team interaction.

IMPLICATIONS OF THE REVISED FRAMEWORK

In this chapter we have attempted to extend the notions regarding team competencies laid out by Cannon-Bowers et al. (1995). In doing so, we have also refined the concept of shared mental models, both by redefining what is meant by *shared* and by expanding the idea to include team skills and attitudes as well as knowledge. Combined with the original framework, this deeper understanding of team competencies has several applied implications. In the following sections we delineate some of these as they relate to training, staffing and team composition, and team management and development.

Training

First, in terms of team training, what we have presented here implies that team members must be trained to share identical knowledge in some cases, while in others, different, but compatible knowledge is required. Moreover, as we have argued, a primary measure of team training effectiveness may be whether team members can draw common expectations for their own and their teammates' behavior. In order for this notion to be applied, research is needed that addresses specifically the nature of the relationship between KSAs and expectations. That is, it must be determined how shared and compatible KSAs allow team members to draw accurate expectations for performance. In this manner, it may be possible to determine what the necessary organizational structure (or structures) of a team mental model must be to yield accurate expectations. This information could then be used as a basis for generating training objectives.

Further, training principles necessary to foster shared KSAs and to foster compatible mental models must be developed (Salas, Cannon-Bowers, & Johnston, 1997). For example, training *shared* mental models may require intact teams, since factors such as the idiosyncrasies of teammates comprise a portion of the shared knowledge base. In the case of compatible KSAs, it may be possible to train team members separately or in ad hoc teams. Here it would seem that the overriding goal is to teach team members a repertoire of effective task strategies (relating to both their own behavior and that of teammates) and when these apply.

What we have said regarding training thus far would seem to indicate that it is impossible to design team training without understanding the difference between shared and compatible KSAs. However, some hypotheses for training can be offered, building on the suggestions of Cannon-Bowers et al. (1995). More particularly, we hypothesize that a number of training strategies may be appropriate for training shared and compatible team competencies (see also Salas & Cannon-Bowers, 1997). These include:

1. *Task Simulations.* Task simulations (particularly with intact teams) can be an effective means to prepare team members for the actual environment. They are particularly well suited for helping team members to develop accurate expectations for each other's performance given varying task demands. It should be noted, however, that in order to be effective, task simulations must incorporate basic tools and methods of training such as feedback, performance measures, well crafted exercises, etc. (Salas & Cannon-Bowers, 1997).

2. *Role plays/behavior modeling.* It may be possible to build shared and compatible KSAs in team members by incorporating role playing and behavior modeling into training. These techniques have been successfully employed in several recent studies (see Salas et al., 1997). They can help foster correct expectations by allowing team members to associate crucial task cues with likely responses.

3. *Team self-correction.* Recently it has been argued that team members should be trained to provide feedback to themselves as a means to clarify expectations (Blickensderfer, Cannon-Bowers, & Salas, 1994). The approach is to train team members to observe their own performance and to provide one another clarifying feedback after task performance. This should be an excellent means to build common knowledge and shared expectations.

4. *Team leader training.* Related to team self-correction, it may be possible to train team leaders to foster frank and open exchange among team members (e.g., see Smith, Salas, & Brannick, 1994). In this way, it is possible for team members to acquire knowledge about one another and likely behavior in the task. By training team leaders to foster self-correction, team development may be accelerated because problems are discussed and remedied and team members can better understand one another.

5. *Cross training.* Cross training is a means of training team members on the important tasks of their teammates. In this context, cross-training is hypothesized to give team members crucial knowledge about the behavioral and information needs of their teammates. Specifically, it has been suggested that cross-training builds shared knowledge by exposing team members to teammates' tasks, thereby helping them to understand when and how best to support fellow team members (Salas et al., 1997). Recently, Volpe, Cannon-Bowers, Salas, and Spector (1996) demonstrated the positive impact of cross-training on team processes and task outcomes; more empirical work is needed in this regard.

6. *Teamwork skills training.* Cannon-Bowers et al. (1995) suggested that general training on generic teamwork skills may be an effective means of introducing team members to team concepts (also see Salas et al., 1997). It is probably not sufficient training for most teams, but may be a necessary

first step especially when team members are called upon to perform together across a variety of tasks.

Staffing and Team Composition

What has been presented thus far suggests that, over time and with proper training, team members develop shared knowledge and expectations. Another means to achieve such commonality may be to identify and select team members who are similar in important respects. In fact, the topic of team member homogeneity has been the subject of some investigation. Briefly, these findings suggest that homogeneity may not always benefit team performance, since team members may become complacent or suffer losses in creativity (see Cannon-Bowers, Oser, & Flanagan, 1992; Tannenbaum et al., 1996). However, in highly dynamic and interdependent tasks, the effects of heterogeneity on performance may be overcome because team members must respond quickly and adapt to changing situations. In such cases, it can be argued that high degrees of shared and/or compatible KSAs probably out weigh gains afforded by heterogeneity (Cannon-Bowers et al., 1993). Therefore, it may be advisable to select team members who have had similar experiences and backgrounds. Empirical data to test this assertion must be collected to test its viability.

It may also be possible to select team members on the basis of their level of particular team competencies. For example, since collective orientation has been shown to be related to effective team performance (Driskell & Salas, 1992), team members could be selected for this attribute. In addition, it may be possible to select team members who hold basic (generic) teamwork KSAs (see Table 7.2). Because these competencies are not hypothesized to be dependent on any particular task or team, organizations may benefit if team members enter the team with these KSAs intact.

Team Management and Development

The line of reasoning expounded here also has implications for team management and development. Implicit in the notion of shared mental models is that, over time, common experience among team members leads them to hold more common and/or compatible mental models. We would argue that such development can be fostered in several ways.

1. *Reward structure and compensation strategies.* The reward structure must reinforce the development of shared or compatible mental models by rewarding effective team behavior (Hackman, 1987; Tannenbaum, Beard, & Salas, 1992;). In particular, it is important to reward effective team behavior, and provide incentives for team members to maximize team outcomes.

2. *Performance measurement.* In order to recognize, reinforce, and reward team behavior, it is necessary to develop mechanisms to track and evaluate performance at the team level (Cannon-Bowers & Salas, 1997; Johnston, Cannon-Bowers, & Smith-Jentsch, 1995). Several of the KSAs presented in Table 7.2 provide a strong basis on which to develop such measures. Interested readers should also consult a recent chapter by Cannon-Bowers and Salas (Cannon-Bowers & Salas, 1997), which covers this topic in some detail.

3. *Climate/culture.* The climate for teamwork set by team leaders and other management must foster frank and open exchange among members so that team members can develop and refine team KSAs and expectations (see Smith et al., 1994; Sundstrom, Perkins, George, Futrell, & Hoffman, 1990). In fact, self-correction in teams, which appears to occur quite naturally in amateur sports teams, may be an effective means for team members to develop and refine their expectations for performance (Blickensderfer et al., 1994). Research is needed to explore this idea further, particularly how such behavior can be incorporated into everyday behavior of team members.

4. *Task design.* It has been argued that the physical and task variables surrounding a team have an impact on team functioning (Tannenbaum, et al., 1992; Urban, et al., 1995). Given what has been presented here, we would expect that the manner in which team tasks are structured may have an impact on the extent to which KSAs are shared or compatible. For example, high task redundancy (between members) requires more overlap in team members' knowledge bases. Likewise, when task interdependence is high, we anticipate that the need to share knowledge is increased.

CONCLUDING REMARKS

Team performance is obviously of great concern in modern society. In order for such performance to be optimized, it is imperative that we achieve a detailed understanding of the competencies required to be an effective team member and also the necessary relationship among team members' competencies. In this chapter, we have provided an attempt to clarify and extend past work in this area, but it is still very much in need of further study. In particular, the following research issues must be resolved:

1. Do the terms *shared* and *compatible* as defined here accurately describe the nature of required relationships among team member KSAs? Are there others that apply; if so, what are they?

2. Which team KSAs must be shared and which must be compatible (our attempt to specify this here requires empirical verification)?

3. How can shared and compatible KSAs be measured? Should we attempt to assess shared knowledge based directly, or concentrate only on the expectations generated by them?

4. Which human resource and training principles apply to teams as a means to foster and maintain shared and compatible KSAs? We have offered some of these above; investigation of these is required.

The answer to these and related questions will allow us to better prepare teams for the challenging tasks for which they are responsible. The thinking laid out in this chapter is still only the beginning. We hope that work continues in this crucial area in the foreseeable future.

ACKNOWLEDGMENT

The opinions expressed herein are those of the authors and do not represent the official positions of the organizations with which they are affiliated.

REFERENCES

Blickensderfer, E. L., Cannon-Bowers, J. A., & Salas, E. (1994). Feedback and team training: Exploring the issues. *Proceedings of the Human Factors and Ergonomics Society 38th Annual Meeting* (pp. 1195–1199). Santa Monica, CA: The Human Factors Society.

Cannon-Bowers, J. A., Oser, R., & Flanagan, D. L. (1992). Work teams in industry: A selected review and proposed framework. In R. W. Swezey & E. Salas (Eds.), *Teams: Their training and performance* (pp. 355–377). Norwood, NJ: Ablex.

Cannon-Bowers, J. A., & Salas, E. (in press). A framework for developing team performance measures in training. To appear in M. T. Brannick, E. Salas, & C. Prince (Eds.), *Team performance assessment and measurement: Theory, methods, and applications.* Mahwah, NJ: Lawrence Erlbaum Associates.

Cannon-Bowers, J. A., Salas, E., & Converse, S. A. (1993). Shared mental models in expert team decision making. In N. J. Castellan, Jr. (Ed.), *Current issues in individual and group decision making* (pp. 221–246). Hillsdale, NJ: Lawrence Erlbaum Associates.

Cannon-Bowers, J. A., Tannenbaum, S. I., Salas, E., & Volpe, C. E. (1995). Defining team competencies and establishing team training requirements. In R. Guzzo & E. Salas (Eds.), *Team effectiveness and decision making in organizations* (pp. 333–380). San Francisco, CA: Jossey-Bass.

Driskell, J. E., & Salas, E. (1992). Can you study real teams in contrived settings? The value of small group research to understanding teams. In R. W. Swezey & E. Salas (Eds.), *Teams: Their training and performance* (pp. 101–124). Norwood, NJ: Ablex.

Duncan, P. C., Rouse, W. B., Johnston, J. H., Cannon-Bowers, J. A., Salas, E., & Burns, J. J. (1996). Training teams working in complex systems: A mental model based approach. In W. B. Rouse (Ed.), *Human/technology interaction in complex systems* (Vol. 8, pp. 173–231). Greenwich, CT: JAI Press.

Goldstein, I. L. (1993). *Training in organizations* (3rd ed.). Belmont, CA: Wadsworth.

Guzzo, R. A., & Salas, E. (Eds.). (1995). *Team effectiveness and decision making in organizations.* San Francisco, CA: Jossey-Bass.

Hackman, J. R. (1987). The design of work teams. In J. Lorsch (Ed.), *Handbook of organizational behavior* (pp. 315–342). New York: Prentice-Hall.

Johnston, J. H., Cannon-Bowers, J. A., & Smith-Jentsch, K. A. (1995). Event-based performance measurement system for shipboard command teams. *Proceedings of the First International Symposium on Command and Control Research and Technology* (pp. 274–276). Washington, DC: The Center for Advanced Command and Technology.

Kleinman, D. L., & Serfaty, D. (1989). Team performance assessment in distributed decision making. *Proceedings of the Symposium on Interactive Networked Simulation for Training* (pp. 22–27). Orlando, FL: University of Central Florida.

Klimoski, S. W., & Mohammed, S. (1994). Team mental model: Construct or metaphor? *Journal of Management, 20,* 403–437.

Kraiger, K., Salas, E., & Cannon-Bowers, J. A. (1995). Measuring knowledge organization as a method for assessing learning during training. *Human Factors, 37,* 804–816.

Lauber, J. K. (1987). Cockpit resource management: Background studies and rationale. In H. W. Orlady & H. C. Foushee (Eds.), *Cockpit resource management training* (Tech. Rep. No. NASA CP-2455). Moffett Field, CA: NASA Ames Research Center.

Minionis, D. P., Zaccaro, S. J., & Perez, R. (1995). *Shared mental models, team coordination, and team performance.* Paper presented at the 10th annual meeting of the Society for Industrial and Organizational Psychology, Orlando, FL.

Orasanu, J., & Salas, E. (1993). Team decision making in complex environments. In G. Klein, J. Orasanu, R. Calderwood, & C. E. Zsambok (Eds.), *Decision making in action: Models and methods* (pp. 327–345). Norwood, NJ: Ablex.

Rentsch, J. R., Heffner, T. S., & Duffy, L. T. (in press). What you know is what you get from experience: Team experience related to team work schemas. *Group and Organization Management.*

Rouse, W. B., Cannon-Bowers, J. A., & Salas, E. (1992). The role of mental models in team performance in complex systems. *IEEE Transactions on Systems, Man, and Cybernetics, 22,* 1296–1308.

Salas, E., & Cannon-Bowers, J. A. (1997). Methods, tools, and strategies for team training. In M. A. Quiñones & A. Ehrenstein (Eds.), *Training for a rapidly changing workplace: Applications of psychological research.* Washington, DC: APA Press.

Salas, E., Cannon-Bowers, J. A., & Johnston, J. H. (1997). How can you turn a team of experts into an expert team?: Emerging training strategies. In C. Zsambok & G. Klein (Eds.), *Naturalistic decision making* (pp. 359–370). Mahwah, NJ: Lawrence Erlbaum Associates.

Smith, K. A., Salas, E., & Brannick, M. T. (1994). Leadership style as a predictor of teamwork behavior: Setting the stage by managing team climate. In K. Nilan (Chair), *Understanding teams and the nature of teamwork.* Symposium presented at the Ninth Annual Conference of the Society for Industrial and Organizational Psychology, Nashville, TN.

Sundstrom, E., Perkins, M., George, J., Futrell, D., & Hoffman, D. A. (1990, April). *Work-team context, development, and effectiveness in a manufacturing organization.* Presented at the Fifth Annual Conference of the Society for Industrial and Organizational Psychology, Miami, FL.

Tannenbaum, S. I., Beard, R. L., & Salas, E. (1992). Team building and its influence on team effectiveness: An examination of conceptual and empirical developments. In K. Kelley (Ed.), *Issue, theory, and research in industrial organizational psychology* (pp. 117–153). Amsterdam: Elsevier.

Tannenbaum, S. I., Salas, E., & Cannon-Bowers, J. A. (1996). Promoting team-effectiveness. In M. West (Ed.), *Handbook of work group psychology* (pp. 503–529). Sussex, England: Wiley.

Urban, J. M., Bowers, C. A., Cannon-Bowers, J. A., & Salas, E. (1995). The importance of team architecture in understanding team processes. In M. Beyerlein, D. A. Johnson, & S. T. Beyerlein (Eds.), *Advances in interdisciplinary studies of work teams* (Vol. 2, pp. 205–228). Greenwich, CT: JAI Press.

Volpe, C. E., Cannon-Bowers, J. A ., Salas, E., & Spector, P. (1996). The impact of cross-training on team functioning. *Human Factors, 38,* 87–100.

Part II

ASSESSMENT OF COMPETENCIES

Chapter 8

Design of Teacher-Scored Measures for Workforce Readiness Competencies[1]

Harold F. O'Neil, Jr.
CRESST/University of Southern California

Keith Allred
Eva L. Baker
CRESST/University of California, Los Angeles

The purpose of this chapter is to provide a context for our assessment work and to suggest a general methodology for measuring workforce readiness competencies. Our approach focuses on measures of two competencies that teachers could administer and score. The chapters by O'Neil, Allred, and Dennis (chapter 9, this volume) and O'Neil, Chung, and Brown (chapter 16, this volume) focus on approaches that are computer-administered and scored.

The methodology that we document consists of 14 steps, from the initial selection of a work environment to the report documenting the process. As seen in Table 8.1, following selection of a work environment, a job and task analysis is conducted to determine the requirements for the job. Then a competency or skill is selected that is assumed or documented to be present in the work environment. Possible competencies would be academic (e.g., reading) or interpersonal skills (e.g., participating as a member of a team), etc. Unfortunately, such molar categories do not map directly onto an assessment measure, and some further level of decomposition is required. Thus, the cognitive analysis step is implemented, and a componential analysis is conducted in order to analyze the competency into its constituent subcompetencies. Next, indicators are created for the subcompetencies.

---

[1]A version of this chapter was presented in the symposium *Workforce Readiness: Competencies and Measures* at the 1992 annual meeting of the American Educational Research Association.

TABLE 8.1
Workforce Readiness Assessment Methodology

Select a work environment
Conduct job and task analysis
Select competency
Conduct componential analysis of competency
Create indicator(s) for subcompetencies
Classify indicator(s) within a cognitive science taxonomy
Create rapid prototypes of measures of indicator(s) test via specifications
Select/develop final measures of indicator(s)
Select experimental/analytical design
Run empirical studies
Analyze statistically
Use/create norms
Report reliability/validity of indicator(s) measure
Report on workforce readiness competency using multiple indicators

---

The indicators are then classified within a cognitive science taxonomy. The purpose of this step is to allow generalization of the findings from an indicator to a high order subcompetency. Then, measures of the competency are selected or developed in two steps: (a) rapid prototypes are developed and tested, and (b) prototypes are refined into final measures. Both process and outcomes are measured. Next, an experimental/analytical design is selected and empirical studies run. The data are statistically analyzed with a focus on psychometric issues (e.g., internal consistency, construct validity), and norms are used or created. A report on the reliability and validity of the indicator is written. Finally, a report on the assessment of the workforce competency using multiple indicators is written.

The application of this general method to a specific case follows in the next section. We use several analytical approaches as "proof of concept" of our methodology, but do not provide an empirical study. The empirical validation of a workforce measure based on this methodology is provided in O'Neil, Allred, and Dennis (chapter 9, this volume) and O'Neil, Chung, and Brown (chap. 16, this volume).

SCANS

The Secretary's Commission on Achieving Necessary Skills (SCANS) was a commission organized by the U.S. Secretary of Labor to determine what is required in tomorrow's workplace and to investigate the extent to which high school students would be able to meet those requirements. SCANS was chosen as a target system for our methodology for two reasons. First, the SCANS approach includes almost all the competencies we are interested in for our assessment approach (for example, SCANS was meant to be a national rather than a state or regional assessment). Second, we had a good, cooperative relationship with the SCANS staff.

As we discussed in an earlier chapter, SCANS identified five competencies: the ability to efficiently use (a) resources, (b) interpersonal skills, (c) information, (d) systems, and (e) technology (see Table 8.2). Additionally, the Commission found that these five competencies are based on a three-part foundation: (a) basic skills, (b) thinking skills, and (c) personal qualities (see Table 8.3).

The Commission also suggested a five-step progression in skills acquisition to define proficiency levels for each of the competencies identified. The five steps in level of skill acquisition are (a) preparatory, (b) work-ready, (c) intermediate, (d) advanced, and (e) specialist. The second level of skill acquisition, *work-ready*, defines the level of proficiency necessary for entry into the workforce. Although the Commission did not specify the precise levels of skill acquisition that would be considered work-ready, it did provide examples of how those proficiency levels might look (U.S. Department of Labor, 1991; see Table 8.4).

CRESST was asked to suggest effective ways of assessing proficiency levels of the competencies identified by the Commission in its June 1991 report. Our chapter documents progress in the development of an assessment model or framework and its instantiation in the assessment of two of the five workforce competencies. Specifically, approaches to the assessment of the information and interpersonal competencies are suggested. It should be noted that while the assessment approaches focus on the two competencies specified, other workforce-readiness components identified by the Commission are also involved. As the Commission explained in its report, "Seldom does one of these eight components stand alone in job performance. They are highly integrated and most tasks require workers to draw on several of them simultaneously" (U.S. Department of Labor, 1991, p. vi). Accordingly, our approach recommends the assessment of competencies in the context of the foundation skills as well.

Developing Rapid Prototypes

The specific approach to developing rapid prototypes for indicators of the two competencies of primary focus (information and interpersonal) will be described in the following manner. First, the specifications for the sources of suggested test content will be elaborated. According to Millman and Greene (1989), when tests are designed to assess future performance in a specified setting, an analysis of the cognitive requirements of that setting includes two steps:

> *First,* the specific cognitive requirements of the criterion setting are identified, through a job analysis for employment settings. . . . *Second,* the content

TABLE 8.2
SCANS: Five competencies

Resources: Identifies, organizes, plans, and allocates resources

    A.   *Time*—Selects goal-relevant activities, ranks them, allocates time, and prepares and follows schedules.

    B.   *Money*—Uses or prepared budgets, makes forecasts, keeps records, and makes adjustments to meet objectives.

    C.   *Material and Facilities*—Acquires, stores, allocates, and uses materials or space efficiently.

    D.   *Human Resources*—Assesses skills and distributes work accordingly, evaluates performance, and provides feedback.

Interpersonal: Works with others

    A.   *Participates as Member of a Team*—Contributes to group effort.

    B.   *Teaches Others New Skills*

    C.   *Serves Clients/Customers*—Works to satisfy customers' expectations.

    D.   *Exercises Leadership*—Communicates ideas to justify position, persuades and convinces others, responsibly challenges existing procedures and policies.

    E.   *Negotiates*—Works toward agreements involving exchange of resources, resolves divergent interests.

    F.   *Works with Diversity*—Works well with men and women from diverse backgrounds.

Information: Acquires and uses information

    A.   *Acquires and Evaluates Information*

    B.   *Organizes and Maintains Information*

    C.   *Interprets and Communicates Information*

    D.   *Uses Computers to Process Information*

Systems: Understands complex interrelationships

    A.   *Understands Systems*—Knows how social, organizational, and technological systems work and operates effectively in them.

    B.   *Monitors and Corrects Performance*—Distinguishes trends, predicts impacts on system operations, diagnoses deviations in systems' performance, and corrects malfunctions.

    C.   *Improves or Designs Systems*—Suggests modifications to existing systems and develops new or alternative systems to improve performance.

Technology: Works with a variety of technologies

    A.   *Selects Technology*—Chooses procedures, tools, or equipment, including computers and related technologies.

    B.   *Applies Technology to Task*—Understands overall intent and proper procedures for setup and operation of equipment.

    C.   *Maintains and Troubleshoots Equipment*—Prevents, identifies, or solves problems with equipment, including computers and other technologies.

Note. U.S. Department of Labor, 1991, p. 12.

TABLE 8.3
SCANS: A Three-Part Foundation

Basic Skills: Reads, writes, performs arithmetic and mathematical operations, listens and speaks.

A.	*Reading*—Locates, understands, and interprets written information in prose and in documents such as manuals, graphs, and schedules.
B.	*Writing*—Communicates thoughts, ideas, information, and messages in writing; and creates documents such as letters, directions, manuals, reports, graphs, and flow charts.
C.	*Arithmetic/Mathematics*—Performs basic computations and approaches practical problems by choosing appropriately from a variety of mathematical techniques.
D.	*Listening*—Receives, attends to, interprets, and responds to verbal messages and other cues.
E.	*Speaking*—Organizes ideas and communicates orally.

Thinking Skills: Thinks creatively, makes decisions, solves problems, visualizes, knows how to learn, and reasons.

A.	*Creative Thinking*—Generates new ideas.
B.	*Decision Making*—Specifies goals and constraints, generates alternatives, considers risks, and evaluates and chooses best alternative.
C.	*Problem Solving*—Recognizes problems and devises and implements plan of action.
D.	*Seeing Things in the Mind's Eye*—Organizes and processes symbols, pictures, graphs, objects, and other information.
E.	*Knowing How to Learn*—Uses efficient learning techniques to acquire and apply new knowledge and skills.
F.	*Reasoning*—Discovers a rule or principle underlying the relationship between two or more objects and applies it when solving a problem.

Personal Qualities: Displays responsibility, self-esteem, sociability, self-management, and integrity and honesty.

A.	*Responsibility*—Exerts a high level of effort and perseveres toward goal attainment.
B.	*Self-Esteem*—Believes in own self-worth and maintains a positive view of self.
C.	*Sociability*—Demonstrates understanding, friendliness, adaptability, empathy, and politeness in group settings.
D.	*Self-Management*—Assesses self accurately, sets personal goals, monitors progress, and exhibits self-control.
E.	*Integrity/Honesty*—Chooses ethical courses of action.

Note. U.S. Department of Labor, 1991, p. 16.

specification of the predictive test is developed . . . [commonly using the] cognitive indicators known or hypothesized to be positively related to the criterion requirements. (p. 341)

The Commission, in its study of the workforce, completed the first step in Millman and Greene's (1989) process by identifying the cognitive re-

TABLE 8.4
Know-How: Work-Ready Level of Proficiency

Competence	Example of Level
Resources	Develop cost estimates and write proposals to justify the expense of replacing kitchen equipment. Develop schedule for equipment delivery to avoid closing restaurant. Read construction blueprints and manufacturer's installation requirements to place and install equipment in the kitchen.[a]
Interpersonal	Participate in team training and problem-solving session with multicultural staff of waiters and waitresses. Focus on upcoming Saturday night when local club has reserved restaurant after midnight for a party. Three people cannot work and team has to address the staffing problem and prepare for handling possible complaints about prices, food quality, or service.[a]
Information	Analyze statistical control charts to monitor error rate. Develop, with other team members, a way to bring performance in production line up to that of best practice in competing plants.[b]
Systems	As part of information analysis above, analyze painting system and suggest how improvements can be made to minimize system downtime and improve paint finish.[b]
Technology	Evaluate three new paint spray guns from the point of view of costs, health and safety, and speed. Vendors describe performance with charts and written specifications. Call vendors' representatives to clarify claims and seek the names of others using their equipment. Call and interview references before preparing a report on the spray guns and make a presentation to management.[b]

Progress in Acquiring Skills

Proficiency Level	Performance Benchmark
Preparatory	Scheduling oneself
Word-ready	Scheduling small work team
Intermediate	Scheduling a production line or substantial construction project
Advanced	Developing roll-out schedule for new product or production plant
Specialist	Develop algorithm for scheduling airline

[a]Competence as demonstrated in a service sector application.
[b]Competence as demonstrated in a manufacturing sector application.
Adapted from U.S. Department of Labor, 1991, pp. 26, 28.

quirements for the criterion setting, namely, specific competencies in the workforce. SCANS derived this information via expert judgment, not job analysis.

In completing the second assessment step, cognitive indicators known or hypothesized to be positively related to the criterion requirements are identified. The research literature is analyzed to identify these cognitive indicators.

Subsequent to the specification of sources of test-item content, sample test items are developed. Finally, specifications for the generation of further test items, such as our sample items, are elaborated. Our format for the test item writing specifications is taken from Baker, Aschbacher, Niemi, Yamaguchi, and Ni (1991) and Millman and Greene (1989).

Unfortunately, the Commission never defined explicitly the work-ready proficiency level. Thus, it is not possible at this point to provide a precise account of how scores on these tests would translate into proficiency levels. However, once a work-ready proficiency level is specified, pilot studies can be conducted to relate the test items to that level.

THE INFORMATION COMPETENCY

Based upon its discussions and meetings with business owners, public employers, unions, and workers and supervisors in shops, plants, and stores, the Commission found that the ability to productively use information is critical to productivity in the workforce. Technological advances have both increased dramatically the amount of information generated and made this information potentially more accessible. This explosion in the amount of information, along with the rapidity of change in today's workplace, has contributed to a heightened need for the efficient use of information. Accordingly, the Commission elaborated the cognitive requirements for the information competency of the workforce as follows (U.S. Department of Labor, 1991, pp. B1–B2).

Acquires and evaluates information. Identifies need for data, obtains it from existing sources or creates it, and evaluates its relevance and accuracy.

Organizes and maintains information. Organizes, processes, and maintains written or computerized records and other forms of information in a systematic fashion.

Interprets and communicates information. Selects and analyzes information and communicates the results to others using oral, written, graphic, pictorial, or multi-media methods.

Uses computers to process information. Employs computers to acquire, organize, analyze, and communicate information.

It is also clear that the information usage competency must be combined with the ability to solve problems—one of the important skills identified by the Commission as part of the thinking skills foundation. That is, information often must be productively used to solve problems.

Researchers have examined the cognitive indicators of the intelligent use of information in problem solving. In particular, we have drawn upon the work of Sternberg (1986) and Mayer, Tajika, and Stanley (1991), who have not only examined the cognitive indicators of the intelligent use of information in problem solving theoretically and empirically, but also developed tests of the indicators they have identified.

Mayer et al. (1991) identified the *integration process* as one cognitive indicator of competence in the intelligent use of information to solve problems. The integration process involves the ability to identify relevant information, distinguish it from irrelevant information, and then integrate the relevant information to solve problems. Table 8.5 provides a suggested test item for this competency.

Sternberg (1986) identified three cognitive processes, similar to Mayer et al.'s (1991) integration process, that indicate competence in the intelligent use of information in problem solving. First, one must *selectively encode* the information available to solve the problem. In other words, one must identify, from the host of information available, which information is relevant to the problem at hand and attend to it. Second, one must *selectively combine* the relevant pieces of information. Sternberg observes that although there are usually several ways of combining information, there is usually a single optimal way for generating a solution to a particular problem. Third, one must *selectively compare* the new information with relevant, previously acquired information in solving problems intelligently. Table 8.6 provides a sample test item for the selective combination aspect of the information and thinking skills cognitive indicators suggested by Sternberg (1986).

TABLE 8.5
Test Item for Integration Process

Lucia had $3.00 for lunch. She bought a sandwich for $.95, an apple for $.20, and a milk for $.45. How much money did she spend?

a.	3.00	.95	.20	.45
b.	.95	.20	.45	
c.	.95	.45		
d.	3			

(Answer: b)

Adapted from Mayer, Tajika, and Stanley (1991). *Journal of Educational Psychology*, 1991, *83*, 69–72. Copyright (1991) by the American Psychological Association, Adapted with permission.

TABLE 8.6
Test Item for Selective Combination

David is a cook in a small restaurant named "Lester's" which specializes in steaks. The restaurant has recently become so popular that the average wait to be seated is one hour. Mr. Lester has therefore asked David to reduce the amount of time needed to cook an order of six steaks to under one hour. The restaurant has a small grill just big enough to broil four steaks at a time. David says to himself, "It takes 30 minutes to broil both sides of one steak because each side takes 15 minutes. Since I can cook four steaks at the same time, 30 minutes will be enough to get four steaks ready. It will take another 30 minutes to cook the remaining two steaks which means a total of one hour." How can David complete cooking all six steaks in just 45 minutes?

Answer:

If one combines the information that there are six steaks that take 15 minutes per side to broil, in other words, 12 sides to be broiled for 15 minutes, with the information that four steaks can be broiled at the same time, one can see that if four steaks are always on the grill, it will take only 45 minutes to grill all six. Assuming the six steaks are grouped into three pairs labeled A, B, and C, David can accomplish keeping four steaks on the grill by first broiling one side of the two A and two B steaks, taking off the B steaks and broiling side 2 of the A steaks and side 1 of the C steaks, and then broiling side 2 of the B and C steaks.

Adapted from Sternberg (1986). Excerpts from Intelligence Applied: *Understanding and Increasing Your Intellectual Skills* by Robert J. Sternberg, copyright 1986 by Harcourt Brace & Company, reprinted by permission of the publisher.

In summary, as seen in Table 8.7, we have used Mayer et al.'s and Sternberg's approaches to assessing the information competency, focusing primarily (P) on the "interpreting and communicating data" aspect of the information competency combined with both the problem-solving skills identified as part of the thinking skills foundation competency and the arithmetic and mathematics skills identified as part of the basic skills foundation competency. A secondary (S) focus in this assessment approach, as also seen in Table 8.7, is the "acquiring and evaluating information" aspect of the information competency as well as the creative-thinking and decision-making skills identified as parts of the thinking skills foundation.

In the next section are item-writing specifications, adapted from Millman and Greene (1989, p. 352), for generating sample items such as those above.

ITEM-WRITING SPECIFICATIONS

Specific SCANS Competencies to be Tested

Competencies of primary focus:

1. The information competency:
The interpreting and communicating information subcompetency, which is the ability to select and analyze information and to communicate the

TABLE 8.7
Information Competency

	Acquiring and Evaluating Information	Organizing and Maintaining Information	Interpreting and Communicating Information	Using Computers
Thinking creatively	S			
Making decisions	S		S	
Solving problems	S		P	
Seeing things in the mind's eye				
Knowing how to learn				
Reasoning				
Reading				
Writing				
Arithmetic	S		P	
Mathematics	S		P	
Listening				
Speaking				

Note. S = secondary focus, P = primary focus.

results to others using oral, written, graphic, pictorial, or multi-media methods (U.S. Department of Labor, 1991, p. B1).

2. The basic skills foundation:

Arithmetic: Performs basic computations; uses basic numerical concepts, such as whole numbers and percentages, in practical situations; makes reasonable estimates of arithmetic results without a calculator; and uses tables, graphs, diagrams, and charts to obtain or convey quantitative information (U.S. Department of Labor, 1991, p. C1).

Mathematics: Approaches practical problems by choosing appropriately from a variety of mathematical techniques; uses quantitative data to construct logical explanations for real world situations; expresses mathematical ideas and concepts orally and in writing; and understands the role of chance in the occurrence and prediction of events (U.S. Department of Labor, 1991, p. C1).

3. The thinking-skills foundation:
The problem-solving subskill, which is the ability to recognize that a problem exists (i.e., there is a discrepancy between what is and what should or could be), includes the ability to identify possible reasons for the discrepancy, and to devise and implement a plan of action to resolve it. It also includes the ability to evaluate and monitor progress, and to revise the plan as indicated by findings (U.S. Department of Labor, 1991, p. C1–C2).

Competencies of secondary focus:

1. The information competency:
The acquire-and-evaluate-information subcompetency is the ability to identify the need for information, to obtain it from existing sources or to create it, and to evaluate its relevance and accuracy (U.S. Department of Labor, 1991, p. B1).

2. The thinking skills foundation:
Creative thinking: Uses imagination freely, combines ideas or information in new ways, makes connections between seemingly unrelated ideas, and reshapes goals in ways that reveal new possibilities (U.S. Department of Labor, 1991, p. C1).
Decision making: Specifies goals and constraints, generates alternatives, considers risks, and evaluates and chooses best alternatives (U.S. Department of Labor, 1991, p. C1).

Rationale

Increased volume of information at work and an increased need for workers at all levels to be able to use this information effectively require that all workers be efficient information managers.

Content Specification

Cognitive indicators, either known or hypothesized, of the ability to perform the cognitive requirements identified by the Commission should be found in relevant research as a source for test content (Millman & Greene, 1989). In our example, test content was drawn from the work of Sternberg (1986) and Mayer et al. (1991). The specific cognitive indicator drawn from Mayer et al. was

Integration: The ability to select and combine information into a coherent representation of the entire problem. (pp. 69–70)

The specific cognitive indicator drawn from Sternberg was:

Selective combination: The ability to combine information into a meaningful whole, in which associations between relevant pieces of information are understood, in order to solve a problem. (pp. 31, 187–189, 210–211)

Context Attributes

The context should include a workplace setting involving a problem that requires the use of information to solve it. The relevant information can be applied in several possible ways but in only one optimal way for the problem involved.

Question Attributes

Questions of the following two types can be asked:

1. Multiple-choice questions in which students must choose the option that accurately identifies all and only the information relevant to the given problem.
2. Questions in which the students must identify the relevant information and then combine it in the optimal way to generate a solution in written form to the given problem.

Response Attributes

Responses to the first type of question are simply identification of the one correct multiple-choice option. Responses to the second type of question must clearly describe in written form the optimal solution to the problem.

Relationship of Scores to SCANS Proficiency Levels

As mentioned, the Commission never explicitly specified proficiency levels. It is therefore not possible, at this time, to specify what levels of performance on these test items would constitute "work-ready" levels of proficiency.

Summary of Instantiation of Assessment Methodology

In the prior section we have instantiated our general methodology in the context of the specific SCANS competency "information." This instantiation is summarized in Table 8.8. An additional step, "Specify basic skills foundation," is added in the SCANS case.

TABLE 8.8
Workforce Readiness Assessment Methodology for SCANS: Example 1

General Methodology	Specific Example
Select a work environment	Analytically derived
Conduct job and task analysis	Analytically derived
Select competency	Information management (SCANS)
Conduct componential analysis of competency	Interpreting and communicating information
Specify basic skills foundation	Arithmetic, mathematics, thinking skills
Create indicator(s) for subcompetencies	Integrating process; Selective combination
Classify indicator(s) within a cognitive science taxonomy	Mayer et al., 1991; Sternberg, 1986
Create rapid prototype of measures of indicator(s) via test specifications	[see examples in Tables 8.5, 8.6]
Select/develop final measures of indicator(s)	To be done
Select experimental/analytical design	Criterion groups
Run empirical studies	To be done
Analyze statistically	To be done
Use/create norms	To be done
Report reliability/validity of indicator(s) measure	To be done
Report on workforce readiness using multiple indicators	To be done

THE INTERPERSONAL COMPETENCY

One of the five competencies the Commission identified as critical to productive performance in the workforce is interpersonal skills. Specifically, interpersonal skills were identified to consist of the ability to participate as a member of a team, to teach others, to serve clients and customers, to exercise leadership, to negotiate, and to work with cultural diversity (U.S. Department of Labor, 1991, p. B1). The identification of the interpersonal competency as critical resulted in part from the Commission's finding that a trend exists toward organizing workers in terms of teams and toward decision making closer to the front line (U.S. Department of Labor, 1991, pp. 3–4).

The Commission's findings that, to be competitive, America needs to organize its workforce in terms of teams that take on problem-solving and decision-making responsibilities formerly left to managers further up the management hierarchy are confirmed by other commissions and task forces examining the skills demands of America's workforce (e.g., Employability Skills Task Force, 1989; National Center on Education and the Economy, 1990). However, as more tasks and responsibilities are shared and fulfilled cooperatively by several persons rather than by individuals acting alone, the potential for interpersonal friction increases.

The Commission went on to argue:

> Interpersonal competence is the lubricant of the workplace, minimizing friction and the daily wear and tear of work. It also undergirds restructured work organizations in factories and provides the "service" in service firms. It is required if teams are to solve problems that they jointly face. All of these competent workers function effectively in quite complicated interpersonal environments. A false step in most of these situations invites resistance from colleagues or clients. .... (U.S. Department of Labor, 1991, p. 13)

In elaborating the cognitive requirements of the interpersonal competency, the Commission defined the following six subcompetencies (U.S. Department of Labor, 1991, p. B1):

1. *Participates as a member of a team.* Works cooperatively with others and contributes to group with ideas, suggestions, and effort.
2. *Teaches others.* Helps others learn.
3. *Serves clients/customers.* Works and communicates with clients and customers to satisfy their expectations.
4. *Exercises leadership.* Communicates thoughts, feelings, and ideas to justify a position; encourages, persuades, convinces, or otherwise motivates an individual or groups, including responsibly challenging existing procedures, policies, or authority.
5. *Negotiates.* Works towards an agreement that may involve exchanging specific resources or resolving divergent interests.
6. *Works with cultural diversity.* Works well with men and women and with a variety of ethnic, social, or educational backgrounds.

The Michigan Employability Skills Employer Survey also identified several skills related to the interpersonal skills identified by the Commission that are critical to the workforce (Mehrens, 1989). The Michigan Employability Skills Task Force administered surveys to a wide variety and large number of employers in the workforce to gain their perceptions of the skills required by all workers. Each employer was asked to rate 86 skills on a 4-point scale, where 1 = critical, 2 = highly needed, 3 = somewhat needed, and 4 = not needed. "Pay attention to the person speaking" received a mean rating of 1.4. "Ask questions to clarify understanding" received a mean rating of 1.5. "Cooperate with others" received a mean rating of 1.6. All three skills just mentioned were in the top 20 of the 86 skills identified on the survey.

Researchers have examined the cognitive indicators of a number of interpersonal competencies that the Commission found critical to productivity in the workforce. Specifically, the cognitive indicators of the "negotiates" subcompetency, which is defined (U.S. Department of Labor, 1991, p. B1) as the ability to work towards an agreement that may involve exchanging specific resources or resolving divergent interests, are examined in the research on integrative negotiation skills (e.g., Komorita & Parks, 1995; Lewicki, Litterer, Minton, & Saunders, 1994; Womack, 1990).

Integrative negotiation skills can be defined in terms of three subskills. The first subskill is the ability to understand and articulate the common goals and interdependency of various colleagues with whom one must work closely and cooperatively. Second, integrative negotiation skills include the ability to communicate the interests, values, knowledge, and priorities underlying one's own suggestions, opinions, and decisions at work, as well as the ability to understand and appreciate similar communications from other colleagues and team members. Third, integrative negotiation skills include the ability to work with others in integrating (thus, the name) the valuable information gained through this effective communication to generate creative solutions to problems and tasks encountered at work.

The three integrative negotiation subskills can also clearly be seen as cognitive indicators of the "participates as a member of a team" and "exercises leadership" interpersonal subcompetencies, as defined by the Commission. The "participates as a member of a team" subcompetency is defined as the ability to work cooperatively with others and to contribute to a group with ideas, suggestions, and effort (U.S. Department of Labor, 1991, p. B1). Certainly proficiency in integrative negotiation skills indicates the ability to cooperatively share ideas and suggestions in group efforts.

The "exercises leadership" subcompetency is defined as the ability to communicate thoughts, feelings, and ideas to justify a position, and to encourage, persuade, convince, or otherwise motivate an individual or group, including responsibly challenging existing procedures, policies, or authority (U.S. Department of Labor, 1991, p. B1). The ability to communicate thoughts, feelings, and ideas to justify a position is explicitly demonstrated by proficiency in integrative negotiation. One of the major advantages of an integrative approach discussed in the literature is to facilitate the responsible challenge of existing policies, procedures, or authority.

As seen in Table 8.9, our analysis of these interpersonal competencies indicates that they also typically draw upon several thinking foundation skills. In particular, creative-thinking, decision-making, and problem-solving skills (U.S. Department of Labor, 1991, pp. C1–C2) are required for, and are indicated by, proficiency in integrative negotiation. In sum, research on integrative negotiation skills documents important cognitive indicators of the interpersonal and thinking skills competencies identified

TABLE 8.9
Interpersonal Skills

	Working on Teams	Teaching Others	Serving Customers	Leading	Negotiating	Working with Cultural Diversity
Thinking creatively	S			S	P	
Making decisions	S			S	P	
Solving problems	S			S	P	
Seeing things in mind's eye						
Knowing how to learn						
Reasoning						

Note. S = secondary focus, P = primary focus.

by the Commission as critical to performance in the workforce. An overview of important research establishing these cognitive indicators follows.

One of the most thorough treatments of research on integrative negotiation skills is offered in Womack's (1990) review of basic and applied research on negotiation of conflict resolution in organizations. Like other researchers (e.g., Brett, Goldberg, & Ury, 1990; Lax & Sebenius, 1986; Lewicki et al., 1994), Womack concludes from her review that an integrative approach is generally more effective than others in achieving solutions that satisfy the interests and concerns of the colleagues involved in problems and disputes, and, therefore, such an approach results in longer-lasting resolutions that better meet organizational and individual needs. She also concludes that integrative negotiation skills contribute to enhanced relations between fellow workers.

Womack (1990) offers an extensive treatment of research on the effectiveness of particular integrative negotiation skills. First, when fellow workers or team members are discussing problems over which there are divergent opinions, the parties concerned should verbally emphasize their interdependence. Each party should also openly and clearly communicate information about its own interests and concerns as well as listen carefully to that information from the other party. The verbal communication should also include exploratory problem solving, expression of arguments in support of the other party's position, and a willingness to accept the other's analysis and proposals as legitimate and reasonable.

Brett et al. (1990) also emphasize the role of integrative communication skills in dealing with work situations involving divergent opinions. Based on their own as well as others' research, they argue the importance of the open exchange of information about each party's interests relevant to the dispute, rather than communication of principles upon which a party thinks a conflict should be resolved. Their argument is that abstract principles and rights imperfectly represent the parties' real interests and concerns, thereby making resolution of the dispute on such grounds virtually impossible.

Also, negotiations should include consideration of more than one interest at a time so that integrative trade-offs, corresponding to the parties' differential priorities, can be discovered. For example, although there may be issues where the parties to the dispute have clearly competitive interests, one issue might be of high priority to one party, while a different issue is of high priority to the other party. If both issues are considered together, each party can compromise on the issue of lesser priority and joint positive outcomes can be increased (Lax & Sebenius, 1986; Lewicki et al., 1994; Neale & Bazerman, 1991). Brett et al. (1990) also emphasize that the mutual dependence of the parties should be articulated so that a cooperative rather than a competitive orientation prevails whenever possible.

Brett et al. (1990) argue that when members of the workforce possess and use these integrative negotiation skills, solutions will be reached that better address the needs, concerns, and interests of the parties within the organization as well as the needs the organization itself. Furthermore, the recurrence of disputes will be diminished. Thus, although an integrative, interest-based approach may result in more extended negotiations initially, those negotiations will lead to agreements which will hold and prevent later conflict.

Tjosvold (1990) has also addressed the issue of integrative communication in the workforce. Tjosvold asked members of a social service organization about recent conflicts they had experienced at work. He found that when cooperative goals prevailed and integrative negotiation skills were used, communication was characterized by assistance and support for the interests and analyses of the other party, by a problem-solving orientation, by brainstorming for creative solutions for the interests of all parties, and by the integration of ideas from the various parties in achieving solutions to work problems. Ratings of the effectiveness of conflict resolution and the degree of trust in the other party were also high when an integrative orientation prevailed. Similar findings were reported by Pruitt and Syna (1983).

In summary, the research on integrative negotiation skills provides clear documentation of cognitive indicators of important aspects of the interpersonal competency identified by the Commission. This stream of research identifies the cognitive indicators of the negotiation subcompetency, which is the ability to negotiate an agreement involving exchanges of specific resources or resolving divergent interests. Integrative negotiation skills also serve as cognitive indicators of the "participates as a member of a team" and "exercises leadership" interpersonal subcompetencies as well as the creative thinking, decision-making, and problem-solving skills. According to the integrative negotiation skills research, three cognitive indicators of these competencies are (a) the ability to identify and articulate the common goals and interdependency of the parties, (b) the ability to effectively communicate the basis for one's own position as well as to understand and appreciate the other party's position(s), and (c) the ability to use the information gained to generate creative solutions.

Drawing upon a workforce scenario (see Table 8.10) described in the SCANS report (U.S. Department of Labor, 1991, pp. 9–10), sample test items for integrative negotiation skills were developed (see Table 8.11). The SCANS report describes three friends' endeavor to open their own restau-

TABLE 8.10
Workforce Scenario

Greg, Anthony, and Kathleen have just embarked upon their entrepreneurial dream—opening their own restaurant (The Three Chefs) in a growing southern town. Each of them independently worked hard to get to this point, spending 10 or more years learning the ropes in the restaurant business, pooling their savings, and borrowing from friends and family to get the start-up capital they needed.

Greg has worked in the restaurant business the longest and has been wanting to start his own restaurant for several years so he could be his own boss and enjoy the benefits of his own labor. He has managed several restaurants and enjoys using his business skills to make restaurants a successful business endeavor. Greg put up 10% more start-up cash than Anthony and Kathleen and also took out a second mortgage on his home to satisfy the local bank's demand for security for operating credit. He serves as manager and "front-of-the-house" shift supervisor during the day.

Anthony loves to combine his creative talent with the skills he gained at a culinary arts school he attended in the Northeast to produce unique and delicious gourmet delights. Anthony trains the staff, does the bookkeeping, and prepares the evening meals, which he loves the most.

Kathleen has always enjoyed the restaurant business for the service it provides of offering a pleasant environment in which family and friends can enjoy a meal together. Kathleen majored in interior design in college and enjoys using the skills she gained to improve the ambiance of a restaurant. Even when not working, Kathleen enjoys going out with friends and family to a restaurant. She is the lunchtime chef and evening manager.

Adapted from U.S. Department of Labor, 1991, pp. 9-10.

TABLE 8.11
Sample Test Items

Item 1: Articulate Common Goals

Write a short essay explaining the goal(s) which Greg, Anthony, and Kathleen have in common. Also explain the ways in which they need each other to accomplish their goals.

Item 2: Understand Others' Positions

The Three Chefs has been in business for one month. It has been moderately busy in that time. During the month, each partner has formed ideas and opinions about how to improve The Three Chefs. They have come together on Monday morning to discuss their various ideas. They have agreed that they will meet weekly at this time for 45 minutes to discuss the business. They have also agreed that they will take turns chairing the meeting, and it was decided that Greg would chair the first meeting.

Greg calls first on Anthony, who is hardly able to restrain his enthusiasm to express his suggestions. Anthony begins by telling Kathleen and Greg about a new commercial food processor he learned of through one of his old classmates at the culinary arts school. As Anthony explains different features of the processor, he comments on all the wonderful dishes he could prepare with it. He believes that the expense of the equipment, which is considerable, will be offset by the volume of business they will do by offering such wonderful food and by being able to charge more for it. He therefore suggests that they purchase the food processor. His discussion of possible new dishes they could offer leads him to a discussion of his other suggestion to order a wider variety of higher quality ingredients in order to serve the truly exquisite food he thinks they should offer.

After listening to Anthony's suggestions, Greg asks Kathleen to share her suggestions. Kathleen says she has received a few comments from their evening customers that The Three Chefs' decor is somewhat barren. Not wanting to expend too much capital without having a sense of how good their business was going to be, the three had decided to initially buy only the essentials for the restaurant's decor. Kathleen suggests that, since they have had an encouraging first month, they should invest in decorating The Three Chefs and offers her ideas of some of the improvements that could be made. Otherwise, she argues, people won't enjoy eating out at The Three Chefs and won't come back.

Pretend that you are Greg. Anthony and Kathleen have expressed their suggestions as described above. It is now your turn, but the 45 minutes allocated for the meeting have been taken up. Knowing that everyone has important preparations to make for the coming day, you suggest that you will write a summary of Anthony's and Kathleen's suggestions as well as your own suggestions rather than explaining them now. It is agreed that you will give this summary to Anthony and Kathleen to read before next Monday when the three of you will continue the discussion. As Greg, write such a summary. You should identify Anthony's and Kathleen's suggestions along with the reasons they offered to support those suggestions. You should try to represent Kathleen's and Anthony's positions as fairly as possible while also representing Greg's position, as if you were Greg. In representing Greg's position, you should keep the following concerns in mind.

First, given that you put up 10% more start-up capital than the others and took out a second mortgage on your home to secure operating credit with the bank, you are quite concerned that costs at The Three Chefs be held to a minimum and profits maximized. Anthony's and Kathleen's suggestions, all of which sound as though they will involve considerable cash outlays, therefore concern you. Second, you have a couple of your

(Continued)

TABLE 8.11
(Continued)

---

suggestions which correspond to your concerns. You have noticed that many of the lunchtime customers are in a hurry. On several occasions you heard customers commenting that they wished the food would be served more promptly so they wouldn't go over their lunch break. You also observed that between 12:00 and 1:00 it was often quite busy, and that customers had sometimes waited for 20 minutes to be seated. Your idea, therefore, is to offer a menu with items that are quicker to prepare to better serve the customers. Furthermore, such a change would allow The Three Chefs to do a higher volume of business and therefore increase profits.

Write a short essay that summarizes everyone's position.

Item 3: Generation of Creative Solutions

With the information given in Item 2, devise a compromise which takes into account Greg's, Anthony's, and Kathleen's mutual and individual concerns and suggestions. Your goal should be to devise a compromise that best incorporates the information provided by the three parties and that best and most fairly accommodates each of their positions. Write an essay describing that compromise and how it takes into account the concerns and issues raised.

---

rant to portray the Commission-identified competencies in the accommodations and food services sector of the economy. Our modified version (Table 8.10) expands on each person's background and responsibilities and indicates how those differences in background and responsibilities led to divergent interests with regard to suggestions for improving the restaurant. In three different questions, students are asked to perform one of the three cognitive indicators. Subsequent to the sample items, the item-writing specifications which were used to generate the items are elaborated.

ITEM-WRITING SPECIFICATIONS

Specific SCANS Competencies to be Tested (see Table 8.9)

Competencies of primary focus:

1. The "negotiates" subcompetency of the interpersonal competency: Works towards an agreement that may involve exchanging specific resources or resolving divergent interests (U.S. Department of Labor, 1991, p. B1).

Competencies of secondary focus:

1. The "exercises leadership" subcompetency of the interpersonal competency: Communicates thoughts, feelings, and ideas to justify position;

encourages, persuades, convinces or otherwise motivates an individual or group(s), including responsibly challenging existing procedures, policies, or authority (U.S. Department of Labor, 1991, p. B1).

2. The "participates as a member of a team" subcompetency of the interpersonal competency: Works cooperatively with others and contributes to group with ideas, suggestions, and effort.

3. The creative-thinking subskill of the thinking skills foundation: Uses imagination freely, combines ideas or information in new ways, makes connections between seemingly unrelated ideas, and reshapes goals in ways that reveal new possibilities.

4. The decision-making subskill of the thinking skills foundation: Specifies goals and constraints, generates alternatives, considers risks, and evaluates and chooses best alternatives.

5. The problem-solving subskill of the thinking skills foundation: Recognizes that a problem exists (i.e., there is a discrepancy between what is and what should or could be), identifies possible reasons for the discrepancy, and devises and implements a plan of action to resolve it. Evaluates and monitors progress, and revises plan as indicated by findings.

Rationale

Developments in the ways people are organized at work require more interpersonal skills to integrate the input of other members of a team in making decisions, thinking creatively, and solving problems to increase effectiveness.

Content Specification

Cognitive indicators, either known or hypothesized, of the ability to perform the cognitive requirements identified by the Commission should be found in relevant research as a source for test content (Millman & Greene, 1989). In this instance, test content was derived from the research on integrative negotiation skills. The cognitive indicators from the integrative research are summarized as:

1. The ability to understand and articulate the common goals and interdependency of various colleagues with whom one must work closely and cooperatively at work.

2. The ability to communicate the interests, values, knowledge, and priorities underlying one's own suggestions, opinions, and decisions at work, as well as the ability to understand and appreciate similar communications from other colleagues and team members.

3. The ability to work with others in integrating the valuable information gained through this effective communication to generate creative solutions to problems and tasks encountered at work.

Context Attributes

The context should include a workplace setting involving a team of three to six members. The context should be presented in the form of a scenario describing the work place, a specific task or problem facing the team, and a description of the goals, knowledge, concerns, priorities, etc. of each team member and the corresponding positions that they take in relation to the given task or problem. In addition to the common task or problem, there should be some diversity and contradiction in the positions taken by the team members.

Question Attributes

Three types of essay questions, corresponding to the three cognitive indicators, should be asked.

1. Students should be asked to write a short essay identifying the ways in which the team members are dependent on each other to accomplish individual and team goals.

2. Students should be asked to fairly represent in essay form, using their own words, the positions of the various team members. Students should include in their essay the reasons behind each member's position.

3. Students should be asked to integrate the information and positions offered by each member into a solution that is optimally effective for the group as a whole as well as for the individual members. The solution, along with an explanation of the ways in which the solution solves the problem and satisfies, to the extent possible, the individual members' interests, should be presented in essay form.

Response Attributes

Because the responses to these test items are in essay form, specifications for response attributes and scoring have been adapted from Baker et al.'s (1991) work in content assessment. Baker et al. (1991) elaborate an approach to content assessment that involves rating essays along several dimensions, with scores ranging from 0 to 5 on each dimension. The relevant dimensions are determined by analysis of expert responses to the same test items. Baker et al. (1991) elaborate methods of determining scoring dimensions as well as methods of training raters and scoring. The dimensions implied by the integrative negotiation literature are described next, along with scoring guidelines. Studies validating these dimensions and guidelines will need to be conducted.

Item 1: Articulate Common Goals

1. General impression—Content quality
How thoroughly, accurately, and persuasively does the student identify and explain the parties' common goals and interdependency?
(0–5 point global rating: 0 = no response, 5 = highest level of thoroughness, accuracy, and persuasiveness)

2. Identification of interdependence of the parties
This is a measure of the number of the parties' common goals accurately identified by the student (e.g., all three want to own and operate their own restaurant successfully) combined with the number of ways identified in which the parties are dependent on each other to achieve common and individual goals. In the sample test item, the parties are dependent upon each other (a) financially—only the capital from all three is sufficient, (b) in scheduling—they need each other to have people to fulfill the various roles, such as manager and chef, during the different shifts, and (c) for their particular expertise and interests, such as Anthony's skills as a chef, Kathleen's skills in creating an ambiance, and Greg's skills as a business manager.

Score point guidelines:
0 = no response
1 = no common goal or point of interdependence identified
2 = one common goal or point of interdependence identified
3 = two common goals or points of interdependence identified
4 = three common goals or points of interdependence identified
5 = four common goals or points of interdependence identified

Item 2: Understand Others' Positions

1. General Impression—Content Quality
How fairly and persuasively does the student represent the positions of the parties?
(0–5 point global rating: 0 = no response, 5 = highest level of fairness and persuasiveness)

2. Number of Positions Identified
This is a measure of the extent to which the student identifies each party's positions accurately.
A *position* in the sample test item is a suggestion for improving The Three Chefs restaurant.

The four basic positions in The Three Chefs scenario are:

a. Anthony's suggestion that they purchase the food processor.
b. Anthony's suggestion that they purchase higher quality ingredients.
c. Kathleen's suggestion that they improve the decor.
d. Greg's position that they offer moderately priced lunches which can be served quickly.

Score Point Guidelines:

0 = no response
1 = no positions
2 = one position
3 = two positions
4 = three positions
5 = four positions

   3. Provides Reasons Behind Each Party's Position(s)
This is a measure of the extent to which the student documents the reasons for the position each party takes.
An example of a *reason* in the sample test item would be Kathleen's argument that if the decor is not improved, people will not enjoy themselves and won't come back.

Score Point Guidelines:

0 = no response
1 = no reasons
2 = one reason for one position
3 = one reason each for two positions
4 = one reason each for three positions
5 = one reason each for four positions

Item 3: Generation of Creative Solutions

   1. General Impression—Content Quality
How thoroughly and persuasively does the student identify the number of positions taken, fairness, and degree of creative integration of positions to generate a solution?
(0–5 point global rating: 0 = no response, 5 = highest number of positions taken and greatest level of fairness and creativity)

   2. Number of Positions Represented
How many of the positions in the conflict are represented in the solution?
(0–5 point global rating: 0 = no response, 1 = no positions, 2 = one position, 3 = two positions, 4 = three positions, 5 = four positions)

   3. Fairness
Of the positions which are represented, how equally are they represented?
(0–5 point global rating: 0 = no response, 5 = highest level of fairness)

   4. Creative Integration of the Positions
This is a measure of the extent to which the solution is integrative as

opposed to distributive. Distributive refers to solutions that are simply compromises on some middle ground between opposing positions. Integrative refers to solutions that attempt to integrate positions in such a way that more of the positions can be served.
(0–5 point global rating: 0 = no response, 5 = highest level of creative integrative solution)

Example:
Distributive: Greg wants to cut costs and offer food that can be served quicker while Anthony wants to buy more efficient equipment and more expensive ingredients to offer fancier food. The fairest solution then is to just stay with the menu they have and not change at all.
Creative Integrative: Greg wants to increase profits by offering food which can be served faster at lunch so they can serve more customers, while Anthony wants to provide fancier food using more expensive ingredients for dinner. The solution is to offer the quick, cheaper kinds of food for lunch and the fancier food for dinner and charge more for it.

Relationship of Scores to SCANS Proficiency Levels

Again, because the Commission's work on specification of proficiency levels was not completed, it is not possible to establish what scores on these items would constitute a "work-ready" level of proficiency. However, it is anticipated that the test will be calibrated so that an average score of 3 or 4 on each of the dimensions would indicate work-ready competence.

Summary of Second Instantiation of Assessment Methodology

In this section we have instantiated our methodology with a second example in the area of interpersonal competence. This instantiation is shown in Table 8.12. As with our first example (see Table 8.8), an additional step, "Specify basic skills foundation," is added in the SCANS case.

The two CRESST prototype assessment approaches relate to the first three major categories of skills found in the five competencies and do not directly assess the personal characteristics and attitudes category. The problem-solving skills prototype assessments are measures of both the mathematics skills identified in the basic skills category and the problem-solving skills identified in the higher order thinking skills category. The negotiation/conflict resolution prototype is a measure of the problem-solving and creativity skills identified as part of the higher order thinking skills category and the negotiation skills identified as part of the teamwork/interpersonal skills category.

TABLE 8.12
Workforce Readiness Assessment Methodology for SCANS: Example 2

General Methodology	Specific Example
Select a work environment	Analytically derived
Conduct job and task analysis	Analytically derived
Select competency	Interpersonal Competency (SCANS)
Conduct componential analysis of competency	Negotiates
Specify basic skills foundation	Thinking creatively, making decisions, solving problems
Create indicator(s) for subcompetency requirement	Articulating common goals, understanding others' positions, creating integrative solutions
Classify indicator(s) within a cognitive science taxonomy	Integrative negotiation (Womack, 1990)
Create rapid prototype of measures of indicator(s) via test specifications	[see examples in Tables 8.10, 8.11]
Select/develop final measures of indicator(s)	To be done
Select experimental/analytical design	Criterion groups
Run empirical studies	To be done
Analyze statistically	To be done
Use/create norms	To be done
Report reliability/validity of indicator(s) measure	To be done
Report on workforce readiness using multiple indicators	To be done

WHERE ARE WE NOW?

We have created a good "first cut" of a general methodology for measuring workforce readiness competencies. Further, we have instantiated this methodology with two prototypic examples. Our plans are to explore the use of technology to administer, score, and interpret our workforce readiness competency measures. Such explorations are documented in chapters by O'Neil, Allred, and Dennis (this volume) and O'Neil, Chung, and Brown (this volume).

ACKNOWLEDGMENTS

The work reported herein was supported under the Educational Research and Development Center Program cooperative agreement R117G10027 and CFDA catalog number 84.117G as administered by the Office of Educational Research and Improvement, U.S. Department of Education. The findings and opinions expressed in this report do not reflect the position or policies of the Office of Educational Research and Improvement or the U.S. Department of Education.

REFERENCES

Baker, E. L., Aschbacher, P. E., Niemi, D., Yamaguchi, E., & Ni, Y. (1991). *Cognitively sensitive assessments of student writing in content areas.* Los Angeles: University of California, Center for Research on Evaluation, Standards, and Student Testing.

Brett, J. M., Goldberg, S. B., & Ury, W. L. (1990). Designing systems for resolving disputes in organizations. *American Psychologist, 45,* 162–170.

Employability Skills Task Force. (1989, October). *Employability Skills Task Force progress report to the Governor's Commission on Jobs and Economic Development and the Michigan State Board of Education.* Lansing, MI: State of Michigan, Office of the Governor.

Komorita, S. S., & Parks, C. D. (1995). Interpersonal relations: Mixed-motive interaction. *Annual Review of Psychology, 46,* 183–207.

Lax, D. A., & Sebenius, J. K. (1986). *The manager as negotiator.* New York: The Free Press.

Lewicki, R. J., Litterer, J. A., Minton, J. W., & Saunders, D. M. (1994). *Negotiation.* Burr Ridge, IL: Irwin.

Mayer, R. E., Tajika, H., & Stanley, C. (1991). Mathematical problem solving in Japan and the United States: A controlled comparison. *Journal of Educational Psychology, 83*(1), 69–72.

Mehrens, W. (1989). *Michigan Employability Skills Employer Survey. Technical Report.* East Lansing, MI: Michigan State University.

Millman, J., & Greene, J. (1989). The specification and development of tests of achievement and ability. In R. L. Linn (Ed.), *Educational measurement* (3rd ed., pp. 335–366). New York: Macmillan.

National Center on Education and the Economy. (1990, June). *America's choice: High skills or low wages!* (Report of the Commission on the Skills of the American Workforce). Rochester, NY: Author.

Neale, M. A., & Bazerman, M. H. (1991). *Cognition and rationality in negotiation.* New York: The Free Press.

Pruitt, D. G., & Syna, H. (1983). Successful problem solving. In D. Tjosvold & D. W. Johnson (Eds.), *Productive conflict management: Perspective for organizations* (pp. 129–148). Minneapolis, MN: Minneapolis Team Media.

Sternberg, R. J. (1986). *Intelligence applied: Understanding and increasing your intellectual skills.* San Diego: Harcourt Brace Jovanovich.

Tjosvold, D. (1990). The goal interdependence approach to communication in conflict: An organizational study. In M. A. Rahim (Ed.), *Theory and research in conflict management* (pp. 15–31). New York: Praeger.

U.S. Department of Labor. (1991, June). *What work requires of schools. A SCANS report for America 2000.* Washington, DC: U.S. Department of Labor, The Secretary's Commission on Achieving Necessary Skills (SCANS).

Womack, D. F. (1990). Applied communications research in negotiation: Implications for practitioners. In M. A. Rahim (Ed.), *Theory and research in conflict management* (pp. 32–53). New York: Praeger.

Chapter 9

Use of Computer Simulation for Assessing the Interpersonal Skill of Negotiation

Harold F. O'Neil, Jr.
CRESST/University of Southern California

Keith Allred
Robert A. Dennis
CRESST/University of California, Los Angeles

In this chapter we describe one computer-based prototype measure of the negotiating subskill of interpersonal competency. Such skills are important in the workplace. For example, management now recognizes a need to have workers take on more responsibility at the points of production, of sales, and of service rendered, if the United States is to compete in rapidly changing world markets. In order to adapt to the need to introduce new products and services quickly with high quality, managers are increasingly emphasizing employee involvement in management decision making, flatter organizational structure, just-in-time management, total quality management, and team work (Blinder, 1990; Cappelli & Singh, 1992; Gerhart, Milkovich, & Murray, 1992; Huselid, 1995; Kochan, Dyer, & Batt, 1992; Pfeffer, 1994; Stasz, Ramsey, Eden, Melamid, & Kaganoff, 1996).

These developments mean that much more is expected of even entry-level members of the American workforce. Beyond generally greater responsibilities, these developments also mean that workers must carry out those responsibilities to greater degrees in cooperation with other workers. Consequently, interpersonal skills are becoming increasingly important to successful performance in the American workforce. All five major studies examining workforce skills that we reviewed in an earlier report (O'Neil, Allred, & Baker, 1992b; O'Neil, Allred, & Baker, chapter 1, this volume) identified interpersonal skills as a major category of job skills critical in today's workforce. Although the studies varied considerably in the particular interpersonal skills they found to be important, most identified negotiation and conflict resolution as priorities. Because of their documented importance in the workforce, we have developed a measure of negotiation skills and have conducted an initial study of its validity. The focus of this chapter is on the use of computer-based simulation as the assessment tool, whereas our other chapter on a rapid prototype used paper-and-pencil measures (see O'Neil, Allred, & Baker, chapter 8, this volume).

DEFINING NEGOTIATION SKILLS
FOR MEASUREMENT

In developing this measure, we followed the general methodology for the development of workforce readiness measures (see Table 9.1) elaborated in an earlier report (O'Neil, Allred, & Baker, 1992a; O'Neil, Allred, & Baker, chapter 8, this volume). That methodology suggests that measures of a performance competency like workforce competencies begin with a job and task analysis to identify the particular skills necessary for performance in the domain of interest. Subsequently, the relevant research literature is surveyed for the cognitive indicators documented to correlate with performance on

TABLE 9.1
Workforce Readiness Assessment Methodology for SCANS: Negotiation Example

General Methodology	Specific Example
Select a work environment	Analytically derived
Conduct job and task analysis	Analytically derived
Select competency	Interpersonal
Conduct component analysis of competency	Negotiates
Specify basic skills foundation	Mathematics, Creative Thinking, Decision Making, Problem Solving, Self-Management
Create indicator(s) for subcompetencies	Proposing and examining possible options and making reasonable compromises
Classify indicator(s) within a cognitive science taxonomy	Carnevale & Pruitt, 1992; Walton & McKersie, 1965; Womack, 1990
Create rapid prototype of measures of indicator(s) via test specifications	Existing simulation modified
Select/develop final measures of indicator(s)	See Methodology section
Select experimental/analytical design	Expert/novice
Run empirical study	This chapter
Analyze statistically	This chapter
Use/create norms	To be done
Report reliability/validity of indicator(s) measure	This chapter
Report on workforce readiness using multiple indicators	To be done

those identified skills. Particular competency measures can then be developed based on those cognitive indicators.

Our analysis was informed by work conducted by the Secretary's Commission on Achieving Necessary Skills (U.S. Department of Labor, 1991, 1992) for the U.S. Department of Labor. The SCANS study defined *negotiate* as working towards an agreement that may involve exchanging specific resources or resolving divergent interests (U.S. Department of Labor, 1991, p. 31). The SCANS analysis further elaborated that the negotiation skills necessary for workforce performance are (a) researching opposition and the history of the conflict, (b) setting realistic and attainable goals, (c) presenting facts and arguments, (d) listening to and reflecting on what has been said, (e) clarifying problems and resolving conflicts, (f) adjusting quickly to new facts or ideas, (g) proposing and examining possible options, and (h) making reasonable compromises (U.S. Department of Labor, 1992, pp. 2–37). Of those eight negotiation skills, we have focused on measuring (g) proposing and examining possible options and (h) making reasonable compromises as the key terminal behaviors. Because setting realistic and attainable goals, presenting facts and arguments, listening to and reflecting on what has been said, clarifying problems and resolving conflicts, and adjusting quickly to new facts or ideas are seen as prerequisites to these terminal behaviors, they are indirectly measured by our assessment. Our measurement environment in simulation (discussed later) did not assess researching opposition and the history of the conflict.

Again following the methodology developed earlier (O'Neil et al., 1992a), we surveyed the negotiation research literature to identify the cognitive indicators known to be associated with the SCANS-identified performance requirements of (g) proposing and examining possible options and (h) making reasonable compromises. In general terms, research on negotiation has long recognized that negotiations take place in the context of mixed-motive, interdependent relationships (e.g., Kelley, 1979; Kelley & Thibaut, 1978; Lax & Sebenius, 1986; Lewicki, Litterer, Minton, & Saunders, 1994; Pruitt & Rubin, 1986; Rubin & Brown, 1975; Walton & McKersie, 1965). The parties to the negotiation must be interdependent in some respect or there would be no reason for them to seek an agreement or resolution with each other. Similarly, their interests must represent a mixture of compatible and incompatible interests, because if their interests were utterly incompatible, there would be no basis for a resolution or agreement. Conversely, if the parties' interests were perfectly compatible there would be no conflict to resolve.

Because negotiations occur in the context of mixed-motive, interdependent relationships, negotiation presents the parties with the challenge of looking out for both their own interests and the interests of the other party (e.g., Blake & Mouton, 1979; Filley, 1975; Kelley, 1979; Kelley & Thibaut, 1978; Pruitt & Rubin, 1986; Rahim, 1986; Rubin & Brown, 1975; Thomas,

1976; see Carnevale & Pruitt, 1992, for an excellent review). One enters into negotiation in order to gain an agreement which protects or serves one's own interests. However, gaining an agreement also requires watching out for the other side's interests to understand what will be acceptable to them. Thus, negotiating is a quintessentially interpersonal activity.

Walton and McKersie's (1965) distinction between distributive and integrative negotiations has been one of the most important guiding theoretical constructs in analyzing the requirements of these dual concerns in negotiations (e.g., Brett, Goldberg, & Ury, 1990; Lax & Sebenius, 1986; Lewicki et al., 1994; Neale & Bazerman, 1991; Pruitt & Rubin, 1986; Pruitt & Syna, 1984; Rahim, 1986; Tjosvold, 1990; Tjosvold & Johnson, 1983; Walton, Cutcher-Gershenfeld, & McKersie, 1994; Womack, 1990). Distributive negotiation focuses on the distribution of the available outcomes to each of the parties. One presumably seeks to gain as high outcomes as one can, but must also see that the other side gets enough to agree to the resolution. Integrative negotiation involves seeking ways in which the outcomes available to the parties can be increased.

Thus, research on negotiation has documented that performance in proposing and examining possible options and in making reasonable compromises requires that one look out for both one's own interests and the interests of the other party. In doing so, an effective negotiator must be effective in both increasing possible joint outcomes (integrative negotiation) and distributing those outcomes (distributive negotiation).

In terms of the first SCANS-identified skill we are trying to assess, the literature indicates that a negotiator is skillful in proposing and examining possible options when he or she considers whether there are options that increase the total outcomes available to the parties (the integrative aspect of proposals). Skill in proposing and examining options also includes considering options regarding how those total available outcomes are to be distributed (distributive aspects of proposals). With respect to the distribution of available outcomes, skillful negotiators must propose and examine options that balance their interest in maximizing their own outcomes with consideration of what will be acceptable to the other party.

In terms of the second SCANS-identified skill targeted for assessment, the literature indicates that negotiators are skillful in making reasonable compromises when those compromises balance the negotiators' concern about securing good outcomes for themselves with their understanding of what the other side needs to agree to the compromise (the distributive aspect of compromises). In other words, a compromise is *un*reasonable either if it sacrifices one's own interests more than would be necessary to gain agreement, or if it does not reflect consideration of what the other side will need to agree. Part of the skill in making reasonable compromises is to not compromise one's own interests where equal or greater gains can be offered to the other side through a proposal which increases both sides' outcomes (the integrative aspect of compromises).

The SCANS performance criteria and the cognitive indicators of those performance criteria found in the negotiation literature guided the development of our negotiation measure. With regard to the SCANS performance criteria, we needed to simulate the activities of proposing and examining options and making reasonable compromises, and the exchange of proposals and counterproposals. With regard to the cognitive indicators identified in the negotiation literature, the exchange of proposals should take place in the context of a situation of mixed-motive interdependence, with both distributive and integrative dimensions. Such a context is provided by our computer simulation.

SIMULATION AS AN ASSESSMENT CONTEXT

O'Neil and Baker (in press) have defined simulation as "a process that imitates a physical or functional situation, thereby providing experience not easily gained otherwise, and that permits realistic problem solving for individuals or teams of students." For example, computer-based instructional simulators can be used to measure performance as a routine by-product of instruction. In turn, simulation may be conceptualized as the third generation of computer measurement (Bunderson, Inouye, & Olsen, 1989). Bunderson et al. define the four generations as follows:

- Generation 1, computerized testing: administering conventional tests by computer; for example, computerized versions of either intelligence tests such as the Slosson Intelligence Test (Hedl, O'Neil, & Hansen, 1973), personality tests such as the Minnesota Multiphasic Personality Inventory (Dunn, Lushene, & O'Neil, 1972), or achievement tests as in computer-managed instruction (CMI) (O'Neil, Hedl, Richardson, & Judd, 1976).
- Generation 2, computerized adaptive testing (CAT): adapting the test to the individual test-taker by selecting each succeeding task on the basis of the test-taker's performance on previous tasks (e.g., computer-adaptive testing).
- Generation 3, continuous measurement: using calibrated measures embedded in a curriculum to continuously and unobtrusively estimate changes in each learner's proficiency (e.g., simulations in this report).
- Generation 4, intelligent measurement: introducing knowledge-based (artificially intelligent) computing to the decision-making processes of computerized measurement; for example, in scoring constructed responses (Braun, 1994; Braun, Bennett, Frye, & Soloway, 1990).

Progress towards these four generations of computerized measurement is documented in Gutkin and Wise (1991). In general, such technology opportunities for assessment can be viewed from a perspective of both presumed advantages and possible problem areas (see Table 9.2). We feel that the advantages outweigh the disadvantages.

Next we describe the negotiation simulation we have developed, with an emphasis on a conceptual explanation of how we simulate and measure the cognitive indicators of the performance criteria identified. The simulation was conducted via computer, and a full description of the logistics of the simulation is found in the procedure section of this report.

The simulated negotiation situation is a job contract negotiation between a graduating Masters of Business Administration student (MBA) and a management consulting firm. In the simulation, the management consulting firm has already interviewed the MBA and decided that it would like to hire the MBA. The MBA is also interested in working for the management consulting firm. Thus, there is some interdependence between the parties, based on the compatible interests of working out an employment relationship. In the simulation, the two parties have come together to negotiate the terms of the contract with respect to three issues: (a) billable hours, (b) severance pay, and (c) salary. The MBA prefers to have to bill fewer hours, to have a longer severance pay period should he or she be laid off, and to have a higher salary. In contrast, the representative for the management consulting firm prefers that the MBA work more billable hours, receive severance pay for a shorter period after a layoff, and receive a smaller salary. Thus, there are also some incompatible or competitive aspects to the interdependence between the two parties.

TABLE 9.2
Technology Opportunities for Assessment

Presumed Advantages	Possible Problem Areas
Provide consistent, high-quality assessment available on a large scale	Cost
	Validity
Provide high-quality assessment at remote sites	Program maintenance
Provide hands-on, performance-oriented instruction/testing	Equity
	Fidelity
Permit individualized testing	Increased security
Permit team assessment	Negative teacher attitudes
More comprehensive domain coverage	
Quicker reporting	
Provide rapid update of testing materials	
Reduce testing time	
Reduce reliance on highly skilled personnel	

To build in both integrative and distributive dimensions to the simulation, the parties also have offsetting priorities regarding the three issues being negotiated. Because the MBA is characterized as being very concerned with having a life outside work, the billable hours issue is most important to him or her. Because of the MBA's concern with the volatility of the consulting industry in the present economy and the common layoffs that result, the severance pay issue is moderately important to the MBA. Because the range of the salary being negotiated is quite satisfactory, the salary issue is least important. Without also elaborating the rationale for the management consulting firm representative's preferences here, the salary issue is most important to the management consulting firm representative, followed by severance pay and billable hours respectively.

The offsetting priorities create some integrative potential in the negotiation simulation. The MBA can compromise on the issue of least importance to him or her (salary) in exchange for a concession from the management consulting firm representative on the issue of most importance to the MBA (billable hours). The management consulting firm representative is likely to be willing to do this because the firm receives a better arrangement on the issue of most importance to it (salary), in return for a concession on the issue of least importance to the firm (billable hours). Besides this integrative aspect of the negotiation, the parties must also consider the total, overall distribution of the good outcomes between the parties to be generated by the conclusion of a job contract.

The subject's task in the simulation is to exchange proposals and messages with the other side to try to reach an agreement. As will be explained in the procedure section, the subject is led to believe that the "other party" is a person also sitting at a networked computer terminal in a computer laboratory. In fact, the "other party" is a computer program designed to reciprocate the subject's proposals in terms of the opposing interests identified above.

Because mixed-motive interdependence is built into the negotiation situation, subjects will be successful in achieving attractive agreements to the extent they exchange offers with the dual concerns in mind as discussed above. Specifically, the subject (who always plays the role of the MBA student) will need to propose options and make reasonable compromises with respect to the MBA's interests as well as the interests of the programmed management consulting firm representative, along both the distributive and integrative dimensions, to elicit favorable offers from the other side. With respect to the distributive aspect of the negotiations, the computer will respond with counterproposals that distribute the outcomes based on the same balance of self and other's interests as the subject's proposal. With respect to the integrative aspect of the negotiations, the programmed management consulting firm representative offers a counterproposal that concedes on the billable hours issue to the same extent that the subject concedes on the salary issue in his or her proposal.

With the negotiation situation so constructed and the management consulting firm representative's counterproposals so programmed, the representative's counteroffer is a measure of the subject's skill in proposing and examining options and in making reasonable compromises. Regarding proposing and examining options specifically, the management consulting firm's counterproposal reflects the same balancing of the interests of both parties as the subject's proposal reflected. In other words, the management consulting firm's counterproposal reflects the subject's skill with respect to the distributive aspect of proposing and examining options. The management consulting firm's counterproposal also reflects the subject's skill with respect to the integrative aspect of proposing and examining options. The counterproposal will offer the same level of concession on the issue of least importance to the management consulting firm representative as the level of concession the subject offered on the issue of least importance to him or her. Thus, subjects elicit counterproposals that are valuable to them by proposing options that reflect their own and the "other party's" interests on both the distributive and integrative dimensions.

Calculating the value of the counteroffer this way is essentially the same as assigning a point value to each subject's final proposal based upon the other party's pay-off matrix. Calculating the value of the final counterproposal, and assigning a point value to the MBA student's final proposal itself, based upon the worth of the proposal to the other party, are the same thing. Additionally, this approach fits neatly with the mechanics of the simulation: The value of the final counteroffer represents the worth of the subject's proposal to the other party.

Similarly, the management consulting firm's counterproposal is also a measure of skill in making reasonable compromises. With respect to the distributive aspect of skill in making reasonable compromises, the management consulting firm's counterproposal reflects the same level of compromise of one's own interests for the other party's interests as the subject's offer. With respect to the integrative aspect of skill in making reasonable compromises, the management consulting firm's counterproposal reflects the level of providing the other side with higher outcomes without an equal sacrifice of one's own outcomes. Specifically, the management consulting firm's counterproposal reflects the same level of concession on the issue of least importance to one's self to provide greater outcomes to the other on the issue of greatest importance to him or her.

With the conceptual aspect of our negotiation skill measure thus defined, we now report on an initial validation study of that measure. In the course of reporting that study, a more detailed description of the actual operation of the simulation will be provided. A more extensive validation study can be found in O'Neil, Allred, and Dennis (chapter 10, this volume).

VALIDITY OF THE SIMULATION ASSESSMENT

The initial validation study we conducted was based on an expert/novice criterion group approach. We assumed that a test that validly measures a given skill should be able to discriminate between experts and novices in that skill. Accordingly, we conducted the simulation with both novices and experts in negotiation. By quantifying the value or quality of the management consulting firm's counterproposal (how we do this is explained in the procedure section), we were able to compare the quality of the counteroffers the experts and novices elicited. Thus, our main hypothesis was that experts would receive better counteroffers than novices.

Further, if the simulation represents a valid negotiation environment, we would expect two critical negotiation biases, the self-serving bias and the fixed-pie bias, to be present in our computer simulation. Based on research by Sillars (1981), we hypothesized that subjects' perceptions of their own behavior and the other party's behavior would be subject to a self-serving bias. The self-serving bias, in the context of negotiations, refers to the tendency for people to see the other party as being less cooperative and reasonable than themselves. The self-serving bias interferes with effective distributive negotiation because one sees oneself as willing to compromise more than the other party and demands greater compromises from the other party than the situation would otherwise dictate. The result is interference with the ability of parties to reach agreements that distribute outcomes to the satisfaction of both parties.

The self-serving bias also tends to inhibit effective integrative negotiation, which requires that the parties mutually exchange information and cooperate to discover ways in which the outcomes of both parties might be increased. However, if the exchange and cooperation is not mutual, the side exchanging less information and being less cooperative will gain advantages in the distributive aspect of negotiations. Therefore, a party will not exchange information and cooperate in the manner necessary to realize integrative potential unless that party believes the other party is being just as forthcoming and cooperative. Because of the self-serving bias, however, parties tend to see the other party as being less forthcoming and cooperative and so are not willing to be forthcoming and cooperative themselves.

Since the other party's behavior was programmed in our computer simulation to be a mirror image of the subject's negotiating behavior, differences in subject ratings of the subject's own behavior and the other party's behavior are a measure of a self-serving bias. Accordingly, we hypothesized that the subjects in general would perceive themselves as more reasonable and fair than the other party.

Finally, the conceptual basis for our computer simulation is the idea that in order to be successful in negotiations, a negotiator must process all aspects of the negotiation in terms of the other party's interests as well as his or her own. The structure of the simulation situation built in mixed-motive interdependence with both distributive and integrative dimensions. Because the management consulting firm representative was programmed to respond to the MBA subject in a particular way, it was necessary, to be successful in the negotiation, for the MBA subject to think of multiple aspects of the negotiation (e.g., options, compromises, messages) in terms of both the other party's and his or her own interests.

As constructed, the simulation facilitates measurement of performance outcomes. However, outcome measures are not direct evidence of some processes we are inferring. Consequently, we were also interested in investigating whether success in terms of outcome measures is associated with cognitively processing facets of the negotiation in terms of consideration of both the other party's and one's own interests. If such cognitive processing is associated with outcome success, we expect subjects who are more successful in the negotiation will be more cognizant of the difference between approaching a negotiation with consideration of only one's own interests and approaching a negotiation with consideration of both one's own and the other party's interests.

METHODOLOGY

Subjects

Two groups of participants, one of novice negotiators and one of experts, took part in the study.

Experts.   Ten experts were drawn from graduate students who had just completed a course in negotiation in a Masters of Business Administration (MBA) program at a prestigious state university business school. The study was conducted in two sessions, with five students participating in one session and four participating in the other (a disk failure during the second data collection session caused the loss of one subject's data). All were second-year MBA students; five were female and four were male. The youngest participant was 26 years old and the oldest 52, with a mean age of 32 years. Six participants were White, two were Hispanic, and one was Asian. Participants had worked an average of four years before entering the MBA program. The mean Graduate Management Admissions Test (GMAT) score was 645, with a lowest score of 580 and a highest score of 720. These graduate students were experts compared to the novice group (high school students) but would be considered intermediate if compared to a "true" expert group of negotiators.

One experimenter made an announcement concerning the study in the class, in which 30 MBA students were enrolled. In addition to $20.00 for participation in the 1-hour study, a $20.00 prize was offered to the top performer in each session. The students were also told that they would receive feedback on their negotiation performance for participating in the study. A total of 10 students participated in the study as a result.

Novices. Twenty-one participants were drawn from a business computer class in a public high school of 1,400 students. Three students were randomly selected and eliminated from the analysis to maintain proportional cell sizes between expert and novice groups. Of the 18 novices, 3 were sophomores, 5 were juniors, and 10 were seniors. Six of the students indicated that they would probably attend college and 12 indicated that they would definitely attend college. Nine of the novices were female, and 9 male; 15 were White, 2 were African-American, and 1 was Hispanic. The average GPA was 3.12 (*SD* .44).

The teacher in the business computer class announced that the researchers would be visiting the class and asking the students to participate. The teacher also informed the students that the study would last 15 minutes longer than the usual class period, which was the last period of the day. Additionally, the teacher told the students that if they chose to participate, they would receive $20.00. They were also told that the top performer would receive an additional $20.00.

Procedure

Each subject was seated in front of an IBM personal computer that presented the instructions, task, and subsequent questionnaire. Our computer program was a modification of Carnevale's program (e.g., Carnevale & Conlon, 1988).[1] Carnevale and Pruitt (1992) viewed negotiation as the resolution of divergent interests by mutual and voluntary decision of the parties to the conflict. They contrasted negotiation with other forms of conflict resolution, that is, resolution by coercion, mediation (a third party facilitates or controls the process leading to the decision), or arbitration (a third party controls the decision itself, decides the outcome).

The Carnevale program was developed as a way of capturing what is known as the "AEI" paradigm. The AEI paradigm is intended to operationalize the integrative/distributive distinction discussed previously. In the paradigm, parties to a conflict face three issues. Integrative potential is operationalized by manipulating the relative priorities of the parties on the three issues. If the parties have exactly offsetting priorities on the issues

---

[1]We thank Dr. Peter Carnevale who provided his program for us to modify for this study.

(i.e., the issue of least importance to one is the issue of most importance to the other), then there is integrative potential which can be achieved by the parties mutually compromising on their respective issues of least importance. The result is that both parties receive a better arrangement on their issue of most importance. Within the paradigm, nine levels (A–I) of possible agreement are specified for each issue. The offsetting priorities are operationalized by assigning different points or values for offer levels on different issues, such as the points represented in Table 9.3. The possible trade-offs on issues create the integrative potential. The distributive aspect is represented by the fact that each side prefers the opposite end of the range on all of the three issues The name AEI derives from the fact that the best solution to both the integrative and distributive dimensions is AEI (see Neale & Bazerman, 1991).

Carnevale operationalized the AEI paradigm with his computer program. Using this program, Carnevale has investigated a variety of factors affecting dispute resolution processes and outcomes, including a number of investigations examining third-party mediation and arbitration scenarios, in addi-

TABLE 9.3
MBA and Management Consulting Firm Representative Issue Chart of Point Values

	Issues				
Billable Hours Per Week		*Severance (Weeks)*		*Salary (Annual)*	
MBA Student					
A	120	A	80	A	40
B	105	B	70	B	35
C	90	C	60	C	30
D	75	D	50	D	25
E	60	E	40	E	20
F	45	F	30	F	15
G	30	G	20	G	10
H	15	H	10	H	5
I	0	I	0	I	0
Management Consulting Firm Representative					
A	0	A	0	A	0
B	5	B	10	B	15
C	10	C	20	C	30
D	15	D	30	D	45
E	20	E	40	E	60
F	25	F	50	F	75
G	30	G	60	G	90
H	35	H	70	H	105
I	40	I	80	I	120

tion to two-party negotiation scenarios (e.g., Carnevale, 1986; Carnevale & Henry, 1989; Carnevale & Mead, 1990). Factors that he has examined include time pressure, perceived expertise, strategy choices, etc. (e.g., Carnevale, 1991; Carnevale & Conlon, 1988; Carnevale & Keenan, 1990).

We have adapted Carnevale's program with one major modification. As developed by Carnevale, the program would not contingently respond to the subject in any dynamic way. Accordingly, we built in an algorithm such that the computer would respond to the subject's negotiating behavior in terms of the issue chart for the management consulting firm representative (see Table 9.3). Specifically, a concession by the subject on a salary issue was answered by an equal concession on the billable hours issue. The management consulting firm representative role was programmed to follow a simple tit-for-tat strategy on the severance pay and billable hours issue. In other words, the computer would concede the same number of proposal levels from its most favored level (I) as the subject would concede from his or her most favored level (A). Finally, the management consulting firm representative's role was programmed so that it would accept any proposal that offered it the points equal to the EEE agreement (120) or better.

Subjects in our study were told they would negotiate with others via the computer in two negotiation sessions. They were also told that they would be negotiating a job contract between an MBA and a consulting firm. The computer, they were instructed, would randomly assign the subject to the role of either a consulting firm representative (the management consulting firm representative) seeking to hire an MBA or an MBA seeking employment with the consulting firm. In fact, the computers were not networked, and all subjects played the role of the MBA in both sessions, while the role of the management consulting firm representative was programmed.

Subjects were instructed that the job contract negotiation centered on three issues: (a) the number of billable hours the MBA would be required to log per week, (b) the number of weeks of severance pay the MBA would receive if fired or laid off, and (c) the annual salary the MBA would receive. Table 9.4 shows the issue chart that was displayed on the computer screen. Each issue had nine proposal levels (listed from "A" to "I"). Subjects were told that the parties would exchange proposals and messages in the negotiation in trying to reach agreement on one proposal level for each issue.

The subjects were also instructed with respect to their relative priorities on the three issues. The subjects were told that the time outside work was highly valuable to the MBA they were playing. The severance pay issue, subjects were told, was moderately important to the MBA because the consulting market was rather volatile at the time. The salary range being discussed, the subjects were further instructed, was quite satisfactory to the MBA. Therefore, the issue of greatest importance to the MBA was billable hours, followed by severance pay and salary respectively. Subjects were

TABLE 9.4
MBA Issue Chart of Real Values

	Issues				
Billable Hours Per Week		Severance (Weeks)		Salary (Annual)	
MBA Student					
A	20	A	9	A	80,000
B	22	B	8	B	77,500
C	24	C	7	C	75,000
D	26	D	6	D	72,500
E	28	E	5	E	70,000
F	30	F	4	F	67,500
G	32	G	3	G	65,000
H	34	H	2	H	62,500
I	36	I	1	I	60,000

subsequently shown another version of their issue chart which reflected these relative priorities on the three issues by assigning point values to each offer level for each issue.

The management consulting firm representative's role was programmed in such a way that the representative's priorities were exactly offsetting, as seen in Table 9.3. Thus, integrative potential was structured into the job-contract negotiation so that if the parties reciprocally conceded on the issue of least importance to them, joint outcomes could be maximized (EEE yields 120 points for each, while AEI yields 160). As was discussed earlier, the management consulting firm representative role was further programmed to reciprocate moves made by the subjects. Specifically, a concession by the subject on the salary issue was answered by an equal concession on the billable hours issue.

Subjects were further instructed that proposals and/or messages chosen from a menu of messages would be exchanged between the parties. Proposals consisted of one of the nine proposal levels for each issue (e.g., BCE). Messages could be chosen from a menu presented on the computer screen, as seen in Table 9.5. Messages, adapted from Carnevale and Conlon (1988), were selected that conveyed either concern for self (e.g., *You are too stubborn—make some concessions*) or concern for self and for the other party's interests (e.g., *Let's consider both of our needs and interests*).

As described, the computer took the subjects through two practice rounds of exchanging proposals and offers before the negotiations began. Using the same scenario, each of the two negotiation sessions continued until an agreement was reached or until the negotiation had proceeded

for 10 rounds, with one exchange of proposals and/or messages constituting one round. After the last round of the second negotiation session,
the subjects answered questions presented on the computer screen. The
questions asked the subjects to rate their own and the other party's negotiating style. Specifically, the subjects were asked to rate how coopera-
tive/competitive they and the other party were in the negotiations on
6-point Likert scales, ranging from *extremely competitive* to *extremely cooperative.*
Subsequently, the students were asked to make pairwise similarity ratings
on a 7-point scale of the messages they had an opportunity to use in the
simulation (see Table 9.5). Following these questions presented on computer, the subjects responded to a paper-and-pencil metacognitive questionnaire developed by O'Neil, Sugrue, Abedi, Baker, and Golan (1992).
The data on the metacognitive instrument will be published elsewhere.

After finishing, the subjects were debriefed. The experimenters explained that the computers were not networked in the simulation and that
the subjects were interacting with a computer program. The experimenters
further explained how the computer was programmed to reciprocate the
subjects' negotiating behavior. Finally, the experimenters explained that
the best agreement possible was AEI and offered lessons that could be
learned from the experience for the real negotiations the subjects might
encounter. No subject appeared to be upset by the deception, and most
found it amusing that they had, in effect, been negotiating with a mirror
image of themselves. The subjects were then thanked for their participation
and paid for their participation.

TABLE 9.5
Message Menu

Msg. #	Concern-With-Self-Only Messages
2	You are too stubborn—make some concessions.
3	Improve your offer if you want an agreement.
7	If we can't reach an agreement, I can easily find another job.
8	This offer is a gift. What more do you want?
10	This is the very best offer that is possible.

	Concern-With-Self-and-Other Messages
1	Here's a proposal for you to consider.
4	Let's consider both of our needs and interests.
5	I hope this offer is good for you.
6	Let's make offers that are good for both of us.
9	I'm interested to see what you think of this offer.

RESULTS

The primary purpose of this study was to test the validity of the simulation as a measure of negotiation skills. Our main hypothesis, therefore, was that the experts, being more skilled in examining and proposing options and in making reasonable compromises, should elicit counteroffers from the programmed management consulting firm representative that represented more points to the subjects. To test this, we calculated the point values (in terms of the subject's point values) of the final counterproposal offered by the programmed management consulting firm representative in each session (see Table 9.6). Since each subject completed two sessions of the negotiation simulation, we treated the value of the final counteroffer of each session as the outcome variable in a repeated measures design. The following repeated measures ANOVA tables summarize the effects for group (expert vs. novice) and session (first negotiation session vs. second negotiation session). The differences between the expert group and the novice group approach statistical significance, $F(1, 25) = 3.74$; $p = .064$. Students in general performed better in Session 2 than in Session 1, $F(1, 25) = 7.46$, $p = .011$. The latter result was expected, as we expected students to learn the implicit rules of the simulation (e.g., there were only 10 rounds in a session) during the initial session.

In addition, the number and quality of final counteroffers that resulted in actual agreements suggested that experts were more successful in the simulation than novices. Twenty-five percent of the novice sessions (9 of 36) were concluded with an agreement. The mean value of those concluded agreements was 136.7 (*SD* 23.6). In contrast, 33% of the expert sessions (6 of 18) were concluded with an agreement. The mean value of the

TABLE 9.6
Means and Standard Deviations for the Value of the Final Counteroffer (in Points)

Group (n)	M	SD
Session 1		
Expert (9)	101.67	37.91
Novice (18)	77.50	42.88
Total (27)	85.56	42.18
Session 2		
Expert (9)	126.67	26.34
Novice (18)	98.33	40.62
Total (27)	101.78	38.44

Note. Maximum value = 160 points.

experts' agreements was 146.7 (*SD* 16.3). However, a χ^2 test indicated that the frequency of agreements reached was not significantly different between groups. Of the agreements that were concluded, expert agreements also more frequently realized some of the integrative potential built into the simulation. We defined *some integrative potential is realized* as whenever an agreement yielded more than the 120 points received for the best purely distributive proposal (EEE) the computer would accept. This offer can only be achieved through some mutual tradeoff of the least important for the most important issue (i.e., AEI, BEH, CEG, DEF). Moreover, the experts' agreements more frequently achieved integrative potential (5 of 6 agreements or 83%) than novices' (4 of 9 agreements or 44%). (The cell sizes were too small for a statistical analysis.) Thus, as predicted, experts achieved both agreements and better agreements more frequently than novices. It should be noted that the statistical evidence for these assertions is weak.

Analysis of the subjects' ratings of their own and the other party's negotiation behavior revealed the hypothesized self-serving bias. Specifically, subjects (experts and novices combined) rated themselves as being more reasonable and fair (*M* = 3.76, *SD* = 1.02) than the "other party" (*M* = 2.85, *SD* = 1.13). A *t* test revealed that difference to be significant (*t* = 3.49; *df* 26; *p* = .002).

Analysis of the similarity ratings of the messages confirmed our expectation that subjects who more clearly distinguished between the messages, according to whether the messages conveyed concern for subjects' own interests or for the other party's interests in addition to their own, would perform better in the negotiations. In our similarity rating task, subjects rated the degree of similarity of all 45 possible pairwise combinations of the messages that were available for use during the negotiation simulation. For example, in Table 9.5, Message 2 would be compared successively with all the remaining messages. The data collected from this task were analyzed using the INDSCAL model option (Carroll & Chang, 1970) of the ALSCAL procedure available in SPSS.

A solution using two dimensions was found. Individual differences were clearly being modeled in terms of the relative influence of the two dimensions on subjects' perceptions of similarity between the statements. In other words, subjects were differentially influenced by these two dimensions. Additionally, one dimension was clearly more salient than the other in subjects' ratings. This primary dimension distinguished the stimulus points in terms of the criterion for which they were selected. As previously explained, the messages were written to represent a distinction between conveying concern with one's own interests only and concern with both the other party's and one's own interests. Furthermore, as expected, greater differentiation of the messages in terms of this dimension was correlated with higher performance in the negotiation task (i.e., elicited final counter-offers of greater value) ($r = 0.51$, $p < .01$).

DISCUSSION

The results were generally supportive of our hypotheses.

Validation of Simulation Assessment

The results provide evidence of the validity of the simulation as a measure of negotiation skills. Content validity was provided by both Carnevale and Conlon (1988) and our analytic tying of the simulation to the SCANS skill *negotiate*. Construct validity was examined in this study by comparing the performance of experts and novices.

Our main hypothesis was supported—that experts would be more success-ful than novices in proposing and examining options and making reasonable compromises as measured by the value of the final counteroffer they were able to obtain from the programmed management consulting firm repre-sentative. Results of the statistical test for this conclusion approached statistical significance ($p = .064$). We feel that in a first study of this series (a feasibility study), our observed probability value ($p = .064$) indicates that there is support for the assertion that experts did better than novices. Although the greater frequency and value of expert agreements was found, it was not statistically significant. In general, the size of our sample is small and requires replication, but we view these results as very promising.

This encouraging initial evidence of validity suggests that further validity tests are warranted. One important question that this study did not examine is whether differences in the two subject groups other than expertise in negotiation might be responsible for the differences in performance. For example, the novice (high school students) and expert (MBA students) groups varied in age, level of education, and familiarity with the MBA negotiation scenarios, in addition to level of expertise in negotiations. Also, the validity of the simulation as a measure of negotiation skills must be further examined by using subjects who are more clearly experts in nego-tiation. Our MBA students had only one course in negotiation. We plan to use professional arbitrators and mediators as expert subjects. Our ex-pectation is that the differences will be more dramatic.

Another issue related to the MBA simulation used in this study concerns the particular form of the scenario. The scenario involved an MBA negoti-ating a job contract with a consulting firm. Clearly, the experts in our study (MBA students) were more familiar with this scenario than the novices (high school students). Although no previous knowledge of job contracts with consulting firms was assumed or necessary to perform in the simulation, the familiarity of the situation for the experts may have aided their ability to process the information provided, or engaged or motivated them more because of the relevance of job contract negotiations for graduating MBAs.

Further evidence for the validity of the measure would be provided by comparing experts and novices across different scenarios involving three issues to be negotiated between parties with mixed-motive interdependence. We are currently analyzing data from a second phase of the present programmatic research that will allow us to investigate this issue in the manner just described. In this second phase, approximately 40 experts and over 200 novices participated in the computer simulation, but with two different scenarios. The experts were third-year law students enrolled in a negotiation course. Novices were drawn from the same high school as the subjects in the present study. Half of the novices and half of the experts played the role of a high school student negotiating a job contract with the personnel manager of a movie theater regarding the number of movie passes received, the number of weekend work hours required, and the hourly wage received for the job. The other half of the novice and expert groups played the role of a third-year law student negotiating a job contract with a law firm regarding the issues of salary, time to partnership, and required billable hours. Data from this second phase will allow us to examine the effect of familiarity with the scenario on negotiation performance within an expert-novice paradigm.

Self-Serving Bias.   Our results provide clear evidence that people in our simulation tended to see themselves as being more reasonable and fair in negotiations than the other party. These results, consistent with the literature, provide further validation of our simulated environment. In addition to the results confirming the validity of simulation, the self-serving bias findings shed light on processes that may explain outcome variables such as the value of the management consulting firm representative's final counteroffer. These findings suggest that attribution processes may be responsible, at least in part, for negotiation behavior. As explained in the introduction, such a bias could have an important impact on how effectively people propose and examine options and make reasonable compromises along both the distributive and integrative dimensions. With respect to our simulation, the self-serving bias probably interferes most with effective distributive negotiation. Because one sees oneself as willing to compromise more than the other party, one tends to demand greater compromises from the other party than the situation might otherwise dictate. Consequently, subjects may have been limited in their ability to elicit more valuable counterproposals from the programmed management consulting firm representative.

Dual Concern. The expected results of the similarity ratings analyses confirm that success in the simulation is associated with cognitively processing different facets of the negotiation in terms of both the other party's and one's own interests. Subjects who more clearly differentiated between messages that conveyed this approach to negotiations and messages that only conveyed concern for one's own interests were more successful in the negotiations. These results also provide some additional evidence of construct validity for our simulation. And, the results support our notion that cognitive processing of various aspects of the negotiations in the simulation in terms of the other party's and one's own interests is related to our outcome measures of performance.

Feasibility

This initial validation study also revealed worthwhile information about the feasibility of a computer simulation approach to measuring negotiation skills. Feasibility is usually viewed as consisting of three issues—time, cost, and performance. With respect to performance, our program is copied onto bootable disks. Thus, setting up for the simulation is as simple as restarting the IBM personal computers with a disk in the "A" drive. Furthermore, the data from the subject's negotiation interaction and the questions following the negotiation are recorded on the disk at the conclusion of the simulation, making data collection and entry into statistical programs quite simple. With respect to cost, because many high schools have IBM computer labs, there probably would be no additional cost for the hardware. Obviously, the simulation would have to be rewritten and recompiled for a Macintosh world. Thus, the hardware/software costs of this approach seem reasonable. Additionally, the time subjects took to read the instructions and complete the first session of negotiations was, on the average, 14 minutes ($M = 840.92$ seconds, $SD = 196.58$ seconds). Thus, the measurement time for an interpersonal skill of negotiation is not excessive. Obviously, some sort of interpretation feature would have to be added to the simulation if feedback to the student was desired. Further, the subjects also seemed quite engaged in the task. A number of subjects told the experimenters that they found the simulation quite interesting. In general, our experience in this initial validation study speaks favorably to the issue of the feasibility of the computer simulation approach to measurement of negotiation skills.

CONCLUSION

In conclusion, our analysis of the research on negotiation has identified important indicators of the negotiation skills identified in the job and task analyses conducted by SCANS (U.S. Department of Labor, 1991, 1992).

First, in terms of skills in proposing and examining possible options, a negotiator must assess possible options with respect to both the negotiator's own interests and the other party's interests, and must do so for both distributive and integrative aspects of the negotiation. With respect to distributive aspects, does an option provide outcomes that are acceptable to both the negotiator and the other party? With respect to integrative aspects, are there options that increase the outcomes for both parties?

Second, in terms of skills in making reasonable compromises, the literature suggests that a negotiator must consider compromises in terms of the other party's and his or her own interests along both the distributive and integrative dimensions. According to the distributive aspects of reasonable compromises, a party must not compromise his or her own outcomes unduly, while making enough of a compromise to make a resolution acceptable to the other party. With respect to the integrative aspect of making reasonable compromises, a reasonable compromise is one that gives up some of one's outcomes for the sake of the other's only where it is not possible to provide a corresponding increase to the other side without giving up on the outcomes one receives him- or herself. In other words, a compromise is not reasonable where the outcomes offered to the other side can somehow be provided without sacrificing one's own outcomes.

Most importantly, the results of our validation study suggest that our simulation can be used to measure these research-identified indicators of negotiation performance in proposing and examining options and making reasonable proposals. The computer simulation approach to measuring negotiation skills appears to be worthy of further investigation. Further exploration along this avenue of assessment holds the promise of yielding an assessment method for other SCANS skills that can assist the United States in its endeavor to develop a workforce with the skills needed to compete successfully in today's world.

ACKNOWLEDGMENTS

The work reported herein was supported under the Educational Research and Development Center Program cooperative agreement R117G10027 and CFDA catalog number 84.117G as administered by the Office of Educational Research and Improvement, U.S. Department of Education. The findings and opinions expressed in this report do not reflect the position or policies of the Office of Educational Research and Improvement or the U.S. Department of Education.

REFERENCES

Blake, R. R., & Mouton, J. S. (1979). Intergroup problem solving in organization: From theory to practice. In W. G. Austin & S. Worchel (Eds.), *The social psychology of intergroup relations.* Monterey, CA: Brooks/Cole.

Blinder, A. S. (1990). *Paying for productivity: A look at the evidence.* Washington, DC: The Brookings Institution.

Braun, H. I. (1994). Assessing technology in assessment. In E. L. Baker & H. F. O'Neil, Jr. (Eds.), *Technology assessment in education and training* (pp. 231–246). Hillsdale, NJ: Lawrence Erlbaum Associates.

Braun, H. I., Bennett, R. E., Frye, D., & Soloway, E. (1990). Scoring constructed responses using expert systems. *Journal of Educational Measurement, 27,* 93–108.

Brett, J. M., Goldberg, S. B., & Ury, W. L. (1990). Designing systems for resolving disputes in organizations. *American Psychologist, 45,* 162–170.

Bunderson, C. V., Inouye, D. K., & Olsen, J. B. (1989). The four generations of computerized educational measurement. In R. Linn (Ed.), *Educational measurement* (3rd ed., pp. 367–408). New York: Macmillan.

Cappelli, P., & Singh, H. (1992). Integrating strategic human resources and strategic management. In D. Lewin, O. S. Mitchell, & P. D. Sherer (Eds.), *Research frontiers in industrial relations and human resources.* Madison, WI: Industrial Relations Research Association.

Carnevale, P. J. (1986). Strategic choice in mediation. *Negotiation Journal, 2,* 241–256.

Carnevale, P. J. (1991). *Cognition and affect in cooperation and conflict.* Presented at the Fourth Annual Meeting of the International Association of Conflict Management, Ernst Sillem Hoeve, Den Dolder, The Netherlands.

Carnevale, P. J., & Conlon, D. E. (1988). Time pressure and strategic choice in mediation. *Organizational Behavior and Human Decision Processing, 42,* 111–133.

Carnevale, P. J., & Henry, R. (1989). Determinants of mediator behavior: A test of the strategic choice model. *Journal of Applied Social Psychology, 19,* 481–498.

Carnevale, P. J., & Keenan, P. A. (1990). *Decision frame and social goals in integrative bargaining: The likelihood of agreement versus the quality.* Presented at the Third Annual Meeting of the International Association of Conflict Management, Vancouver, BC.

Carnevale, P. J., & Mead, A. (1990). *Decision frame in the mediation of disputes.* Presented at the annual meeting of the Judgment Decision Making Society, New Orleans, LA.

Carnevale, P. J., & Pruitt, D. J. (1992). Negotiation and mediation. *Annual Review of Psychology, 43,* 531–582.

Carroll, J. D., & Chang, J. J. (1970). Analysis of individual differences in multidimensional scaling via an *N*-way generalization of Eckart-Young decomposition. *Psychometrika, 35,* 283–319.

Dunn, T. G., Lushene, R., & O'Neil, H. F., Jr. (1972). The complete automation of the Minnesota Multiphasic Personality Inventory. *Journal of Consulting and Clinical Psychology, 39,* 381–387.

Filley, A. C. (1975). *Interpersonal conflict resolution.* Glenview, IL: Scott, Foresman.

Gerhart, B., Milkovich, G. T., & Murray, B. (1992). Pay, performance, and participation. In D. Lewin, O. S. Mitchell, & P. D. Sherer (Eds.), *Research frontiers in industrial relations and human resources.* Madison, WI: Industrial Relations Research Association.

Gutkin, T. B., & Wise, S. L. (Eds.). (1991). *The computer and the decision-making process* (Buros-Nebraska Symposium on Measurement and Testing). Hillsdale, NJ: Lawrence Erlbaum Associates.

Hedl, J. J., Jr., O'Neil, H. F., Jr., & Hansen, D. N. (1973). The affective reactions toward computer-based intelligence testing. *Journal of Consulting and Clinical Psychology, 40,* 310–324.

Huselid, M. A. (1995). The impact of human resource management practices on turnover, productivity, and financial performance. *Academy of Management Journal, 38,* 635–672.

Kelley, H. H. (1979). *Personal relationships: Their structures and processes.* Hillsdale, NJ: Lawrence Erlbaum Associates.

Kelley, H. H., & Thibaut, J. W. (1978). *Interpersonal relations: A theory of interdependence.* New York: Wiley-Interscience.

Kochan, T. A., Dyer, L., & Batt, R. (1992). International human resource studies: A framework for future research. In D. Lewin, O. S. Mitchell, & P. D. Sherer (Eds.), *Research frontiers in industrial relations and human resources.* Madison, WI: Industrial Relations Research Association.

Lax, D. A., & Sebenius, J. K. (1986). *The manager as negotiator.* New York: Free Press.

Lewicki, R. J., Litterer, J. A., Minton, J. W., & Saunders, D. M. (1994). *Negotiation.* Burr Ridge, IL: Irwin.

Neale, M. A., & Bazerman, M. H. (1991). *Negotiator cognition and rationality.* New York: Free Press.

O'Neil, H. F., Allred, K. G., & Baker, E. L. (1992a). *Measurement of workforce readiness competencies: Design of prototype measures* (CSE Tech. Rep. No. 344). Los Angeles: University of California, Center for Research on Evaluation, Standards, and Student Testing (CRESST).

O'Neil, H. F., Allred, K. G., & Baker, E. L. (1992b). *Measurement of workforce readiness: Review of theoretical frameworks* (CSE Tech. Rep. No. 343). Los Angeles: University of California, Center for Research on Evaluation, Standards, and Student Testing (CRESST).

O'Neil, H. F., Jr., & Baker, E. L. (in press). A technology-based authoring system for assessment. In S. Dijkstra & N. Seel (Eds.), *Instructional design: International perspectives. Vol. II: Solving instructional design problems.* Mahwah, NJ: Lawrence Erlbaum Associates.

O'Neil, H. F., Jr., Hedl, J. J., Jr., Richardson, F. C., & Judd, W. A. (1976). An affective and cognitive evaluation of computer-managed instruction. *Educational Technology, 16,* 29–34.

O'Neil, H. F., Jr., Sugrue, B., Abedi, J., Baker, E. L., & Golan, S. (1992). *Report of experimental studies on motivation and NAEP test performance* (Report to National Center for Education Statistics, Contract No. RS90159001). Los Angeles: University of California, National Center for Research on Evaluation, Standards, and Student Testing.

Pfeffer, J. (1994). *Competitive advantage through people: Unleashing the power of the workforce.* Boston, MA: Harvard Business School Press.

Pruitt, D. G., & Rubin, J. Z. (1986). *Social conflict: Escalation, stalemate, and settlement.* New York: Random House.

Pruitt, D. G., & Syna, H. (1984). Successful problem solving. In D. Tjosvold & D. W. Johnson (Eds.), *Productive conflict management: Perspective for organizations* (pp. 129–148). Minneapolis, MN: Minneapolis Team Media.

Rahim, M. A. (1986). *Managing conflict in organizations.* New York: Praeger.

Rubin, J., & Brown, B. R. (1975). *The social psychology of bargaining and negotiation.* New York: Academic Press.

Sillars, A. L. (1981). Attributions and interpersonal conflict resolution. In J. H. Harvey, W. Ickes, & R. F. Kidd (Eds.), *New directions in attribution research* (Vol. 3, pp. 279–305). Hillsdale, NJ: Lawrence Erlbaum Associates.

Stasz, C., Ramsey, K., Eden, R. A., Melamid, E., & Kaganoff, T. (1996). *Workplace skills in practice* (MR-722-NCRVE/UCB). Santa Monica, CA: RAND.

Thomas, K. W. (1976). Conflict and conflict management. In M. Dunnette (Ed.), *Handbook of industrial and organizational psychology* (pp. 889–935). Chicago: Rand McNally.

Tjosvold, D. (1990). The goal interdependence approach to communication in conflict: An organizational study. In M. A. Rahim (Ed.), *Theory and research in conflict management* (pp. 15–31). New York: Praeger.

Tjosvold, D., & Johnson, D. W. (1983). *Productive conflict management.* New York: Irvington Publishers.

U.S. Department of Labor. (1991, June). *What work requires of schools. A SCANS report for America 2000.* Washington, DC: U.S. Department of Labor, The Secretary's Commission on Achieving Necessary Skills (SCANS).

U.S. Department of Labor. (1992). *Skills and tasks for jobs: A SCANS report for America 2000.* Washington, DC: U.S. Department of Labor, The Secretary's Commission on Achieving Necessary Skills (SCANS).

Walton, R. E., Cutcher-Gershenfeld, J. E., & McKersie, R. B. (1994). *Strategic negotiations.* Boston: Harvard Business School Press.

Walton, R. E., & McKersie, R. E. (1965). *A behavioral theory of labor negotiations.* New York: McGraw-Hill.

Womack, D. F. (1990). Applied communications research in negotiation: Implications for practitioners. In M. A. Rahim (Ed.), *Theory and research in conflict management* (pp. 32–53). New York: Praeger.

Chapter 10

Validation of a Computer Simulation for Assessment of Interpersonal Skills

Harold F. O'Neil, Jr.
CRESST/University of Southern California

Keith Allred
Robert A. Dennis
CRESST/University of California, Los Angeles

In this chapter, we describe the results of two validation studies of our negotiation skill measure. We begin with a description of our prior work in negotiation skills assessment to provide the context and rationale for the present research.

In creating our computer-based measure, we adopted the methodology for the assessment of workforce readiness that we developed in an earlier report (O'Neil, Allred, & Baker, 1992; chapter 8, this volume). As shown in Fig. 10.1, that methodology dictates that in developing measures of workforce readiness skills competencies, one begins in the work environment. First, a job and task analysis should be conducted (see O'Neil et al., chapter 8, this volume, for logic). The creation of indicators for subcompetencies is a critical next step. The use of cognitive science taxonomies allows for generalization of the findings. Then a set of empirical studies are conducted. Finally, the results are documented.

Based on the assessment of the performance criteria by the Secretary's Commission on Achieving Necessary Skills (SCANS; U.S. Department of Labor, 1991, 1992) and the cognitive indicators of those performance criteria, we developed a rapid prototype of negotiation simulation. With regard to the SCANS performance criteria, we needed to simulate the activities of proposing and examining options and making reasonable compromises. Accordingly, the negotiation task is the exchange of proposals and counterproposals. With regard to the cognitive indicators identified in the negotiation literature (e.g., Kelley, 1979; Kelley & Thibaut, 1978; Lax &

General Methodology	Specific Example
Select a work environment	Analytically derived
Conduct job and task analysis	Analytically derived
Select competency	Interpersonal
Conduct component analysis of competency	Negotiate
Specify basic skills foundation	Mathematics, Creative thinking, Decision making, Problem solving, Self-management
Create indicator(s) for subcompetencies	Proposing and examining possible options and making reasonable compromises
Classify indicator(s) within a cognitive science taxonomy	Carnevale & Pruitt, 1992; Walton & McKersie, 1965; Womack, 1990
Create rapid prototype of measures of indicator(s) via test specifications	Existing simulation modified
Select/develop final measures of indicator(s)	See Methodology section
Select experimental/analytical design	Expert/Novice
Run empirical studies	This report
Analyze statistically	This report
Use/create norms	To be done
Report reliability/validity of indicator(s) measure	This chapter
Report on workforce readiness using multiple indicators	To be done

FIG. 10.1. Workforce readiness assessment methodology for SCANS: negotiation example.

Sebenius, 1986; Lewicki, Litterer, Minton, & Saunders, 1994; Pruitt & Rubin, 1986; Rubin & Brown, 1975; Walton, Cutcher-Gershenfeld, & McKersie, 1994; Walton & McKersie, 1965), the exchange of proposals should take place in the context of a situation of mixed-motive interdependence, with both distributive and integrative dimensions (e.g., Brett, Goldberg, & Ury, 1990; Lax & Sebenius, 1986; Lewicki, Litterer, Minton, & Saunders, 1994; Neale & Bazerman, 1991; Pruitt & Rubin, 1986; Pruitt & Syna, 1984; Rahim, 1986; Thompson, 1990a, 1990b, 1991; Tjosvold, 1990; Tjosvold & Johnson, 1983; Walton et al., 1994; Walton & McKersie, 1965; Womack, 1990).

Distributive negotiations concern the zero-sum or win/lose elements of a negotiation. Integrative negotiations involve variable sum or win/win aspects of a negotiation. To ensure a resolution to the negotiation in either case, one must ensure that the other party to the negotiation gets enough to agree.

In our work, we use computer simulations for assessment. The logic for the use of simulation as an assessment context is documented elsewhere (O'Neil, Allred, & Dennis, 1992; see also chapter 9, this volume). Our use of a computer simulation for assessment purposes is consistent with the guidelines for computer testing (Green, 1991). As may be seen in Fig. 10.2, domain specifications (e.g., Baker, 1992; Millman & Greene, 1989) are embedded in the software.

General Domain Specification	Specific Example
Scenario	Role play a job contract negotiation by exchanging proposals in mixed motive context
Players	One student and one manager (computer software)
Student	Either expert or novice, individual or team
Manager	Computer software (Carnevale & Conlon, 1988; O'Neil, Allred, & Dennis, 1992)
Priorities	Offsetting
Moves	Reciprocal
Rounds	Offer from student and counteroffer from manager
Subcompetencies	Propose options; make reasonable compromises
Negotiation issues	Three in number (e.g., salary) with offsetting priorities
Negotiation measures	Agreement (yes/no), type of agreement (distributive vs. integrative), final counteroffer
Cognitive processes (domain-dependent)	Fixed-pie bias, self-serving bias
Cognitive processes (domain-independent)	Metacognitive skills
Affective processes (domain-independent)	Effort, anxiety

FIG. 10.2. Domain specifications embedded in the software.

The negotiation scenario is a job contract negotiation. The parties to the negotiation are a representative of the potential employer and the potential employee. Two scenarios entailing different employers and employees were used in the validation studies reported here. The conceptual framework for the simulation situation is described in terms of the movie theater scenario used with half the subjects in the first study and all subjects in the second study. In this scenario, the contract negotiation takes place between a high school student and the manager of a movie theater. In the simulation, the high school student is seeking employment and the movie theater manager is looking to hire. The two parties negotiate the terms of the contract with respect to three issues: (a) free movie passes, (b) weekend hours, and (c) hourly wage. The high school student prefers to have more movie passes, to work fewer hours on the weekend, and to have a higher hourly wage. In contrast, the movie theater manager prefers that the high school student receive fewer passes, work more weekend hours, and receive a smaller hourly wage.

To build in both integrative and distributive dimensions to the simulation, the parties also have offsetting priorities regarding the three issues being negotiated. The high school student is characterized as being a big movie buff, and the free passes issue is most important to him or her as the passes are worth $7.50 each. Because the high school student likes time free on the weekends, the weekend hours issue is moderately important to him or her. The hourly wage issue is least important as the range of various hourly wages is quite small. In contrast, the movie theater manager's most important issue is the hourly wage, followed by the weekend hours and free passes issues, respectively. The offsetting priorities of the high school student and the manager create the integrative potential in the simulation.

COMMON HYPOTHESES OF THE PRESENT STUDIES

We conducted a preliminary validation study of our negotiation skill measure based on an expert/novice criterion group approach (O'Neil, Allred, & Dennis, 1992; chapter 9, this volume). One would expect that a test with construct validity would discriminate between experts and novices. Thus, we validated the simulation with both novices and experts in negotiation. Eighteen high school students served as novices and nine graduating MBA students who had just completed a course on negotiation served as experts. We wanted to conduct a preliminary study to document whether further validation studies were warranted.

In the O'Neil, Allred, and Dennis study (1992), the mean value of the counterproposals the experts exhibited was higher than that of novices. These differences approached statistical significance ($p = .064$). Additionally, in terms of actual agreements, the experts concluded more and higher quality agreements than novices. The expected differences did indicate that larger scale validation studies were warranted. Based again upon an expert/novice criterion group approach and the results from the initial validation study, we expected experts in the present study to perform better than novices.

In addition to performance, we were interested in examining domain-specific cognitive processes known to be related to negotiation outcomes. Evidence that cognitive processes related to negotiation performance in other contexts were also related to performance in our simulation would further support the construct validity of our simulation. According to the cognitive perspective in psychology, people manage interdependence by accurately processing and interpreting both (a) the negotiation *situation* and (b) the goals and attitudes of the *other party* with whom one is negotiating (Thompson, 1990b). The decision-analytic perspective (e.g., Neale & Bazerman, 1991; Raiffa, 1982) has proved a powerful framework for understanding how negotiators process and understand negotiation situations. Research conducted within the decision-analytic framework has identified, among other biases, a fixed-pie bias, which prevents people from realizing the potential that often exists for increasing joint gains (Bazerman, Magliozzi, & Neale, 1985; Thompson, 1990a, 1991; Thompson & Hastie, 1990). The fixed-pie bias is the tendency people have to assume there is a fixed amount of outcomes to be distributed that cannot be expanded. Accordingly, a gain for one side means an equal loss for the other side and vice versa. The fixed-pie bias thus prevents people from pursuing integrative potential or seeking to expand the pie. In particular, the fixed-pie bias has been reported in the same type of negotiation context that we use. For example, Bazerman, Thompson, and their colleagues have documented that people tend to assume that what is most important to them is also most important to the other party. Consequently, the potential for joint gain is often unrealized. Based upon this research we hypothesized the following for the present study: Experts will exhibit less fixed-pie bias than novices.

Although it has received less attention from negotiation researchers, the attributional perspective in social psychology also suggests processes by which negotiators interpret the other party in a negotiation. According to the attributional perspective, how we interact with another person is partly a function of the causal attributions we make about how and why the people with whom we are interacting are behaving as they are (e.g., Heider, 1958; Weiner, 1986, 1992). In the context of negotiation specifically, behavior will be dependent in part on how a person interprets the behavior of the parties with whom he or she is negotiating (Sillars, 1981).

We suggest that a particularly important aspect of attribution in negotiation concerns the judgments of the relative degree of reasonableness, cooperativeness, and concern for the opposite party exhibited by one's self and by the other party. Specifically, we predicted that people would exhibit a self-serving attributional bias such that they would tend to perceive themselves as more concerned for the other party, more cooperative, and more reasonable than the other party (Brandstatter, Kette, & Sageder, 1982; Kramer, Newton, & Pommerenke, 1993; Sillars, 1981; Turnbull, Strickland, & Shaver, 1976).

The implications of a self-serving bias for negotiator performance seem clear. To the extent that one perceives oneself to be more reasonable or concerned for the other party's welfare, one is likely to feel exploited and reciprocate by exhibiting less cooperation and concern subsequently. As a result of a self-serving bias, one is likely to demand greater compromises from the other party than the situation would otherwise dictate. The probable result is to interfere with the ability of parties to reach agreements satisfactory to both parties.

In our simulation, the "other party" with which subjects negotiated was always the computer program—however, this was not revealed to them until they were debriefed at the end of the negotiation. Because the other party's behavior was programmed to be a mirror image of the subject's negotiating behavior, differences in subject ratings of the subject's own behavior and the other party's behavior are a measure of a self-serving bias. Thus, although the empirical support for the link between self-serving biases and negotiator performance is limited, we hypothesized that the relationship would emerge in our simulation. Specifically, we hypothesized the following: In this study, experts will exhibit less self-serving bias than novices.

One aspect of our research on assessing negotiation skills was focused on self-regulating processes (Glaser, Raghavan, & Baxter, 1992). We view self-regulating processes as consisting of metacognition and effort. In turn, metacognition consists of planning, self-monitoring, cognitive strategies, and awareness. We have developed state measures of these constructs (O'Neil & Abedi, 1996; O'Neil, Sugrue, Abedi, Baker, & Golan, 1992). We reasoned that experts engaged in our computer-based simulation of negotiation would exhibit more metacognitive activity and effort with less anxiety than novices.

SPECIFIC HYPOTHESES OF THE VALIDATION STUDY

In the validation study, in addition to the common hypotheses identified above, we were also interested in examining the effects of different scenarios of the same basic simulation. Thus, in addition to the movie theater scenario described above, we employed a scenario in which a third-year law student was negotiating a job contract with a law firm. We anticipated there would be no significant effects for scenario in terms of performance, fixed-pie bias, self-serving bias, or self-regulatory skills.

METHOD

Participants

One group of expert and one group of novice negotiators participated. Thirty-seven second- and third-year law students (24 males, 13 females) who were nearing completion of a course on negotiation participated as experts. The average age of the law students was 28. Their law negotiation course focused on integrative and distributive aspects of effective negotiation. In addition to lectures and discussions, the course included simulations designed to provide the students with practical experience with the principles discussed. However, no simulation was based on the exact paradigm employed in our simulation. The students participated in our study during regular class time as part of the requirements for the course, in one of two sessions conducted in the law school's computer laboratory, which had 20 IBM personal computers.

Two hundred forty-eight novice participants (102 males, 146 females) were drawn from various classes of college-bound students in a public high school with a total of approximately 1,400 students. Sixty-eight participants were sophomores, 86 were juniors, and 94 were seniors. All students participated as part of the classroom activity for that day. The students came to one of the two high school computer laboratories, each of which had 20 IBM personal computers.

Procedure

Each subject was seated in front of an IBM personal computer that presented the instructions, task, and subsequent questionnaire. The computer program used was a modification of Carnevale's program (e.g., Carnevale & Conlon, 1988).[1]

Subjects were instructed that they would negotiate with others via the computer. Subjects were randomly assigned to either the movie theater scenario described above or the parallel scenario involving employment negotiations between a graduating law student and a law firm. The procedure will be described in terms of the movie theater scenario. (For more details on the law firm scenario, see Appendix A.) The computer, the subjects were instructed, would randomly assign the subject to the role of either a movie theater manager seeking to hire a high school student or a high school student seeking employment with a movie theater. In fact, the computers were not network-aware, and all subjects played the role of the high school student (or law student in the law firm scenario), while the role of the movie theater manager (or law firm representative in the law firm scenario) was programmed.

Subjects were instructed that the job contract negotiation centered on three issues: (a) the number of free, transferable passes per month that the high school student would receive, (b) the number of weekend hours the student would work of the 10 total hours worked per week, and (c) the hourly wage the high school student would receive. Subjects were told that they preferred more free passes, fewer weekend hours, and higher hourly wages, whereas the personnel manager preferred fewer passes, more weekend hours, and lower hourly wages. Subjects were also instructed with respect to their relative priorities on the three issues and told that the parties would exchange proposals in the negotiation in trying to reach agreement on one proposal level for each issue. The computer presented the subjects with the issue chart shown in Table 10.1. Subjects did not have a paper copy of this chart. (See O'Neil, Allred, and Dennis, chapter 9, this volume, for more extensive discussion of subject priorities.)

The manager's role was programmed so that it would accept any proposal as an agreement that offered it the points equal to EEE (120) or

---

[1]We thank Dr. Peter Carnevale who provided his program for us to modify for this set of studies.

TABLE 10.1
High School Student Issue Chart of Real Values

		Issues			
Passes (Per Month)		Weekend Hours (Per Week)		Wage (Per Hour)	
A	9	A	2	A	5.05
B	8	B	3	B	4.95
C	7	C	4	C	4.85
D	6	D	5	D	4.75
E	5	E	6	E	4.65
F	4	F	7	F	4.55
G	3	G	8	G	4.45
H	2	H	9	H	4.35
I	1	I	10	I	4.25

better. Consequently, the programmed negotiator mirrored the subject's negotiating behavior in terms of whether proposals moved toward realizing the integrative potential or not. Additionally, it was not possible for the subject to conclude an agreement valued at more than 120 points without engaging in logrolling. (Pruitt and Rubin [1986] refer to the trading off of issues of different priority as a "logrolling" integrative strategy; for example, a subject in the role of the high school student can compromise on the issue of least importance to him or her [hourly wage] in exchange for a concession from the movie theater manager on the issue of most importance to the high school student [free passes].)

The computer presented the subjects with two practice rounds of exchanging proposals. Subjects were instructed that the negotiations would continue until an agreement was reached or until the negotiation had proceeded for 12 rounds, with one exchange of proposals constituting one round. After the last round of the negotiations, the subjects completed a fixed-pie and self-serving bias questionnaire using a 7-point Likert scale. This questionnaire was presented on the computer screen. Following the questionnaire, the subjects responded to a paper-and-pencil self-regulation questionnaire (O'Neil, Sugrue, et al., 1992; O'Neil & Abedi, 1996).

After finishing, the subjects were debriefed. The experimenters explained that the computers were not network aware and that the subjects were interacting with a computer program. The experimenters further explained how the computer was programmed to reciprocate the subjects' negotiating behavior. Finally, the experimenters explained that the best agreement possible was AEI and offered lessons that could be learned from the experience for real negotiations the subjects might encounter.

No subjects appeared to be upset by the deception, and most found it amusing that they had, in effect, been negotiating with a mirror image of themselves. The subjects were then thanked for their participation.

Results

The results were generally supportive of the hypotheses. With respect to hypothesis 1, that experts would perform better than novices, we measured performance in three ways. First, as previously described, the value of the final counteroffer the subject elicited from the programmed employer is a measure of negotiation skill. As seen in Table 10.2, experts elicited final counteroffers of greater value to themselves than did novices in both the movie theater (theater) and law firm (law) scenarios. An analysis of variance showed this main effect for criterion group to be significant, $F(1, 284) =$ 9.28, $p < .001$. As expected, there was no main effect or interaction for scenario. However, the interaction approached statistical significance, $F(1, 128) = 2.99$, $p = .085$.

A second measure of performance was the frequency of actual agreements. In the final counteroffer measure all subjects are included, whether they actually concluded an agreement or not. Thus, we thought it also important to examine actual agreements concluded. As seen in Table 10.3, experts more frequently concluded agreements than novices. As also seen in Table 10.3, however, the chi-square analyses showed that this difference only approached significance ($p = .059$). With respect to scenario effects, contrary to our expectations, subjects in the theater scenario concluded agreements significantly more frequently than subjects in the law scenario (see Table 10.4). We will say more about this result later.

A third measure of performance was the frequency of integrative versus distributive agreements reached. In other words, of those agreements ac-

TABLE 10.2
Mean Final Counteroffer Performance Measure (Entire Data Set)

	Scenario	
Criterion Group	Theater	Law
Novice		
Mean *(SD)*	111.77 (29.14)	111.29 (28.89)
n	124	124
Expert		
Mean *(SD)*	136.11 (19.67)	118.16 (31.06)
n	18	19

TABLE 10.3
Frequency of Agreement by Criterion Group
(Entire Data Set)

| | Criterion Group | | |
	Novice	Expert	n
Agreement	155 (62.5%)	29 (78.4%)	184
No Agreement	93 (37.5%)	8 (21.6%)	101
Total N	248 (100.0%)	37 (100.0%)	285

Note. $\chi^2 = 3.55$. $df = 1$. $p < .059$.

tually reached, did experts tend to achieve integrative agreements more frequently? Our results offer strong evidence that the answer is yes. As described before, subjects had to engage in integrative logrolling strategy to achieve an agreement of greater value than 120. Thus, we examined the frequencies of agreements above 120 versus those at or below 120. As seen in Table 10.5, experts' agreements were integrative almost two thirds of the time, while novices' agreements were integrative less than one third of the time. The chi-square analysis (Table 10.5) showed this difference to be extremely significant. As seen in Table 10.6, with respect to scenario effects, subjects achieved integrative agreements with about the same frequencies as they achieved distributive agreements. The differences were not significant.

These results provided strong support for our main hypothesis that experts would perform better in the simulation than novices. However, the

TABLE 10.4
Frequency of Agreement by Scenario
(Entire Data Set)

| | Scenario | | |
	Theater	Law	n
Agreement	101 (71.1%)	83 (58.0%)	184
No Agreement	41 (28.9%)	60 (42.0%)	101
Total N	142 (100.0%)	143 (100.0%)	285

Note. $\chi^2 = 5.33$. $df = 1$. $p < .02$.

TABLE 10.5
Frequency of Integrative Versus Distributive Agreements by Criterion Group
(Entire Data Set)

	Criterion Group		
	Novice	Expert	n
Distributive	99 (63.9%)	7 (24.1%)	106
Integrative	56 (36.1%)	22 (75.9%)	78
Total N	155 (100.0%)	29 (100.0%)	184

Note. $\chi^2 = 15.79$. $df = 1$. $p < .001$.

prediction that there would be no scenario effects was confirmed in two of the three measures, but was contradicted in the frequency of agreement analyses (see Table 10.4). After collecting the data, we discovered a programming error in the simulation program in the law scenario version. This error may explain the inconsistent results. For reasons that are yet unclear to us, the program responded inaccurately in the law firm scenario when a subject offered the other party a proposal which included an "F" on the billable hours issue, which was the issue of least importance to the subject. Rather than responding with a counterproposal that included a "D" on the subject's most important issue, as it should have, the program countered with a proposal that included "F" on the most important issue as well as an "F" on the least important issue. Thus, in this particular case, the program did not reciprocate a subject's move toward an integrative solution, although it did in all other cases. Because the program would always respond to an "F" this

TABLE 10.6
Frequency of Integrative Versus Distributive Agreements by Scenario
(Entire Data Set)

	Scenario		
	Theater	Law	n
Distributive	54 (53.5%)	52 (62.7%)	106
Integrative	47 (46.5%)	31 (37.3%)	78
Total N	101 (100.00%)	83 (100.0%)	184

Note. $\chi^2 = 1.57$. $df = 1$. $p = .20$.

way in the law scenario, but always responded appropriately in the theater scenario, we have a confound between the scenario and the programming error. Thus, it seems possible that the observed lower frequency of agreement in the law scenario may be due to this computer program error.

To further examine this issue, we have conducted the same analyses with a subset of the total sample. This second set of analyses was conducted for law scenario subjects who did not encounter the program error and all theater scenario subjects (no theater scenario subjects actually encountered the error; the error was not present in the theater scenario).

Means and standard deviations for the final counteroffers for this subset of the data are presented in Table 10.7. Experts outperformed novices, $F(1, 212) = 9.082$, $p < .001$, but no other main effect or interaction was significant. The results for the frequency of agreement of expert versus novice analyses for the subset of the data are presented in Table 10.8. Experts reached agreement significantly more often than novices. The significant difference in frequency of agreement for law scenario versus theater scenario subjects found for the entire data set (Table 10.4) does not go away for these subjects, who did not encounter the computer error (see Table 10.9). As shown in Table 10.9, there were significantly more agreements in the theater scenario. The results for the frequency of integrative versus distributive agreements in this subset of the data are presented in Table 10.10. Experts achieved significantly more integrative solutions than novices. Finally, there was no effect of scenario on integrative versus distributive agreements (see Table 10.11).

To summarize the performance results, it seems clear that experts perform better than novices. However, the programming error makes it difficult to know whether there is an effect of scenario on performance.

The hypothesis relating to fixed-pie bias was also strongly supported by the statistical results. The measure of fixed-pie bias is discussed in terms of

TABLE 10.7
Mean Final Counteroffer Performance Measure (Subset of Data)

	Scenario	
Criterion Group	Theater	Law
Novice		
Mean *(SD)*	111.77 (29.14)	109.44 (31.08)
n	124	63
Expert		
Mean *(SD)*	136.11 (19.67)	115.00 (35.15)
n	18	8

TABLE 10.8
Frequency of Agreement by Criterion Group
(Subset of Data)

	Criterion Group		
	Novice	Expert	n
Agreement	117 (62.6%)	23 (88.5%)	140
No Agreement	70 (37.4%)	3 (11.5%)	73
Total N	187 (100.0%)	26 (100.0%)	213

Note. $\chi^2 = 6.79$. $df = 1$. $p < .01$.

TABLE 10.9
Frequency of Agreement by Scenario
(Subset of Data)

	Scenario		
	Theater	Law	n
Agreement	101 (71.1%)	39 (54.5%)	140
No Agreement	41 (28.9%)	32 (45.1%)	73
Total N	142 (100.0%)	71 (100.0%)	213

Note. $\chi^2 = 5.51$. $df = 1$. $p < .02$.

TABLE 10.10
Frequency of Integrative Versus Distributive Agreements by Criterion Groups
(Subset of Data)

	Criterion Group		
	Novice	Expert	n
Distributive	73 (62.4%)	5 (21.7%)	78
Integrative	44 (37.6%)	18 (78.3%)	62
Total N	117 (100.0%)	23 (100.0%)	140

Note. $\chi^2 = 12.88$. $df = 1$. $p < .001$.

TABLE 10.11
Frequency of Integrative Versus Distributive Agreements by Scenario
(Subset of Data)

	Scenario		
	Theater	Law	n
Distributive	54 (53.5%)	24 (61.5%)	78
Integrative	47 (46.5%)	15 (38.5%)	62
Total N	101 (100.00%)	39 (100.0%)	140

Note. $\chi^2 = 1.74.$ $df = 1.$ $p < .38.$

the movie theater scenario. The movie theater manager's role was programmed with the assumption that his or her most important issue was hourly wage and least important issue was passes. The issue of weekend hours was moderately important to the programmed theater manager. Thus, if subjects answered that the other party's least important issue was passes and most important issue was hourly wage, they received a zero for each response, indicating no bias. If subjects answered that the weekend hours issue was most important, they received a 1 for that item; or if they answered that the weekend hours issue was the least important issue, they received a 1, reflecting a moderate mistake in the perception of the other party's priorities. If subjects answered that the manager's most important issue was passes (when actually passes were least important to the manager), they received a 2 for that item, indicating a major mistake in the perception of the other party's priorities. Similarly, subjects received a 2 if they answered that the hourly wage was most important to the manager. The subjects' scores for the two items were summed to create a fixed-pie bias measure in which a zero indicates no bias and a 4 indicates the most extreme form of the bias.

As seen in Table 10.12, experts, as predicted, were significantly less biased than novices, $F(1, 284) = 6.84$, $p < .01$. However, there was also a main effect for scenario, $F(1, 284) = 35.12$, $p < .001$, so that subjects in the law scenario had more fixed-pie bias than subjects in the theater scenario. Table 10.13 shows the results of the analyses for the subset of the data according to the programming error. The fixed-pie bias analyses reveal a significant effect for criterion group, $F(1, 212) = 4.35$, $p < .05$. There was also a significant effect of scenario, $F(1, 212) = 33.44$, $p < .001$.

The results do not support the self-serving bias hypothesis. Self-serving bias scores were computed as the difference in the ratings, on 7-point Likert scales, of the self and the other party on questions of concern for

TABLE 10.12
Mean Fixed-Pie Bias (Entire Data Set)

Criterion Group	Scenario	
	Theater	Law
Novice		
Mean *(SD)*	1.51 (1.50)	2.57 (1.55)
n	124	124
Expert		
Mean *(SD)*	0.78 (1.40)	1.89 (1.70)
n	18	19

opposite party, cooperativeness/competitiveness, and reasonableness/un-reasonableness, with higher numbers indicating greater self-serving bias. Experts were not significantly less biased with respect to ratings of concern for the opposite party (see Table 10.14), nor was there any main effect for scenario. As seen in Table 10.15, novices actually perceived the other party to be more cooperative, while experts exhibited a clear self-serving bias with respect to cooperativeness, $F(1, 284) = 11.36$, $p < .001$. There was again no effect for scenario. As seen in Table 10.16, the results of the subsample are consistent in that only the main effect of expertise was significant, $F(1, 212) = 13.30$, $p < .001$. Experts were also somewhat more biased than novices with respect to ratings of reasonableness of self and other, as seen in Table 10.17, which also reveals that subjects in the law scenario were not more biased on the reasonableness dimension than

TABLE 10.13
Mean Fixed-Pie Bias (Subset of Data)

Criterion Group	Scenario	
	Theater	Law
Novice		
Mean *(SD)*	1.51 (1.50)	2.75 (1.50)
n	124	63
Expert		
Mean *(SD)*	0.78 (1.40)	2.25 (1.91)
n	18	8

TABLE 10.14
Mean Self-Serving Bias in Ratings of Concern for Other Party (Entire Data Set)

	Scenario	
Criterion Group	Theater	Law
Novice		
Mean *(SD)*	0.55 (1.52)	0.65 (1.42)
n	124	124
Expert		
Mean *(SD)*	0.06 (0.87)	0.84 (1.39)
n	18	19

TABLE 10.15
Mean Self-Service Bias in Ratings of Cooperativeness (Entire Data Set)

	Scenario	
Criterion Group	Theater	Law
Novice		
Mean *(SD)*	-.07 (1.64)	-.11 (1.81)
n	124	124
Expert		
Mean *(SD)*	0.89 (1.37)	1.00 (2.06)
n	18	19

TABLE 10.16
Mean Self-Serving Bias in Ratings of Cooperativeness (Subset of Data)

	Scenario	
Criterion Group	Theater	Law
Novice		
Mean *(SD)*	-.07 (1.64)	-.08 (1.79)
n	124	63
Expert		
Mean *(SD)*	0.89 (1.37)	2.00 (2.73)
n	18	8

TABLE 10.17
Mean Self-Serving Bias in Ratings of Reasonableness (Entire Data Set)

Criterion Group	Scenario	
	Theater	Law
Novice		
Mean *(SD)*	1.00 (1.90)	1.26 (2.01)
n	124	124
Expert		
Mean *(SD)*	1.22 (1.63)	1.84 (2.01)
n	18	19

subjects in the theater scenario. There was no effect of scenario. As seen in Table 10.18, with respect to the subsample, results on the self-serving bias of reasonableness on the main effect of scenario were significant, $F(1, 212) = 5.13$, $p = .025$. It is also interesting to note that the experts seemed to be more biased in the law scenario than in the theater scenario on each of the three dimensions. Basically, with respect to bias, the same patterns found for the entire data set hold across the subset.

Self-Regulation Results

In this study we viewed self-regulation as consisting of metacognition (self-checking and planning) combined with effort. We also measured worry. As seen in Tables 10.19 and 10.20, although experts exhibited more self-checking and planning activity, these differences only approached

TABLE 10.18
Mean Self-Serving Bias in Ratings of Reasonableness (Subset of Data)

Criterion Group	Scenario	
	Theater	Law
Novice		
Mean *(SD)*	1.00 (1.90)	1.57 (2.01)
n	124	63
Expert		
Mean *(SD)*	1.22 (1.63)	2.38 (2.07)
n	18	18

TABLE 10.19
Metacognition: Self-Checking

	Scenario	
Criterion Group	Theater	Law
Novice		
Mean *(SD)*	2.66 (0.67)	2.68 (0.72)
n	124	124
Expert		
Mean *(SD)*	2.86 (0.59)	2.91 (0.53)
n	18	19

significance, $F(1, 281) = 3.19$, $p = .075$ and $F(1, 281) = 3.26$, $p = .072$ respectively. There was no effect of scenario in either analysis. With respect to effort (see Table 10.21), experts exhibited significantly more effort, $F(1, 281) = 3.99$, $p = .047$. There was no effect of scenario. With respect to worry (see Table 10.22), experts exhibited less worry than novices, $F(1, 281) = 9.91$, $p = .002$. There was no effect of scenario. Thus, in general, experts exhibited more self-regulatory behavior than novices. However, with respect to metacognition, these differences only approach significance ($p < .10$).

Discussion

The purpose of the study was to investigate the construct validity of the negotiation skill measure we have developed. Our main test of validity was to see if our simulation approach would discriminate between expert and

TABLE 10.20
Metacognition: Planning

	Scenario	
Criterion Group	Theater	Law
Novice		
Mean *(SD)*	3.07 (0.64)	3.02 (0.66)
n	124	124
Expert		
Mean *(SD)*	3.26 (0.57)	3.24 (0.40)
n	18	19

TABLE 10.21
Effort

| Criterion Group | Scenario | |
	Theater	Law
Novice		
Mean *(SD)*	3.19 (0.61)	3.15 (0.68)
n	124	124
Expert		
Mean *(SD)*	3.27 (0.62)	3.51 (0.36)
n	18	19

novice criterion groups. The results clearly indicated that the simulation does in fact discriminate between experts and novices. The mean final counteroffer experts elicited was significantly higher than the mean for novices. Furthermore, when an actual agreement was reached, experts concluded integrative agreements much more frequently than novices.

We also examined construct validity by investigating whether cognitive processes associated with negotiation performance in the negotiation literature were also associated with performance in the simulation. With respect to the fixed-pie bias, novices were clearly more biased than experts, further supporting the validity of our simulation. The results did not support our predictions concerning the self-serving biases, however. Experts generally exhibited no less self-serving bias than novices.

One possible explanation of the failed self-serving bias results is the computer setting. The self-serving bias is an explicitly interpersonal bias,

TABLE 10.22
Worry

| Criterion Group | Scenario | |
	Theater	Law
Novice		
Mean *(SD)*	1.66 (0.50)	1.67 (0.53)
n	124	124
Expert		
Mean *(SD)*	1.34 (0.26)	1.44 (0.35)
n	18	19

while the fixed-pie bias regards the negotiation situation rather than the other party. The self-serving bias may exert its influence on negotiation performance primarily through emotions that are generated in face-to-face interpersonal negotiations when one perceives a concrete other person who is showing less concern, cooperation, and reasonableness. It should also be repeated that the empirical evidence linking self-serving biases to negotiation performance is as yet quite small. Few empirical studies suggest a relationship between self-serving bias and negotiation performance (Kramer et al., 1993; Sillars, 1981), while there is clear empirical support for the link between the fixed-pie bias and performance (Neale & Bazerman, 1985; Thompson, 1991). It may be that the self-serving bias is simply not as strongly associated with negotiation performance.

The secondary question concerning scenario effects could not be adequately answered because of the programming error. However, for both the entire data set and the subset of data, our results indicated little effect for scenario. Where there was an effect, one other factor should be noted: Attorneys' formal training outside of negotiation classes is in competitive, adversarial approaches to resolving disputes. The competitive orientation of this training for lawyers may have been cued more by role-playing in the law scenario than by role-playing a high school student. Thus, the law students may be a somewhat problematic population for testing negotiation skills in a setting requiring integrative negotiation. This adversarial approach to negotiation may also explain the unexpected results regarding the self-serving bias, particularly the finding that the law students were more biased than novices in terms of ratings of cooperation.

In summary, our data offered clear evidence for the construct validity of our simulation approach to measuring negotiation skills. However, the question of effects for different scenarios, and their interaction with different populations, will require further investigation. The use of explicit domain specification in the form of software parameters appears promising.

PILOT STUDY FOR TEAM ASSESSMENT

Our R&D plans next involved developing assessment measures for groups of students working together in our negotiation context. However, in that context, we did not know the feasibility of students working together to solve the computerized negotiation problem. We were also aware of issues due to group composition (Slavin, 1990; Webb, 1993). Our strategy to deal with differences in group composition is to randomly assign students to a group. We believe that the random assignment in an accountability scenario is fair. Moreover, there is some evidence that the number of students affects productivity (Hagman & Hayes, 1986). Thus, we were also interested in the impact of two-person versus three-person groups. In our second validation study, we expected three-person groups to perform better than two-person groups.

METHOD

Subjects

Subjects participating as teams were drawn from the same high school as the participants in the first study. Fifty-one students (25 males, 26 females) from business computer courses aimed for noncollege-bound students participated. Four of the participants were sophomores, 23 were juniors, and 24 were seniors.

Procedure

The procedure followed was the same as that in our first study with one alteration. Rather than working at a computer alone, the subjects were randomly assigned to groups of two or three. The participants were instructed to work together as a team in the negotiation. They were organized in 15 groups of three and 14 groups of two. However, for data analysis, due to missing data, there were a total of 19 groups (6 groups of two; 13 groups of three).

Results

The results offered limited support of the hypotheses in that most differences between groups were in the expected direction but not significant. With respect to the final counteroffer measure of performance, three-person ($n = 13$) groups achieved a greater mean value ($M = 126.53$, $SD = 6.25$) than two-person groups ($n = 6$) ($M = 115.00$, $SD = 32.09$), but the analysis of variance revealed this difference was not significant. Three-person teams also more frequently concluded the negotiation with actual agreement than did two-person teams (100% of three-person teams reached an agreement whereas 66.7% of two-person teams reached agreement). This difference was significant, ($\chi^2 = 4.84$, $df = 1$, $p = .03$). Three-person teams and two-person teams did not differ significantly in terms of the quality (integrative vs. distributive) of reached agreements.

The results did not support the hypothesis with respect to the fixed-pie bias. The mean bias for three-person teams was 0.62, whereas the mean bias for two-person teams was 1.67. However, analysis of variance revealed that this difference approached significance, $F(1, 17) = 4.53$, $p = .056$. The results did not support the hypotheses with respect to the self-serving biases (or concern, cooperativeness, or reasonableness).

Self-Regulation Results

With respect to metacognition, the two-person and three-person groups were equivalent. The mean self-checking score for the groups of three ($n = 12$) was 2.65 ($SD = 0.46$), whereas the mean for the groups of two ($n = 5$) was 2.74 ($SD = 0.37$). With respect to cognitive strategy, the mean score was 2.83 ($SD = 0.46$) for the groups of three ($n = 12$) and 2.82 ($SD = 0.67$) for the groups of two ($n = 5$). Further, there were also no significant differences between two-person ($n = 5$) and three-person ($n = 12$) groups for worry: mean for three-person groups was 1.41 ($SD = 0.30$); the mean for two-person groups was 1.57 ($SD = 0.36$). With respect to effort, the mean for three-person groups was 3.00 ($SD = 0.52$) and for two-person groups 2.96 ($SD = 0.43$). Thus, with respect to self-regulation in general, there were no significant differences between two-person and three-person groups. In summary, for the group assessment pilot study there were minimal or no effects of group size. However, it is clear that a collaborative environment is feasible.

GENERAL DISCUSSION

This validation study revealed worthwhile information about the feasibility of the computer simulation approach to measuring negotiation skills.

The primary purpose of our studies has been to determine whether the computer simulation we have developed reliably and validly measures the negotiation subskill of the interpersonal workplace competency (U.S. Department of Labor, 1991, 1992). The most direct test of the simulation's validity has been to see whether the simulation discriminates between expert and novice negotiators. Across all phases and versions of the simulation tested so far, experts' performance has been clearly superior to novices' performance on all measures of negotiation performance in the simulation.

In addition to measures of performance outcome, we have also examined several process variables. We reasoned that if our scenarios were capturing the negotiation context, then students should display cognitive processes similar to those that the literature indicates occur in "real" negotiations. If this was true, then such process information would add to the construct validity of our assessment. Research on negotiation has documented a number of cognitive biases that present obstacles to negotiator performance. In other words, superior negotiators are less susceptible to the cognitive biases that commonly beset people in negotiation situations. Thus, we have examined the validity of our simulation by assessing the degree to which the absence of known cognitive biases with performance implications is associated with higher negotiation performance on our simulation. Although not as robust as the expert/novice findings, we have consistently found that (a) such biases are associated with lower negotiation performance in the simulation, and (b) experts exhibit more self-regulation skills than novices.

Several other properties of the negotiation simulation as an assessment tool have also been examined. It appears that the simulation is promising in various negotiation scenarios. Specifically, in two scenarios of negotiating a job contract, whether as a high school student with a movie theater manager or as a law student for a job with a law firm, subject performance on the simulation did not vary significantly as a function of scenario. In effect, we may have parallel forms of our test.

We plan to continue to measure self-regulation for both individuals and teams. However, our current performance measures do not capture the group process well. Our major new effort will be to develop domain-independent measures of teamwork skills so that a score can be assigned to an individual as well as to a team. This work is documented in O'Neil, Chung, and Brown (chapter 16, this book). In summary, our experience in conducting these studies suggests that the simulation approach we have taken is a valid and feasible method of assessing negotiation skills for both individuals and teams.

ACKNOWLEDGMENTS

The authors would like to thank Dr. Randy Lowry, Dr. Cheryl McDonald, and Dr. Peter Robinson of the Institute for Alternative Dispute Resolution, Pepperdine University, for their assistance. We also wish to thank the principal, students and teachers at Twin Falls High School, Twin Falls, ID, for their assistance.

The work reported herein was supported under the Educational Research and Development Center Program cooperative agreement R117G10027 and CFDA catalog number 84.117G as administered by the Office of Educational Research and Improvement, U.S. Department of Education. The findings and opinions expressed in this report do not reflect the position or policies of the Office of Educational Research and Improvement or the U.S. Department of Education.

REFERENCES

Baker, E. L. (1992, February). *The role of domain specifications in improving the technical quality of performance assessment* (Project 2.2, Final Deliverable to OERI). Los Angeles: University of California, Center for Research on Evaluation, Standards, and Student Testing.

Bazerman, M. H., Magliozzi, T., & Neale, M. A. (1985). Integrative bargaining in a competitive market. *Organizational Behavior and Human Decision Processes, 35*, 294–313.

Brandstatter, H., Kette, G., & Sageder, J. (1982). Expectations, attributions, and behavior in bargaining with liked and disliked partners. In R. Tietz (Ed.), *Aspiration levels in bargaining and economic decision making.* Berlin: Springer-Verlag.

Brett, J. M., Goldberg, S. B., & Ury, W. L. (1990). Designing systems for resolving disputes in organizations. *American Psychologist, 45,* 162–170.

Carnevale, P. J., & Conlon, D. E. (1988). Time pressure and strategic choice in mediation. *Organizational Behavior and Human Decision Processes, 42,* 111–133.

Carnevale, P. J., & Pruitt, D. J. (1992). Negotiation and mediation. *Annual Review of Psychology, 43,* 531–582.

Glaser, R., Raghavan, K., & Baxter, G. P. (1992). *Design characteristics of science performance assessments* (CSE Tech. Rep. No. 349). Los Angeles: University of California, Center for Research on Evaluation, Standards, and Student Testing (CRESST).

Green, B. F. (1991). Guidelines for computer testing. In T. B. Gutkin & S. L. Wise (Eds.), *The computer and the decision-making process* (pp. 245–273). Hillsdale, NJ: Lawrence Erlbaum Associates.

Hagman, J. D., & Hayes, J. F. (1986). *Cooperative learning: Effects of task, reward, and group size on individual achievement* (ARI Tech. Rep. No. 704). Alexandria, VA: Army Research Institute for the Behavioral and Social Sciences.

Heider, F. (1958). *The psychology of interpersonal relations.* Hillsdale, NJ: Lawrence Erlbaum Associates.

Kelley, H. H. (1979). *Personal relationships: Their structures and processes.* Hillsdale, NJ: Lawrence Erlbaum Associates.

Kelley, H. H., & Thibaut, J. W. (1978). *Interpersonal relations: A theory of interdependence.* New York: Wiley-Interscience.

Kramer, R., Newton, E., & Pommerenke, P. (1993). Self-enhanced biases and negotiator judgment: Effects of self-esteem and mood. *Organizational Behavior and Human Decision Processes, 56,* 110–133.

Lax, D. A., & Sebenius, J. K. (1986). *The manager as negotiator.* New York: The Free Press.

Lewicki, R. J., Litterer, J. A., Minton, J. W., & Saunders, D. M. (1994). *Negotiation.* Burr Ridge, IL: Irwin.

Millman, J., & Greene, J. (1989). The specification and development of tests of achievement and ability. In R. L. Linn (Ed.), *Educational measurement* (3rd ed., pp. 335–366). New York: Macmillan.

Neale, M. A., & Bazerman, M. H. (1985). The effects of training and negotiator overconfidence on bargainer behavior. *Academy of Management Journal, 28,* 34–44.

Neale, M. A., & Bazerman, M. H. (1991). *Cognition and rationality in organizations.* New York: Free Press.

O'Neil, H. F., Jr., & Abedi, J. (1996). Reliability and validity of a state metacognitive inventory: Potential for alternative assessment. *Journal of Educational Research, 89,* 234–245.

O'Neil, H. F., Allred, K. G., & Baker, E. L. (1992). *Measurement of workforce readiness competencies: Design of prototype measures* (CSE Tech. Rep. No. 344). Los Angeles: University of California, Center for Research on Evaluation, Standards, and Student Testing (CRESST).

O'Neil, H. F., Jr., Allred, K., & Dennis, R. (1992). *Simulation as a performance assessment technique for the interpersonal skill of negotiation* (Deliverable to OERI, Contract No. R117G10027). Los Angeles: University of California, Center for Research on Evaluation, Standards, and Student Testing (CRESST).

O'Neil, H. F., Jr., Sugrue, B., Abedi, J., Baker, E. L., & Golan, S. (1992). *Report of experimental studies on motivation and NAEP test performance* (Report to National Center for Education Statistics, Contract No. RS90159001). Los Angeles: University of California, Center for Research on Evaluation, Standards, and Student Testing (CRESST).

Pruitt, D. G., & Rubin, J. Z. (1986). *Social conflict: Escalation, stalemate, and settlement.* New York: Random House.

Pruitt, D. G., & Syna, H. (1984). Successful problem solving. In D. Tjosvold & D. W. Johnson (Eds.), *Productive conflict management: Perspective for organizations* (pp. 129–148). New York: Irvington.

Rahim, M. A. (1986). *Managing conflict in organizations.* New York: Praeger.

Raiffa, H. (1982). *The art and science of negotiation.* Cambridge, MA: Belknap Press/Harvard University Press.

Rubin, J., & Brown, B. R. (1975). *The social psychology of bargaining and negotiation.* New York: Academic Press.

Sillars, A. L. (1981). Attributions and interpersonal conflict resolution. In J. H. Harvey, W. Ickes, & R. F. Kidd (Eds.), *New directions in attribution research* (Vol. 3, pp. 279–305). Hillsdale, NJ: Lawrence Erlbaum Associates.

Slavin, R. E. (1990). *Achievement effects of ability grouping in secondary schools: A best-evidence synthesis.* Madison: University of Wisconsin, National Center on Effective Secondary Schools.

Thompson, L. L. (1990a). The influence of experience on negotiation performance. *Journal of Experimental Social Psychology, 26,* 528–544.

Thompson, L. L. (1990b). Negotiation: Empirical evidence and theoretical issues. *Psychological Bulletin, 108,* 515–532.

Thompson, L. L. (1991). Information exchange in negotiation. *Journal of Experimental Social Psychology, 27,* 161–179.

Thompson, L. L., & Hastie, R. M. (1990). Social perception in negotiation. *Organizational Behavior and Human Decision Processes, 47,* 98–123.

Tjosvold, D. (1990). The goal interdependence approach to communication in conflict: An organizational study. In M. A. Rahim (Ed.), *Theory and research in conflict management* (pp. 15–31). New York: Praeger.

Tjosvold, D., & Johnson, D. W. (1983). *Productive conflict management.* New York: Irvington Publishers.

Turnbull, A., Strickland, L., & Shaver, K. (1976). Medium of communication, different power, and phasing of concessions: Negotiating success and attributions to the opponent. *Human Communications Research, 2,* 262–270.

U.S. Department of Labor. (1991, June). *What work requires of schools: A SCANS report for America 2000.* Washington, DC: U.S. Department of Labor, Secretary's Commission on Achieving Necessary Skills.

U.S. Department of Labor. (1992). *Skills and tasks for jobs: A SCANS report for America 2000.* Washington, DC: U.S. Department of Labor, Secretary's Commission on Achieving Necessary Skills.

Walton, R. E., Cutcher-Gershenfeld, J., & McKersie, R. B. (1994). *Strategic negotiations: A theory of change in labor-management relations.* Boston, MA: Harvard Business School Press.

Walton, R. E., & McKersie, R. E. (1965). *A behavioral theory of labor negotiations.* New York: McGraw-Hill.

Webb, N. (1993). *Collaborative group versus individual assessment in mathematics: Processes and outcomes* (CSE Tech. Rep. No. 352). Los Angeles: University of California, Center for Research on Evaluation, Standards, and Student Testing (CRESST).

Weiner, B. (1986). *An attributional theory of motivation and emotion.* New York: Springer-Verlag.

Weiner, B. (1992). *Human motivation: Metaphors, theories, and research.* Newbury Park, CA: Sage.

Womack, D. F. (1990). Applied communications research in negotiation: Implications for practitioners. In M. A. Rahim (Ed.), *Theory and research in conflict management* (pp. 32–53). New York: Praeger.

Subjects were instructed that the job contract negotiation centered on three issues: (a) the annual salary the law student would be paid, (b) the number of months to become a partner in the firm, and (c) the number of billable hours the law student would be required to log per year. Subject were told that they preferred a higher salary, fewer months to make partner, and fewer billable hours required, whereas the law firm preferred a lower salary level, a longer time to partnership, and more billable hours. Subjects were also instructed that the parties would exchange proposals in the negotiation in trying to reach agreement on one proposal level for each issue.

The subjects were also instructed with respect to their relative priorities on the three issues. The subjects were told that because the law student had incurred substantial student loans, salary was most important to him or her. Because the law student was willing to work hard and to earn a higher salary, billable hours was least important to him or her. The subjects were also instructed that the months to partnership was of intermediate importance. The issue chart presented on the computer screen, as seen in Table A.1, represented these relative preferences in that the highest points attainable were on the salary issue, followed by the months to partnership and the billable hours issues respectively. The law firm representative's priorities were offsetting, also shown in Table A.1.

TABLE A.1
Law Student and Law Firm Representative Issue Chart of Point Values

		Issues			
Salary		*Months to Partnership*		*Billable Hours*	
		Law Student			
A	120	A	80	A	40
B	105	B	70	B	35
C	90	C	60	C	30
D	75	D	50	D	25
E	60	E	40	E	20
F	45	F	30	F	15
G	30	G	20	G	10
H	15	H	10	H	5
I	0	I	0	I	0
		Law Firm Representative			
A	0	A	0	A	0
B	5	B	10	B	15
C	10	C	20	C	30
D	15	D	30	D	45
E	20	E	40	E	60
F	25	F	50	F	75
G	30	G	60	G	90
H	35	H	70	H	105
I	40	I	80	I	120

Chapter 11

Assessing Foundation
Skills for Work

Thomas G. Sticht
Applied Behavioral & Cognitive Sciences, Inc.

The Organization for Economic Co-operation and Development (OECD) recently reported that in many industrialized nations, the educational system is not as effective as it should be in producing young adults with the competence needed to meet the demands of contemporary society, particularly the world of work (Benton & Noyelle, 1992). While acknowledging the paucity of trustworthy data, the OECD estimated that about one third of workers could do their jobs better if they were more literate. In one survey, about one third of Canadian firms reported serious difficulties in introducing new technology and increasing productivity because of the poor skills of their workers.

In the United States, the U.S. Secretary of Labor's Commission on Achieving Necessary Skills (SCANS) reported that many of the young adults entering, and many adults already in the workforce are lacking in what they considered key "foundation skills" for work, namely, basic literacy— reading and writing (U.S. Department of Labor, 1991).

ISSUES IN WORKFORCE LITERACY COMPETENCE

The OECD and SCANS reports raise a number of issues regarding workforce readiness and workplace literacy. The major issue is the concept of the "literacy gap." This refers to the idea that the literacy requirements of work in a society can be considered in terms of supply and demand. On the supply side, the home, community, and schools are supposed to continuously produce new generations of adults who form a workforce with a supply of literacy in the society. On the demand side, the roles of citizen, family member, community service provider, and worker are considered to demand certain types and amounts of literacy for satisfactory performance. For the role of worker, a given workplace is considered to pose particular demands for literacy. The problem of workplace literacy, then, is to have the supply of literacy in the workforce meet the demands for literacy in the workplaces of the society.

These supply and demand issues raise other issues. One of these is how to understand the nature of literacy, so that the home, community, and school can promote the development of the literacy skills of its new adults. This is not a straightforward problem. For instance, in a major policy-oriented report on literacy in the United States, one that focuses on work-related needs for literacy, the authors include a chapter that discusses ". . . the fact that literacy is a continuum of skills" (Brizius & Foster, 1987, p. 24).

Yet on the very same page they discuss the U.S. Department of Education's National Assessment of Educational Progress (Kirsch & Jungeblut, 1986) assessment of adult literacy and the fact that it tested literacy skills in three areas of competence: prose literacy, document literacy, quantitative literacy. In this case, the test developers considered that these three domains were different kinds of literacy, and three different scales were developed for measuring these three different kinds of literacy. Literacy was considered not as *a continuum* but rather as *three continua*.

Literacy in the Larger Context of Cognitive Competence and Work

A further problem in understanding workplace competence issues is that *literacy*—the SCANS *foundation skills* of reading and writing (U.S. Department of Labor, 1991)—is frequently thought of as either something separate and distinct from other human cognitive abilities, or, at other times, as something synonymous with *all* cognitive ability (e.g., cultural literacy, family literacy, computer literacy; geography literacy; science literacy, workplace literacy, etc.). From the former point of view it is widely believed that intelligence is something that people possess in greater or lesser amounts, and that intelligence influences whether and to what extent people can develop high levels of literacy. Further, it has been assumed by some that one could be highly developed in one of these aspects of cognition (intelligence) while being poorly developed in the other (literacy; Sticht & Armstrong, 1994; see the discussion of the Army Alpha and Beta tests of World War I that follows). What is not so frequently debated, however, is whether it is likely, or even possible, that one could be quite highly developed in literacy but poorly developed in intelligence.

This issue was broached by Carroll (1987) in his discussion of the National Assessment of Educational Progress of young adults. He made the point that, although it is probably not popular to say, at the higher levels of difficulty, many of the adult literacy tasks resemble the kinds of tasks found on tests of verbal intelligence or aptitude. In his discussion he raises the possibility that because high levels of verbal intelligence may be needed to achieve high levels of literacy, and because, in his judgment, we have not had much success in raising the performance of adults on measures of verbal intelligence, it might not be possible, except with extraordinary efforts, to raise large numbers of less literate adults to the very high levels of workplace literacy that many feel are needed for work in high performance, high wage workplaces (U.S. Department of Labor, 1991, p. 3).

The concept of more or less fixed levels of verbal and quantitative aptitudes is prevalent among all of the military establishments of the industrial powers. These nations maintain psychological testing groups who develop and use tests of mental aptitudes to screen military recruits (Gal & Mangelsdorf, 1991). Most colleges in the United States include tests such as the Scholastic Achievement Tests (previously called aptitude tests) in their selection procedures, and the use by business and industry of tests such as the Differential Aptitude Tests (DAT) and the U.S. Employment Service's General Aptitude Test Battery (GATB) is widespread (Fremer, 1989, pp. 61–80).

Assessing the Supply and Demand for Cognitive Competence

Implicit in workforce and workplace literacy initiatives by human resource specialists or educators is the notion that adult cognitive competence can be adequately defined, developed, measured, used for counseling and job placement, and modified through instruction to improve productivity in the workplace. Further, it is anticipated that as workplaces change, either through the introduction of new technology or new organizational structures (the *high performance* organization), demand for literacy skills may change, generally increasing, and these changes can be identified and the workforce can be educated to meet the new demands for literacy (National Alliance for Business, 1995).

At the present time, there is only one workplace organization in the world that has conducted research and development into adult cognitive skills that addresses the full range of these issues and beliefs. That organization is the U.S. Department of Defense with its associated military services: the Army, Navy, Marine Corps, and Air Force. This chapter provides a review of research and development projects concerned with assessing workforce readiness conducted by the military services of the United States over the last 75 years.

Assessing Cognitive Skills. The military's experiences in mental testing from World War I to the present are summarized. As in the civilian context, the military has displayed considerable confusion about the nature of cognitive abilities, including literacy. Nonetheless, it has operationalized cognitive assessments because of their empirical utility in selecting adults for military service and assigning them to training for particular career fields.

In this chapter, the military's mental testing procedures will be interpreted within a conceptual framework that includes a simple model of the architecture of the human cognitive system, an information processing interpretation of the various aptitude and literacy tests that the military has studied and utilized over the years, and a developmental model of literacy that military researchers have used in developing both literacy assessments and literacy training programs.

Assessing the Cognitive Demands of Jobs. The military services have pioneered the study of the literacy and other cognitive skills demands of jobs (Sticht, 1995). This research has included traditional studies of the validity of cognitive tests such as the Armed Forces Qualification Test (AFQT), various reading, arithmetic, and listening tests, and other cognitive tests for predicting job performance. It has used a variety of criterion indicators of job performance, such as success in training, completion of the first term of military service, supervisor's ratings of job performance, job knowledge tests, and job sample tests. In this *predictive validity* approach to identifying cognitive skills requirements of jobs, the goal has been to state the cognitive demands of work in terms of scores on the predictor tests associated with scores on one or more of the job performance indicators.

A second approach to the study of the knowledge and skills demands of work has involved various methods of job and task analysis. The most extensively used methods are those based on the Air Force's Comprehensive Occupational Data Analysis Program (CODAP) (Payne et al., 1992). The purpose of the job and task analysis procedures such as the CODAP is to identify the knowledge and skills that personnel need to be trained in on entry to the job and at advanced stages of career progression. Special analytical procedures have been developed for identifying literacy requirements of jobs. This "literacy task analysis" approach involves observing workers at their job sites, interviewing them to identify the types of reading tasks they perform, and collecting and analyzing reading materials to identify the reading demands of the job (U.S. Departments of Labor and Education, 1988).

COGNITIVE SCIENCE FRAMEWORK

During World War I, military psychologists developed two tests of intelligence (Sticht & Armstrong, 1994). The Army Alpha test was developed to be used by literates. The Army Beta test was designed to be used by illiterates and non-English speakers. It was administered in pantomime with a minimum of language use. The basic idea was that intellectual processes could be exhibited with or without the use of oral or written language skills.

In the mid-1960s, the military services lowered the aptitude requirements for military service and admitted hundreds of thousands of personnel with low literacy skills. Studies were initiated to determine whether it would be feasible in many training situations to substitute learning by listening for learning by reading. The basic idea was that much knowledge could be exhibited and learned either by written or spoken language. Therefore, persons low in literacy skills might be taught using oral language.

The Human Cognitive System

Both the World War I and the 1960s projects were based implicitly upon a simple model of the human cognitive system (see Fig. 11.1). In this model, there are two bodies of knowledge, one inside and the other outside the person's head. The internal knowledge is stored in the person's long term memory and is operated on by information processing procedures in the short term or working memory. These procedures may include searching for, locating, and retrieving information from the long term memory knowledge base. The information may be represented internally in either images or language. It may be thought about and then dismissed from working memory. It may be represented outside the person's head in

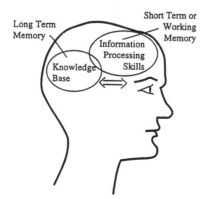

Simple Model of Human Cognitive System: Long Term Memory contains the knowledge base with language and various information processing knowledge. When thinking is taking place, information processing occurs in Short Term or Working Memory. The information processing skills operate on information in the knowledge base and information picked up from "knowledge bases outside the head," such as books, speech, etc.

FIG. 11.1. Model of a human cognitive system used to interpret studies on the assessment of literacy and other cognitive abilities.

pantomime, as in administering the Army Beta test in World War I, it may be represented in spoken language, as in the studies on learning by listening in the 1960s, or in written language as in the Army Alpha tests of World War I and all aptitude and reading tests used by the military services since then.

From the point of view of the simple model of the human cognitive system, whenever someone wants to know something about another person's cognitive make-up, he or she addresses the person's knowledge base and information processing abilities by using some form of language, either written, spoken, or gestural. For instance, in a reading vocabulary test, the written words on the test are presented to the examinee who is supposed to read the words into working memory while also looking up the meanings in long term memory. The person is then supposed to make a response that indicates the outcome of his or her information processing activity.

A Developmental Model of Literacy

Social groups with shared cultural experiences form normative expectations for what people at different ages can and cannot be expected to do. In a typical case, one will not use a form of addressing a person's cognitive system that the person is not expected to be able to process. For instance, one does not typically hand a written note to a two-year-old that asks for personal information such as a name. But one would ask for the child's name using oral language. Here the normative assumption is that, typically, a two-year-old can talk and comprehend simple oral language but generally cannot read. Similarly, one does not ask an infant for information because it is assumed that typically infants cannot comprehend and speak oral language well enough to communicate. But facial expressions, gestures, laughing sounds, etc., might be used in a communicative manner to play with the infant and receive responses such as laughing, smiling, or hand and arm movements. In short, our literate society has expectations for how infants, two-year-olds, six-year-olds, adolescents, and adults develop their cognitive systems over time. In short, we all have an implicit developmental model of literacy that guides our choice of communication methods in different circumstances.

This implicit developmental model of literacy was made explicit by Sticht, Beck, Hauke, Kleiman, and James (1974) and Sticht and Hickey (1991). According to the developmental model, in the typical case, people develop a fair amount of competence in oral language before they are exposed to formal instruction in reading in elementary grades (though informal learning of literacy may begin in the home and community in literate cultures). Written language skills build upon the earlier developed oral language skills and add new vocabulary and concepts, as well as special knowledge about how to represent information in the graphic medium, to the person's knowledge base. In turn, learning new vocabulary and conventions of language through reading and writing enlarges the person's oral language abilities. The development of oral and written language ability may continue indefinitely as the person studies and develops new knowledge domains.

A major component in the developmental model is the person's long term memory or "knowledge base." The long term memory contains all the knowledge developed by the person in interaction with the environment. Much of the knowledge acquired by the person will not be understood in consciousness (for example, the rules of grammar). Rather, it will be unconsciously used to accomplish tasks such as developing language competence and comprehending the events of the world. In addition to the general world knowledge and processes that are in the mind, though not necessarily accessible to conscious understanding without considerable analysis, the memory also contains the language knowledge (words and grammar) that can be used to represent information that arises from experience in the world such as bodies of knowledge about machines, parts of the body, houses, neighborhoods—sometimes called "schema" (Rumelhart, 1980) or "mental models" (Rouse & Morris, 1985), and from didactic instruction, as in training programs.

The model holds that the development of the oracy skills of speaking and auding are built upon the prior development of prelinguistic knowledge through information processing activities. It is important that it be understood that this early, prelinguistic cognitive content, or knowledge, will form the foundation for the acquisition of new knowledge over the person's lifetime, including that knowledge known as "literacy."

Much of this knowledge will remain personal, and will not be explicitly represented in language for communication to others. Nonetheless, such personal, tacit knowledge, which includes perceptual learnings and general knowledge of "how the world works," will be absolutely necessary for learning to comprehend the spoken, and later the written, language. This reflects the fact that language is selective in the features and concepts chosen to be represented. We may think of language as producing a verbal *figure*, which can be comprehended only in terms of its relationship to a nonlinguistic conceptual *ground* of "world knowledge." A simple illustration of the role of "world knowledge" in literacy training is seen in the recommendation to give students experience with objects and events in the world through field trips, demonstrations, movies, etc., before they read about them. This approach provides an experiential base or a "world knowledge" that will permit a deeper comprehension of the words and concepts the students read, and greater "access" to prior knowledge via perceptual learning (Bransford, Sherwood, Vige, & Rieser, 1986).

A final aspect of the model is that it recognizes that, on one hand, the literacy skills of reading and writing utilize the same knowledge base that is used in auding and speaking, plus the special decoding and encoding skills of reading and writing. On the other hand, the very nature of the written language display—characterized by being more or less permanent, being arrayed in space, and utilizing the features of light (color, contrast)— makes possible the development of skills and knowledge entirely different from those involved in oral language.

The model incorporates the role of prelinguistic looking and marking abilities as contributors to later utilization of the visual display of written language in conjunction with graphic marks such as lines, white space, and color to develop graphic tools for thinking and problem solving like matrices, flow charts, color coded graphs, and so forth. These tools combine with written language and nonlanguage graphic symbols, such as arrowheads and geometric figures, to produce analytical products beyond those obtainable through the fleeting, temporal, oral language.

A point to be emphasized is the fact that much of the acquisition of literacy is not simply learning to read; that is, it is not just learning a graphic language system that can be substituted for the oral language system. Rather, a large part of learning to be literate, and perhaps the most important part for acquiring higher levels of literacy, is learning how to perform the many tasks made possible by the unique characteristics of printed displays—their permanence, spatiality, and use of light—and using that knowledge to develop large amounts of new knowledge (Sticht, Hofstetter, & Hofstetter, 1995).

In summary, the theoretical framework used to interpret military research on literacy includes the concept of a human cognitive system comprised of a long term memory with it's knowledge base and a short term working memory that operates through information processing activities on the information in the long term memory and the external world of information. Cognition includes both knowledge and the processes for developing knowledge.

The long term memory develops over the life of a person as he or she undergoes the transitions from pre-language, to oral language, to written language-based information processing. As a major developmental information processing outcome in our literate society, the person comes to acquire the information processing skills involved in a wide variety of cognitive acts involving graphic symbol systems for language processing and for performing various cognitive tasks involved in problem solving and reasoning.

This developmental progression of the person occurs as the result of active, constructive information processing activities that represent and re-represent knowledge to forge new learning from old. The new knowledge adds to the knowledge base in the long term memory and is activated in various contexts through constructive processes that are sensitive to the different contexts. This means, for instance, that even though one possesses certain knowledge and skills, they may not be accessed if the context in which the person is immersed does not activate them. This conceptual framework is used to reinterpret various research projects in the military services concerned with the assessment of cognitive skills, including literacy.

ASSESSMENT OF COGNITIVE COMPETENCE

For over 75 years, the armed forces have pursued a policy of assessing the mental ability of adults who are eligible for military service. In World War I, some 1.9 million men were tested on the Army Alpha test of intelligence for literates and the Army Beta test of intelligence for illiterates and non-English speakers (Yerkes, 1921).

The initial use of mental tests for literates and illiterates in World War I reflected the military's need to rapidly mobilize a force that was the first to fight within the era of modern mechanization. The goal of those who introduced mental testing into the military was to provide a supplement to traditional personnel procedures to assist in assigning men to crafts and jobs of varying complexity (Yerkes, 1921, p. 153).

From World War I through the Vietnam War, there was a steady decline of military/combat occupational groups and a rise in "blue collar" and "white collar" occupations in the military services. Since around 1965, there has been a roughly stable distribution of personnel in the three occupational groups: around 15% general military/combat; 40% "blue collar"; 45% "white collar" (Eitelberg, 1988). Because of the changes in the types of jobs in the military, it has been argued that the modern military is so technologically complex that high mental quality personnel are needed to use, maintain, and repair the tools of war (Binkin, 1986).

This is an argument similar to that heard more and more in the civilian world in discussions of workforce skills. It is argued that technologically complex societies demand higher levels of literacy than in the past (Venezky, Kaestle, & Sum, 1987; Wells, 1995). Interestingly, when discussed in the military context, the need is for personnel of higher "aptitude." Yet, when discussed in the civilian context, the need is for citizens of higher "literacy." These perspectives appear to reflect mainly the differences in the origins of mental tests in the military, which were construed to be measures of innate intelligence, and not measures of achievement per se. Even today, these ideas appear commingled in military management documents:

> "Aptitude tests are designed to reveal an individual's innate intellectual abilities, interests, and experiences in a wide variety of activities, as influenced by education and training. Achievement tests are designed to measure skill

and knowledge accomplishment in specific subject areas." (Assistant Secretary of Defense, Force Management and Personnel, 1987, II-18–II-19)

These distinctions between "aptitude" and "achievement" tests are drawn by the military even though the Armed Forces Qualification Test (AFQT), the major screening test of aptitude for "trainability," consists of reading (word knowledge and paragraph comprehension) and arithmetic (mathematics knowledge and arithmetic reasoning) tests, both "basic skills" that are generally thought of as skills achieved through education and training, not indicators of innate intellectual abilities. Yet the AFQT is considered by the military as an "aptitude" test, not an "achievement" test.

Over the years, the mental ability tests used by the armed services have changed in content and in the definition of what they measure (see Table 11.1), but they still represent attempts to assess the cognitive skills of adults and to use that information to select people for military service and to assign them to work for which their "aptitudes" suit them.

Four major periods in the history of military mental testing are outlined in Table 11.1. The first witnessed the introduction of mental testing, mostly as a developmental activity, during World War I. The second occurred during World War II, when the first large-scale, operational use was made of mental tests for classifying recruits into job assignments. Separate tests were used by the Army and the Navy. The third major period occurred in the 1950s, when the Armed Forces Qualification Test (AFQT) was specially designed and developed to serve as a test for screening out low mental ability persons for all military services. The AFQT subtest scores were then combined with other tests, which differed for each service, to classify recruits into job assignments. The fourth period in military mental testing began in 1976, during the All Volunteer Force, when the Armed Services Vocational Aptitude Battery (ASVAB) was introduced as the single test battery to be used by all military services for both screening and job classification.

In Table 11.1, only the four subjects of the ASVAB that make-up the AFQT that is used for screening for military service are shown. Additionally, the ASVAB includes six other subtests that assess special knowledge or skill (Electronics Information; Mechanical Comprehension; Automotive and Shop Information; Coding Speed; General Science; Numerical Operations). The special knowledge subtests are combined with subtests from the AFQT to form ASVAB composites for classifying military applicants into job fields for which the military has determined that their aptitudes suit them best. For instance, all four military services use an electronics composite consisting of Arithmetic Reasoning + General Science + Mathematics Knowledge + Electronics Information. The different services weight each subtest score differently in combining them into one composite score (Wigdor & Green, 1991, p. 51).

TABLE 11.1
Tests Used by the Military for Assessment of cognitive Skills From World War I to the Present

Test	Historical Period Used	Dates Used	Content
Army Alpha	WW I	1918	Oral directions, arithmetical problems, practical judgment, synonyms-antonyms; disarranged sentences, number series, analogies, information
Army Beta	WW I	1918	Maze, cube analysis, O-series, digit symbol, number checking, pictorial completion, geometrical construction
Army General Classification Test (AGCT)	WW II	1941-1945	Verbal, arithmetic computation, arithmetic reasoning, pattern analysis
Armed Forces Qualification Test (AFQT)	Post WW II	1950-1952	Word knowledge, arithmetic reasoning, spatial relationships
	Korean War/ Vietnam War	1953-1973	Word knowledge, arithmetic reasoning, spatial perception, tool recognition
Armed Services Vocational Aptitude Battery (ASVAB-AFQT)	All-Volunteer Force	1975-1980	Word knowledge, arithmetic reasoning, spatial relationships, mechanical comprehension
		1980-1988	Word knowledge, arithmetic reasoning, paragraph comprehension, numerical operations
		1988-present	Word knowledge, paragraph comprehension, arithmetic reasoning, mathematics knowledge

Note. From Sticht, 1995.

Each of the four periods in Table 11.1 is discussed with regard to the theoretical position regarding what it was believed the tests measured, the contents of the tests, and the uses to which the tests were put.

World War I

The first mental tests designed to be used for mass, group testing were developed by psychologists for the U.S. Army in 1917–1918. The group tests were modeled after intelligence tests designed for individual use in one-on-one assessment. In developing the mental tests, the psychologists subscribed to the position that one could be quite intelligent, but illiterate or not proficient in the English language. Based on this reasoning, two major tests were developed; the Army Alpha for literate groups, and the Army Beta for illiterates, low literates or non-English speaking (Yerkes, 1921). Both tests were based on the theoretical position that intelligence was an inherited trait, and the assumption was made that *native intelligence* was being assessed. Each test was made up of a number of subtests (Table 11.2), the contents of which differed depending on whether the test was for literates or illiterates, low literates, or non-English speakers.

The Alpha Test. As indicated in Table 11.2, the Alpha test battery for literates included a wide range of tests of knowledge and various cognitive skills. Using the simple model of the human cognitive system given in Fig. 11.1, the Alpha test can be reinterpreted not as a test of native intelligence but as a sampling of a wide variety of cognitive abilities by addressing the person's knowledge base by both oral language and written language. Test 1 involved auding and comprehending simple or complex oral language directions and looking at and marking in the appropriate places on the answer sheet. To a large extent, this test was a test of the ability to hold information in working memory and to combine earlier instructions with later ones to determine the correct marking responses.

The role of special bodies of knowledge in the performance of information processing activities is clearly illustrated in the remaining tests. Test 2 required both the ability to read and comprehend the stated problem and the knowledge of arithmetic to perform the computations called for. Again, working memory was stressed by having to hold more than one phrase in it that was information bearing, then combining the phrases and performing the required computations. Mathematics knowledge was also required for Test 6.

Test 3, Practical Judgment, clearly required reading and comprehending language. Additionally, however, it required knowledge of culturally normative expectations to make the "correct" choice. In terms of the developmental model of literacy, this means that the person's mind would have

ARMY ALPHA	ARMY BETA
Test for Literates	*Test for Illiterates and Foregn Language Speakers*

Test 1 Following Oral Directions.

"When I say 'go,' make a cross in the first circle and also a figure 1 in the third circle."

⊕ ◯ ⓵ ◯ ◯ ◯

Test 2 Arithmetical Problems

Ex: If it takes 6 men 3 days to dig a 180-foot drain how many men are needed to dig it in half a day. Answer: (36)

Test 3 Practical Judgment

If a man made a million dollars, he ought to
[ ] Pay off the national debt
[x] Contribute to various worthy charities
[ ] Give it all to some poor man

Test 4 Synonyms-Antonyms

Samples: good - bad   same - <u>opposite</u>
              little - small   <u>same</u> - opposite

Test 5 Disarranged Sentences

leg flies one have only   true - <u>false</u>

Test 6 Number Series Completion

2 3 5 8 12 17 <u>23</u> <u>30</u>

Test 7 Analogies

gun - shoots : : knife - run <u>cuts</u> hat bird

Test 8 Information

The Wyandotte is a kind of:
horse <u>fowl</u> cattle granite

Test 1 Maze. Credit for correct tracing of mazes.

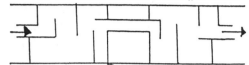

Test 2 Cube Analysis. Correct count of cubes.

Test 3 X - O Series. The series is to be carried out to the end of the line.

X O X O X O X O **X** <u>**O**</u> **X** <u>**O**</u>

Test 4 Digit Symbol. The appropriate symbol is to be written under each number.

1 2 3 4 5 6 7 8 9
⌐ и ⅃ L ∪ O ∧ ✕ =

3 1 2 1 3 2 1 4 2 3 5 2 9
⅃ ⌐ и ⌐ ⅃ и ⌐ L и ⅃ ∪ и =

Test 5 Number Checking. Correct response indicated.

699310 . . . .X . .699310
251004818 251004418

Test 6 Picture Completion. Identify missing parts.

   (mouth missing)

Test 7 Geometrical Construction. Construct a square (on right) out of figures on left.

Note. From Sticht (1995).

had to develop in an external context or environment in which the information needed to make the normatively "correct" response would be presented in some form. Test 8, too, was heavily loaded with cultural knowledge requirements. It was a probe of the person's knowledge base to discover the extent to which it included both very familiar and less familiar declarative knowledge available in the United States' culture.

Specific vocabulary knowledge was required for Tests 4 and 7, in addition to the knowledge of "same" and "opposite" (Test 4) and the content knowledge and information processing skills (Sternberg, 1977) needed to perform analogical reasoning (Test 7).

Based on a person's total Alpha score he was assigned a letter grade of A (superior intelligence), B, C+, C (average intelligence), C−, D, or D− (inferior intelligence). The letter grade became the person's mental category, and was taken as a general indicator of the person's native intelligence. This position was held even though there was a clear relationship of Alpha scores to years of schooling, in which much of the special knowledge, vocabulary and cultural knowledge would have been developed. Generally, the correlations of Alpha *total* test scores with education ranged from .65 if the low literates and non-English speaking were excluded to .75 when the latter were included (Yerkes, 1921, pp. 779–780).

The Beta Test. In determining who should take the Beta test, decisions were made frequently in terms of the number of years of education reported. Generally, those with fewer than four, five, or six years of education were sent to Beta testing. Additionally, men who were non-English speakers, or very poor in speaking English were sent for Beta testing. In some cases, men who tried the Alpha tests but were subsequently judged to be poor readers were readministered the Beta tests. The procedures were not uniform across the testing locations.

As shown in Table 11.2, like the Alpha test, the Beta test battery for illiterates, low literates, or non-English speakers also used a number of subtests. However, unlike the Alpha test in which instructions were given in oral and written language, the Beta subtest instructions were executed in pantomime by the test administrators. The examinees marked their responses on paper using pencils, but they were not required to use written language (though reading of number symbols was required in some subtests). As with the Alpha test, the Beta subtest scores were combined into one score, and that score was used to assign letter grades indicating general intelligence.

Though the attempt was to use the Beta test as an intelligence test comparable to the Alpha but freed of influences of literacy and the English language, examination of the subtests in Table 11.2 reveals major differences between the Alpha and Beta tests both in terms of the knowledge called for and the information processing skills involved in processing graphically presented information. As noted in the discussion of the developmental model of literacy, there are two main aspects to literacy. On one hand, literacy involves the use of graphics technology to produce a second signaling system for speech. That is, the written language is a graphical representation of the spoken language to a large degree.

However, the second major aspect of literacy is the use of the elements of graphics technology—light, space, and permanence—to produce graphic devices to be used in information processing for problem solving, reasoning, and communicating. In the subtests of the Beta test, as illustrated in Table 11.2, it is clear that literacy as the use of graphics technology for problem solving and reasoning is included in every subtest. In Test 1, the maze requires looking at the graphically represented maze while reasoning about the path to be taken. In Test 2, counting cubes in the graphic representation combines the use of graphics information with knowledge of the language of arithmetic for counting, while in Test 7, the graphics information on the left must be studied and mentally rearranged to construct the figure on the right. Reading graphics displays in left to right sequences, as in the written language, is a requirement of Test 3, and a useful processing skill for Tests 4 and 5, as is knowledge of the language of arithmetic.

From the foregoing analysis, it becomes that clear that both the Alpha and Beta tests assessed cognitive skills with the concomitant use of literacy. That is, the ability to utilize graphic marks arrayed in various designs for information processing is common to both tests. The primary difference between the two sets is that the Alpha requires extensive reading defined in the developmental model as looking while languaging, while this is not required to any significant degree in the Beta test (though reading the letters and numbers of Tests 3, 4, and 5 permits some reading while languaging).

In turn, languaging of sentences requires the retrieval of semantic information from the knowledge base stored in long term memory to be used in working memory for comprehending the information being picked up from the graphic display in the text of the test. In the Beta test, on the other hand, most of the information processing could be done in working memory without the need to locate and retrieve semantically encoded knowledge from the long term memory.

The correlation of Beta total scores with schooling ranged from .45 for a sample of over 11,000 native born men with education levels ranging from none to college, to .67 for a sample of 653 native-born draftees (Yerkes, 1921, p. 781, tables 327, 328). For a sample of 5,803 foreign born, the correlation of Beta total score with schooling was .50. In general, then, the correlations of Beta scores with years of schooling were lower than the correlations of Alpha scores and education (.75) when the full range of Alpha test takers (including those subsequently sent for Beta testing) was included.

Intelligence, Literacy, and Occupations. In a special study by the Surgeon General's Office, psychologists studied the intelligence distributions in various occupations as represented by the mental categories derived from the Alpha and Beta total scores (Yerkes, 1921, p. 829, Figure 57). Figure 11.2 summarizes data from that study and presents along with it information on the literacy competence of young adults in similar occupational categories as determined in the 1985 survey of literacy skills using the Document scale (Barton & Kirsch, 1990).

As Fig. 11.2 illustrates, there is a remarkable similarity between the data on intelligence and occupations in 1919 and literacy and occupations in 1985. Interestingly, however, in 1919, 32% of the scores for laborers were from the use of the Beta test for low literates or non-English speaking. All of the 1985 scores are for literates. Apparently, the laborers of today are more literate and English speaking than 75 years ago. But the social distribution of cognitive skills in the workforce is the same: the least skilled cognitively occupy the lowest rung on the occupational ladder. The best educated and most skilled occupy the top rungs of the ladder.

World War II

As the United States moved into World War II, new mental tests were developed by the Army and Navy to aid in the classification of recruits into jobs (Sticht & Armstrong, 1994). As in World War I, the tests were originally

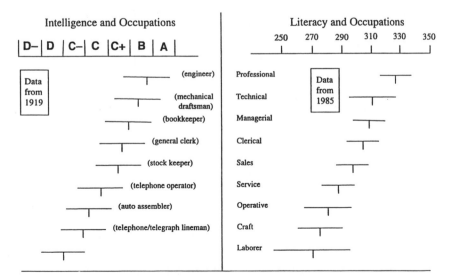

FIG. 11.2. Occupational distributions of intelligence as measured by the Army Alpha and Beta tests in World War I, and literacy as measured by the National Assessment of Educational Progress's survey of young adult literacy skills in 1985.

not used to screen civilians for service. The purpose of the Army General Classification Test (AGCT) test was to serve "as a test of 'general learning ability' and was intended to be used in basically the same manner as the Army Alpha (i.e., an aid in assigning new recruits to military jobs)" (Eitelberg, Laurence, Waters, & Perelman, 1984, p. 15). The Navy General Classification Test (NGCT) served similar purposes. However, beginning in June, 1943, the AGCT was used to screen out people with limited ability for succeeding in military training.

By the time the AGCT was developed, psychologists in the military personnel research sections considered that "intelligence tests do not measure native mental capacity. They measure actual performance on test questions. A test is a fairly valid measure of the native capacities which underlie the abilities tapped by its questions when every one tested has had equal opportunity and equal incentive to develop the abilities measured" (Zeidner & Drucker, 1983, p. 34). As interpreted by Zeidner and Drucker (1983), both long time members of the Army's personnel research activity, "the Army psychologists' World War II position was that the test scores represented nothing more than an index of measured abilities at the time the test was taken" (p. 35).

Because the Army of World War II had to draw upon a primarily inexperienced young adult population and train recruits in a wide variety of technical and administrative fields, as well as mechanized combat jobs, the AGCT was validated as a classification instrument by correlating scores on AGCT with grades in training courses. This seemed consistent with the interpretation of the AGCT as a measure of "general learning ability." Indeed, in content, the AGCT (Table 11.1) resembled measures of schooling, such as reading vocabulary ("verbal" ability) and arithmetic computation and word problem solving. Only a measure of "pattern analysis" was included that differed from school-based achievement tests. Given that the AGCT so much resembled a measure of past school learning, it is not surprising that it was found to correlate reasonably well with achievement in Army schools.

Korea, Vietnam, and the All-Volunteer Force

In 1950, the Armed Forces Qualification Test (AFQT) was introduced as the single test that would be used to screen draftees and volunteers for entry into any of the armed services. With some changes in content (Table 11.1), the AFQT has remained as the primary screening test for military service through the Korean and Vietnam Wars, and up to the present (Sticht & Armstrong, 1994). As noted by Eitelberg, et al. (1984):

> "Unlike the AGCT and the aptitude tests of World War I, the AFQT was specifically designed to be used as a *screening device.* Thus, the AFQT was established for the purpose of both (a) measuring the "examinee's general

mental ability to absorb military training within a reasonable length of time, so as to eliminate those who do not possess such ability;" and (b) providing a "uniform measure of the examinee's potential general usefulness to the service, if qualified on the tests." (pp. 16, 17)

In developing the AFQT, care was taken to make certain that speed was not emphasized, so that slow workers would not be penalized, and that the verbal instructions were not so difficult as to obscure the test items themselves. In these ways, the test was designed to be especially useful for distinguishing among the least able. On June 30, 1951, Congress passed Public Law 51 establishing the minimum acceptable standard for entry into military service at the 10th percentile, hence excluding the lowest scoring persons from service (Maier & Sims, 1986, A-14).

The All-Volunteer Force. In 1973, the draft was ended and the nation entered the contemporary period in which all military recruits are volunteers. Three years later, in 1976, the Armed Services Vocational Aptitude Battery (ASVAB) was introduced as the official mental testing battery used by all services (Table 11.1). The ASVAB combined the AFQT and special aptitude tests that differed for the services into one battery that today includes 10 subtests: word knowledge, paragraph comprehension, numerical operations, arithmetic reasoning, mathematical knowledge, coding speed, general science, mechanical comprehension, electronics information, and automotive-shop information.

Four subtests on the ASVAB are combined to form the AFQT. From 1976 to 1980 the AFQT was made up of the subtests of word knowledge, arithmetic reasoning (word problems), spatial relationships (for example, determining what kinds of boxes might result from folding two dimensional drawings), and mechanical comprehension (for example, if a lever is moved on a piece of equipment, what cogs and gears would move and produce a change in the equipment). From 1980 to 1988, the AFQT consisted of word knowledge, paragraph comprehension, arithmetic reasoning, and numerical operations (rapidly adding, subtracting, multiplying, and dividing up to two digit numbers; no one completes this speeded test). In 1988 numerical operations was dropped and mathematics knowledge was added. Today, the AFQT and special aptitude subtests of the ASVAB are used both to screen out lower aptitude applicants from military service, and to classify applicants according to occupations for which they qualify.

Discussion

In the 70 years since the military first introduced mental testing on a mass scale, the concept of just what it is that the tests measure has changed. At first, the Army Alpha and Beta tests were considered to measure "innate intelligence." In World War II, the AGCT was considered a measure of "general learning ability." For the last 40 years, the AFQT has been considered as a measure of "trainability."

The contents of the military mental tests changed most dramatically from the Army Alpha and Beta tests to the AGCT. From World War II to the present, the content has always included vocabulary and arithmetic knowledge. From 1941 through 1980, the aptitude tests always included a test of pattern or spatial perception. For 20 years the AFQT included a test of tool recognition, and for 5 years the tool recognition test was replaced with a test of mechanical comprehension. Since 1980, the AFQT has included only reading and mathematics components.

Since World War II, the usefulness of mental tests has been established by determining the correlation between scores on the tests and various indicators of performance in the military. The earliest and most frequently used measure of performance was success in training. The highest correlations have typically been found between the mental tests and course grades. This is why the AFQT is called a measure of "trainability."

Intelligence, Aptitude, and Basic Skills. Figure 11.3 shows relationships among the AGCT of World War II, AFQT mental categories and percentiles, IQ scores as measured by the Wechsler Adult Intelligence Scale (WAIS), five standardized, grade school normed tests of reading, the Comprehensive Adult Student Assessment System (one version of the tests that can be configured from the more than 5,000 items in the CASAS item bank), and the Reading Proficiency Scale administered by the National Assessment of Educational Progress (NAEP) to students in the 4th, 8th, and 11th grades. In 1985 the NAEP Reading Scale was administered to a representative sample of young adults 21 to 25 years old in a survey of the literacy skills of young adults, and those results are depicted in Fig. 11.3.

The figure addresses the issue of whether it is likely that one can be highly literate (high in the SCANS foundation skills of reading and writing) but low in intelligence or aptitude. The data suggest that this is not a likely occurrence. In fact, in studies of the interrelationships of six major reading tests (including the TABE, ABLE, G-M, and N-D tests) with the AFQT, one of the military's measures of aptitude ("trainability"), all of the correlations were in the range from .80 to .95 (Waters, Barnes, Foley, Steinhaus, & Brown, 1988, p. 46, Table 15). Given the range of the test-retest reliabilities of the reading tests (from .77 to .92), (Waters et al., 1988, p. 30, Table 6) these intercorrelations of AFQT and reading test scores are about as high as they can get.

From the developmental model of literacy it is apparent that literacy is a cognitive capability that develops on the basis of the earlier basic adaptive processes, and includes aspects of information processing involved in rea-

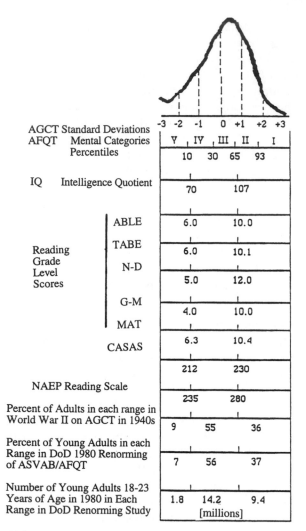

FIG. 11.3. A composite figure showing the distribution of scores on the World War II Army General Classification Test in relation to a variety of aptitude, academic reading, and functional literacy tests. The figure illustrates the similarity of scores on various tests referred to as measures of intelligence, aptitude, or literacy. AGCT = Army General Classification Test; AFQT = Armed Forces Qualification Test; IQ = Intelligence Quotient on the full scale of the Wechsler Adult Intelligence Scale; ABLE = Adult Basic Learning Examination; TABE = Test of Adult Basic Education; N–D = The Nelson–Denny Test of Reading; G–M = The Gates–MacGinitie Reading Test; MAT = Metropolitan Achievement Test; CASAS = The Comprehensive Adult Student Assessment System (a functional literacy test); NAEP = The National Assessment of Educational Progress Reading Proficiency Scale administered to a sample of young adults age 21 to 25 years old in 1985. Reading, CASAS, and NAEP scores are all approximations.

soning about the world in a nonliterate manner, making judgments about experiences in the past, present, and future, and languaging, including learning of words, their meanings, and the rules of syntax. All these information processing abilities are incorporated into the thinking about information picked up by reading as that skill develops from the earlier looking and marking abilities.

An additional factor of importance from the developmental model is the bodies of knowledge about the world that are drawn on in responding to questions on intelligence, aptitude, and then literacy tests. The interaction of knowledge and information processing skills is common in all these types of assessments. For this reason, it is to be expected that correlations of intelligence, aptitude, and literacy with schooling should be moderately high. As one progresses through school, larger bodies of knowledge are developed, along with the vocabulary (lexicon) needed to understand and express aspects of this expanding knowledge base. The data from World War I for both the Alpha and Beta tests, the AGCT in World War II, the AFQT over the last 40 years, and numerous reading assessments all show a direct, positive relationship to schooling such that the greater the number of years of school completed, the higher the scores on the various cognitive tests.

The Learning Abilities Measurement Program (LAMP).

In 1986 the Air Force Human Resources Laboratory (AFHRL—now called the Armstrong Laboratory) initiated a program of research into the cognitive skills that comprise general learning ability (Payne, Christal, & Kyllonen, 1986).

A major outcome of the first 5 years of LAMP research was the formulation and validation of an architecture of the human cognitive system that includes three basic components: a *perceptual system* that picks-up information from outside the head; a *cognitive system* consisting of a working memory that differs in capacity for different individuals, and two long term memory stores consisting of a declarative knowledge base and a procedural knowledge base; and a *motor system* for transforming thought into action (Armstrong Laboratory, 1991, pp. 5–6).

In addition to the architectural components in the LAMP model of the mental apparatus, the framework includes the idea that there is an *information processing speed* for the various processing systems (perceptual, cognitive, and motor) that leads to differences in the learning abilities of individuals. Additionally, individual differences are predicted by the capacity of working memory, the breadth of declarative knowledge, and the breadth of procedural knowledge. In a variety of learning tasks, including reading comprehension, this model of the cognitive system and its operating characteristics (capacity; speed) has accounted for 40% to 50% of the individual differences variance.

One of the conclusions from the LAMP is that to a large extent, reasoning ability is predictable from tests of working memory capacity and vice versa. This means that improved reasoning ability may increase working memory capacity in a given domain of knowledge. In several studies, estimates of the correlation between working memory capacity and reasoning ability ranged from .80 to .90 (Kyllonen & Christal, 1990).

This relationship between reasoning and working memory capacity may underly the ability of the Beta test used in World War I to assess "intelligence." As mentioned earlier, the Beta tests appear to have drawn very little on declarative (language mediated vocabulary, facts, semantic information) knowledge but rather to have involved attentive processes in looking and reasoning about the sequence of graphic displays. This would emphasize perceptual and working memory processes, and some access and use of procedural knowledge of reasoning, but not much processing of information into or out of the declarative knowledge base.

Reaching a Consensus on a Model of the Human Cognitive System Model. It is clear that there is a high degree of similarity between the model of the human cognitive system given in Fig. 11.1 and the model of the information processing system established in the LAMP. A similar model has emerged in the work of various military and civilian laboratories that study human cognition. A military-sponsored conference in 1984 resulted in the recognition that while there is not a uniform acceptance of a human cognitive system model amongst all cognitive scientists, most research does, in fact, involve the components described in Fig. 11.1 (Sticht, Chang, & Wood, 1986, p. 308). Most cognitive system architectural models include a long-term memory store or knowledge base, a short-term or working memory which operates on information from the knowledge base and on information picked up from the external world by means of information sending and recording processes, such as speaking and writing (sending) and auding and reading (recording).

According to the LAMP, a major factor contributing to the learning ability of individuals is the *breadth* of their declarative and procedural knowledge. Broadly literate people, those with high levels of the SCANS foundation skills of reading and writing (U.S. Department of Labor, 1991), know a lot about both general and specific domains of knowledge, while narrowly literate individuals are not likely to possess knowledge in numerous special domains. However, the narrowly literate person may have a depth of knowledge in a special domain, such as automotive and shop information, that permits fairly high scores in that domain. Discussions of "general" workforce readiness rarely take into account these special domains of competence that a person may possess (U.S. Department of Labor, 1991).

ASSESSING THE COGNITIVE DEMANDS OF JOBS

The determination of the cognitive demands of military jobs has been approached following two different general methods: the *predictive validity* method and the *job and task analysis method*. The various studies that the military services have conducted using these different methods have revealed that there is no one best way to identify the cognitive skill requirements of jobs, in or outside of the military. Generally speaking, different methodologies produce different results. So the choice of methodology should reflect the purposes of the inquiry.

For the most part, military research has focussed on two purposes for conducting an inquiry into the cognitive skills requirements of jobs. One purpose is to determine the correlation among scores on predictor and criterion assessment instruments that can be used to determine selection and classification cut-off scores on aptitude or other cognitive tests that predict some standard of success on the job. Typically, research for this purpose uses predictive validity methods.

A second military purpose for identifying the cognitive skills demands of jobs is to use the information to design and develop training programs for teaching job skills. Task analysis methods are generally followed to identify the contents and materials to be included in training programs.

Predictive Validity Studies of the Cognitive Skills Demands of Jobs

Typically, the predictive validity approach is followed to develop information to be used in the selection of people to work in a given job or career field (e.g, clerical). Methods to certify people as *workforce ready* might be validated using a predictive validity approach. This approach usually produces scores on a predictor test battery that are correlated with scores on some criterion assessment. The latter may be success in training, including completion of training, grades, or instructor's ratings. Additionally (or alternatively), indicators of job proficiency after training may be used as criterion measures to which predictor scores are related. These indicators may include supervisor ratings, peer ratings, job knowledge tests, job sample tests, etc.

The assessment of the cognitive skills of personnel reviewed this way has typically been one side of the predictive validity approach for determining the cognitive skill demands of military jobs. Typically, the ASVAB and other such tests are used to develop scores made up of various subtests (called composite scores) that are used to assign people to work that their skills match. These composites are validated by determining the extent to which scores on the tests are correlated with performance either in training or on the job. Cut-off scores on the assignment composites are then chosen, to increase the likelihood that a given person will be successful in training and on the job. A prediction is made from a person's scores on the cognitive tests as to the probability that the person will be successful in that field of work. The prediction is checked, or validated, by following up on people and finding out whether they do or do not succeed in training or on the job. This general approach to determining the cognitive skills demands of work illustrates the *predictive validity* method for determining the cognitive skills requirements of jobs. Those skills or skill levels that discriminate those who are from those who are not successful on the job are established as the cognitive skills requirements of the job.

The predictive validity method is the method the armed services have followed in developing their selection and classification tests since World War II. The U.S. Congress has established a law that excludes from military service those scoring below the 10th percentile on the Armed Forces Qualification Test (AFQT) because hundreds of predictive validity studies have indicated that there is a positive correlation between scores on the AFQT and success in both job training and on the job (Wigdor & Green, 1991). As a general rule, those who score below the 10th percentile on the AFQT will not be as successful in training or on the job as those who score above that cut-off score. Therefore Congress has stipulated that time and money should not be spent trying to accommodate the least cognitively developed in the armed services.

As indicated in Fig. 11.3, the 10th percentile on the AFQT translates roughly into a general reading level of the 4th–6th grade level, a CASAS score of around 212, and a NAEP score in the vicinity of 235. The data of Fig. 11.3 confirm that the least literate are restricted from military service by the Congressionally-set limits on the AFQT. The significance of this is that, to the author's knowledge, this constitutes the only Congressionally-mandated, national standard for literacy or, in SCANS terms, "foundation skills" for entry level work in the United States.

The JPM Study of the Cognitive Skills Required by Military Jobs.   The largest study of the cognitive skills requirements of military jobs that uses the predictive validity approach is the Job Performance Measurement/Enlistment Standards Project (JPM; Wigdor & Green, 1991). The JPM was initiated in 1980 and completed in 1992. In this study, some $36 million dollars was invested in developing measures of job performance to serve as criterion indicators of job proficiency. More than 15,000 troops from all four military services were involved over the decade, and dozens of professionals in psychological measurement participated in one way or another in studies to determine if better measures of job-performance could be developed and used to validate the enlistment standards set by the Department of Defense.

The results, after a decade of research, are summarized in the volume produced by the National Academy of Science's, National Research Council, Committee on the Performance of Military Personnel (Wigdor & Green, 1991). In a summary statement, the National Research Council states that, "The [job] performance measures provide a credible criterion against which to validate the ASVAB, and the ASVAB has been demonstrated to be a reasonably valid predictor of performance in entry-level military jobs" (p. 5).

Four ASVAB tests are combined to form the composite called the Armed Forces Qualification Test—Word Knowledge, Paragraph Comprehension, Arithmetic Reasoning, and Mathematics Knowledge. Other subtests are combined to form special Aptitude Area (AA) scores. For instance, the Arithmetic Reasoning (AR), Word Knowledge, and Paragraph Comprehension (PC) scores are combined to form the Army's General Technical (GT) composite and the Air Force's General (G) composite.

In the JPM project, the AFQT and various ASVAB composite scores were correlated with different indicators of job proficiency. The central indicator of job proficiency was taken to be hands-on job sample tests (Wigdor & Green, 1991, p. 147). Additional indicators of job proficiency that were given special attention in the JPM project were paper-and-pencil job knowledge tests. These are much easier to develop, administer, and score, and to the extent that they provide similar information as hands-on test performance, they represent cost-effective surrogates of the latter tests.

The JPM study provides data that permit the determination of the extent to which specific job knowledge (paper-and-pencil tests of knowledge), special job-related knowledge (aptitude composites derived from ASVAB subtests), and general knowledge (the AFQT) relate to the ability to perform actual job tasks.

In the JPM study, it was found that correlations of paper-and-pencil tests of job knowledge that required reading and hands-on job-sample test performance in nine Army, four Marine Corps, and three Navy jobs ranged from .35 to .61, with a median correlation of .47 (approximately .59 when corrected for limitations in the reliabilities of both tests) (Wigdor & Green, 1991, p. 151, Table 8-5). These data include some substantial relationships between these two types of job proficiency indicators, but the relationship is far from perfect and one indicator cannot substitute for the other. In other words, a reading test that assesses specific job knowledge is not necessarily a good predictor of one's ability to perform job tasks.

The effects of using special aptitude composites, that typically utilize one or more measures of "general literacy" (e.g., word knowledge, paragraph comprehension, arithmetic reasoning) combined with one or more measure of "specific literacy" (e.g., auto and shop information, electronics information) was also studied. These composites represent measures of job-related knowledge accessed via reading. For thirteen jobs, the median correlation of the special aptitude area score with hands-on job sample test performance was .50 (corrected for restriction of range) (Wigdor & Green, 1991, p. 161, Table 8-10).

In a set of studies involving 23 jobs, the predictability of hands-on job sample test performance increased from 21% when "general" knowledge was the predictor, to 25% when both "general" and "job-related" knowledge were combined into Aptitude Area composites, and to 35% when "job-specific" knowledge was used to predict hands-on test performance (Wigdor & Green, 1991, p. 160). In all these assessments, reading in the sense of using the written language as a second signaling system for speech was common to the predictor tests. What appears to have made a difference, then, is the relationship between the knowledge base in the human cognitive system being accessed via reading and the knowledge needed to perform the hands-on job tests.

Predicting Job-Related Reading Task Test Performance. Most jobs have a requirement that employees must be able to read the technical manuals, procedural job aids, safety instructions, and other job-related reading material used in performing their jobs. Research sponsored by the U.S. Army investigated relationships among the AFQT, a standardized reading test (the Survey of Reading Achievement, Junior High Level, California Test Bureau), and a specially developed job-related reading task test (JRTT) in three Army jobs: Cooks, Supply Clerk, and Vehicle Repairman (automobile mechanics) (Sticht, 1975, Chapter 3).

Correlations for over 200 men who had completed job technical training were obtained in each job. It should be noted (see Table 11.1) that the AFQT in use at the time of this study included two subtests—spatial perception and tool recognition—that would lower the relationship between the AFQT and a standardized reading test. These correlations were in fact, .66 for Cooks, .75 for Clerks, and .78 for Repairmen.

For the AFQT and JRTT, the correlations of AFQT with JRTT performance were .54, .58, and .65 for Cooks, Clerks, and Repairmen, respectively. For the standardized reading test, these correlations were .73, .70, and .65. Corrected for restriction of range, each of these coefficients would likely be increased by some .10. This would mean that the AFQT predicted 40% of the variance in JRTT performance, while the standardized reading test predicted 64% of the variance in JRTT.

All of the foregoing correlations involve predictor and criterion *tests*. When the criterion indicator of job proficiency is changed to the widely-used supervisor's ratings, the correlations with literacy (paper-and-pencil) tests drop precipitously, to levels ranging from .06 to .17 in work reported by Sticht (1975, p. 67, Table 24) and .12 to .15 in Zeidner's extensive review of the predictive validity of the GATB and ASVAB (see data for Global Ratings in Zeidner, 1987, pp. 44–48, Tables 13 & 14).

What this means is that general literacy tests are good, but far from perfect, predictors of the ability to perform job reading tasks. But the farther one gets from correlating reading-based, knowledge assessments with similar criterion measures of job performance, the less accurate is the predictive validity of the cognitive assessments. In order of magnitude, when the JPM studies are considered, the predictive validity of general literacy tests like the AFQT, Aptitude Area composites, and nationally, normed standardized reading tests are best at predicting job-related reading task performance (40–64% of the criterion variance), next best at predicting job knowledge, paper-and-pencil test performance (35% of the criterion variance), next best at predicting hands-on job-sample test performance (21% of the criterion variance), and least best at predicting supervisor's ratings of performance (less than 2–3% of the criterion variance).

Establishing Reading Requirements of Jobs. The predictive validity approach was used by Sticht and colleagues (Sticht, 1975) to establish reading requirements for Army jobs that would enlist large numbers of lower ability personnel. In this work, standardized reading tests were administered to some four hundred personnel in each of three Army jobs: Cook, Supply Clerk, and Automotive Repairman. Additionally, data were obtained on the three indicators of job performance previously discussed: job-related reading task tests (JRTT), paper-and-pencil job knowledge tests, and hands-on job sample tests.

The goal of the study was to establish reading requirements for Army jobs that could be used to set standards for military literacy training. A major point to be made here is that the setting of such standards is essentially a judgmental process: There is no technical methodology for setting cut-off scores on what are essentially continuous relationships among predictor and criterion tests.

For instance, Fig. 11.4 shows the relationships among reading grade levels, as measured by the California Test Bureau's Survey of Reading Achievement tests for the middle grades, and job-related reading task tests (JRTT) for Cooks, Repairmen, and Supply Specialists. The figure shows the relationship of general reading to three different criterion levels of performance on the JRTT: 50%, 60%, and 70% correct. Clearly, there is no one general reading level below which all are unsuccessful on the JRTT and above which all are successful. Indeed, the very definition of success is not specified by the empirical data. That is, should the criterion of success be 50% correct? If so, then how many people should have to reach that level—100%?

If that were the case, then for Cooks, an 8th grade level or general reading would permit 100% of personnel to reach a criterion of 50% correct on the Cook's JRTT. That would seem to be true for the Repairman's JRTT also, if we assume that the curve for the 50% correct criterion

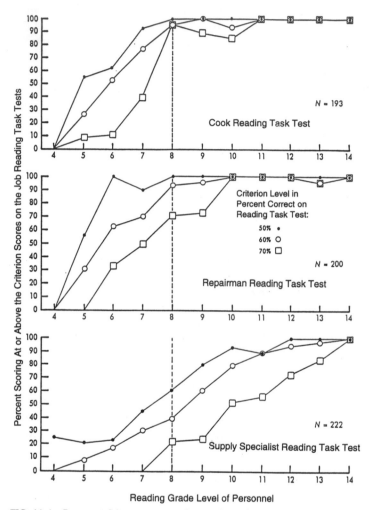

FIG. 11.4. Percent of Army personnel at each reading grade level reaching
different criterion levels of performance on Job-Related Reading Task Tests
(JRTT).

between the 5th and 8th grades increases gradually, and the score for the
6th grade general readers is spurious. Following this same decision rule
for setting standards, those entering the Supply Specialist's field would
have to read at the 12th grade level. This would permit 100% of personnel
to reach the criterion of 50% correct on the JRTT.

Regarding the relationships among reading and job knowledge and
hands-on job performance tests, Fig. 11.5 shows the quartile distributions at
each reading grade level of those who took these three tests. The correlation

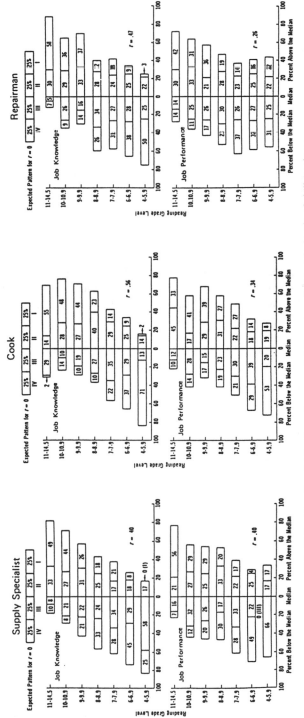

FIG. 11.5. Quartile distributions for men at different reading grade levels of skill in three Army jobs. The figure shows that for Cooks reading at the 4–5.9 grade level, 71% were in the bottom fourth of performers on the paper-and-pencil job knowledge tests while 53% were in the bottom fourth of those taking the hands-on, job performance tests. In general, the figure indicates that job performance, whether indexed by job knowledge or job sample tests, improves as reading ability improves.

coefficients are also given. As with the JRTT, there is no point along the reading scale below which all score in the bottom quarter or bottom half of performers on the job knowledge or job performance tests. Given that there is no obvious empirical or technical solution to the problem of setting standards of performance on the job knowledge and job performance tests as criterion for identifying the general reading level needed to reach those criteria, a judgmental rule must be established and followed.

In the study under consideration (Sticht, 1975, Chapter 4), the rule was established that the level of reading that should be the aim of a literacy training program is one in which those having that level of reading skill are not overrepresented in the bottom quartile of job performers. Applying this rule to the Cook's data, it is noted that over 70% of those reading at the 4–5.9 grade level score in the bottom quartile on the Cook's job knowledge test. It is not until one reaches the 7th grade general reading level that one finds fewer than the expected 25% in the bottom quartile. Therefore, following the specified decision rule, literacy training would be targeted at the 7th grade level for Cooks. Similar applications of the decision rule to the Repairman's and Supply Specialist's data, leads to the designation of an 8th grade level for literacy training for Repairman, and a 9th grade level for Supply Specialists. (Note that this ignores some rather egregious deviations from continuity, as in the case of finding only the expected 25% in the bottom quartile of those reading at the 4–5.9 grade level on the Supply Specialist's job knowledge test. Such deviations were considered as sampling errors.)

It is important to note that these decision rules were aimed at providing recommended standards for literacy training, *not* for screening people into or out of the military. For one thing, all of the people tested were "employed" in the military at the time, including the least literate. And most were performing satisfactorily (as determined by other methods, including supervisor's ratings). Clearly, setting standards for selection purposes would dictate different decision rules, and standards would not be set as high. That would result in rejecting too many people who could, in fact, learn and perform jobs very well.

Discussion of the Predictive Validity Approach to Identifying Cognitive Skills Demands of Jobs. The military research reviewed on the predictive validity approach to identifying cognitive skills demands of jobs is only a very small and narrowly selected sampling of such work in the armed services. Literally thousands of such studies exist, correlating almost all cognitive (and hundreds of noncognitive) tasks imaginable to a wide variety of indicators of job proficiency.

The research reviewed in this chapter focused on the largest project yet undertaken to relate cognitive and noncognitive predictors to well-developed and thoroughly critiqued measures of job performance (the Department of Defense's JPM project, Wigdor & Green, 1991). Additionally, the work by the author and various colleagues summarized here forms the largest corpus of research to directly investigate the relationships of literacy variables to job performance within the framework of a conceptual model of a human cognitive system, a model of literacy and its development, and an information processing approach to cognitive performance, including learning.

Both the JPM and the earlier work by Sticht and associates indicates that:

1. The relationship of literacy to job performance is continuous, not dichotomous. That is, there is no empirical "literacy gap" to be found in these studies. Rather, these studies show that productivity on the job is a continuous function of literacy and other cognitive abilities incorporated within the literacy assessments.

2. Generally speaking, the greater one's literacy ability, the more productive one is on the job. But this is a function of the criterion for job performance. The more the criterion incorporates literacy directly, as in performing job-related reading tasks, the greater will be the predictive validity of measures of reading.

3. Establishing reading requirements of jobs following the predictive validity approach requires human judgment regarding the types of predictor variables to be used, the criterion variables that serve as indicators of job proficiency, and the standards of performance desired on these criterion variables; there are no completely empirical or technical methods that can substitute for human judgment in setting these standards of performance. Wigdor & Green (1991, pp. 207–210) comment on a variety of very complex models for assessing human resources and using such assessments to set standards for allocating individuals to jobs. They state, "The solutions provided by such models are not intended to and will not supplant the overarching judgment that policy officials must bring to bear ..." (p. 209).

4. The purpose of the decision making determines the rules followed in setting standards for cognitive skills requirements. For instance, because of the relatively low predictive validities of literacy and other cognitive tests for predicting success on the job, decisions for setting standards of literacy that will exclude people from job opportunities should follow different rules from those used to establish standards for literacy training. Generally, higher standards may be set for training because it will probably not harm people to receive training beyond what they might need as a minimum. But as the data reviewed indicate, unreasonably high standards for entry into job training or employment may exclude from work thousands who could, in fact, perform satisfactorily in both the classroom and on the job.

Job and Task Analysis Studies of the Cognitive Skills Demands of Jobs

Recent reports provide extensive reviews of job and task analysis methods for identifying cognitive skills requirements of military jobs (Payne et al., 1992; Wigdor & Green, 1991, Chapter 4). The Payne et al. report focuses on task analysis methods for identifying basic skills (reading, mathematics) requirements of jobs to be used in the development of training programs. Wigdor and Green discuss job and task analysis to identify tasks to be used in the development of job performance measures, such as the hands-on tests used in the Job Performance Measurement/Enlistment Standards Project (JPM) discussed earlier.

In addition to their discussion of methods for identifying tasks job holders perform, Wigdor and Green describe several approaches to *trait* and *task-by-trait* analyses of jobs. Traits are relatively fixed characteristics of individuals, such as verbal and quantitative aptitudes, psychomotor skill, information processing speed, interpersonal skills, and so forth. In trait analysis, analysts estimate the types of general abilities workers should have to perform various jobs or job tasks.

In task-by-trait analysis, two approaches are typically followed: a generic task-by-trait approach and a local task-by-trait approach. In the generic task-by-trait analysis method, the assumption is made that all tasks can be broken down into a limited set of general human abilities. Analysts decompose tasks into generic skills. For instance, in one widely studied approach, the generic traits (abilities) that are rated for their presence or absence in performing job tasks include a general category called *Written Comprehension*. Examples of tasks within various jobs that may fall under this category are: Understand an instruction book on repairing a missile guidance system. Understand an apartment lease. Read a road map (Wigdor & Green, 1991, p. 83, Table 4-2).

In the local task-by-trait approach to task analysis, the lists of generic skills are developed as an emergent property of the task analysis. The methodology followed by the developers of the Army's Job Skills Education Program (JSEP; Hoffman, 1989) offers an example of the local task-by-trait approach to task analysis. In that work, job tasks were identified through interviews with job holders and supervisors, and a review of documentation about the job. Lists of job tasks were identified and then analyzed to identify the "basic skills" or "prerequisite skills" embedded within the tasks. This resulted in extensive lists of "decontextualized" basic skills, such as "Numbering and Counting," "Vocabulary," "Reference Skills," "Tables & Charts," "Outlining," and so forth. Each of these categories contained additional categories of basic skills decontextualized from the tasks in which they were found. Altogether, 97 subskills were identified in the quantitative domain and 56 in the verbal domain (Hoffman, 1989, p. 6, Table 3).

In this approach, for instance, a skill such as "outlining" is considered to be a general ability trait a person can acquire that will then permit him or her to operate effectively in any context that requires the ability to outline. In the generic task-by-trait approach, lists of "traits," "abilities," or "skills" are developed and then applied as an inventory to determine which skills are found in which tasks. In the JSEP approach, the list of skills was only partially determined at the outset. Then, during the task analysis additional skills were added to the list as they became evident to the analysts. In some cases, the JSEP developers added lessons not directly derived from task analyses (e.g., "learning strategies"), but which were considered as useful skills for soldiers to develop.

A difficulty with the task analysis method for identifying literacy (or other cognitive) skills demands of jobs is that the task analysis is likely to produce extensive lists of tasks, their associated skills, and even a monumental amount of materials obtained from the job sites. While such information and material is necessary for curriculum developers, policymakers and managers find such lists and stacks of materials overwhelming as indicators of the literacy requirements of jobs. They prefer some summary index number that expresses the general level of skills needed so that they can set policies for selection and program outcomes. One example is the U.S. Congress's use of the 10th percentile on the AFQT as the minimum basic skills level for entrance into the armed services. Another (civilian) example is given in a recent study of the federally sponsored Job Opportunities and Basic Skills (JOBS) program for welfare recipients. This report notes that the policy regulations governing the program emphasize a general goal for basic skills education comparable to those of a student in the 9th month of the 8th grade (the 8.9 grade level; Chisman & Woodworth, 1992, p. 5).

The only project that the author is aware of in either civilian or military settings that has approached both the problem of developing a general method for deriving summary index numbers stating a general level of literacy (reading) required of jobs and the identification of reading tasks and materials for curriculum development was the Navy's Project SEAREAD (Sticht, 1977; Sticht, Hauke, Fox, & Zapf, 1977).

In the SEAREAD work, job-related reading task test items were administered to large samples of servicemembers, and the reading grade level needed to achieve 80% correct on the items was determined. Then, the items were used in job analysis inventories in which the test items were shown to job incumbents and they were asked to rate whether or not they read that kind of material to answer the types of questions asked on their jobs. A weighted average reading grade level of tasks performed was derived as the summary index number for the general level of reading needed for the job. The task items themselves then provided materials for designing literacy training programs for teaching job-related reading.

Aspects of the SEAREAD methodology was used by Mikulecky and Diehl (1980) in an adapted form to study the reading requirements of some 100 civilian jobs, and it served as the forerunner to what the U.S. Departments of Labor and Education (1988) called "literacy task analysis" or "literacy audits."

IMPLICATIONS FOR POLICY AND PRACTICE

Properly setting standards for entry into or remaining productive in the workforce requires some method for identifying the cognitive competence that will render youths and adults members in good standing of the workforce of the United States in the coming decades. This will generally entail methods for assessing the workforce readiness of youths and adults, and the latter will, in turn require some way of analyzing and characterizing the cognitive demands of work. This chapter has emphasized that there are two general approaches for identifying the literacy or other cognitive skills requirements of work that is even moderately complex. The predictive validity approach relates literacy skills to measures of job proficiency. This requires methods for identifying what will be measured in the literacy assessments and what will be the criteria of job proficiency. Typically, literacy measures of competence will better predict measures of job-related literacy task performance than they will job sample tests or supervisor ratings. In fact, literacy and other cognitive assessments typically account for only some 10% to 20% of the differences in the latter types of indicators of job performance.

The predictive validity approach can provide a useful indication of the degree of variation in job performance that is associated with variations in literacy levels. But, significantly, the military research reveals that there is no technical solution to establishing standards for literacy or other cognitive competence. There is no noticeable literacy gap in the numerous studies done in the military using literacy and/or other types of cognitive assessments as the obvious point to set standards. Rather, productivity on the job appears to be a continuous function of cognitive ability. Generally, the more a person knows and the better the person can reason with his or her knowledge, the more productive the person is likely to be. For this reason, *setting standards is a judgmental process* that must be performed by those involved.

Another limitation to the predictive validity approach to identifying the literacy demands of jobs is that it does not provide curriculum materials needed for developing job-related literacy programs (Mikulecky, 1995). The discussion of the *task analysis* approach to identifying literacy demands of jobs notes that there is one way to conduct a literacy task analysis. To a large extent, how one goes about conducting such an analysis will reflect one's understanding of work, literacy, and the aims and context of the development effort. An important conclusion from the review of the methods for identifying literacy requirements of jobs is the observation that policymakers and management generally wish to express these requirements in an index number that establishes a general goal for the literacy program. Curriculum developers, on the other hand, need numerous examples of the types of tasks and materials used in performing literacy tasks on the job. An example is cited of a research project that combined the predictive validity and task analysis approaches to provide both a summary index number (e.g., grade level) that policymakers could use to set general levels of literacy to qualify for and complete literacy education, and a sample of materials that could be used by curriculum developers to prepare job-related literacy programs. That study suggested that such an approach is feasible. But a great deal more research is needed to evaluate the utility of such an approach.

Finally, and perhaps most importantly, the conceptual framework developed in this chapter and the review of the assessment of cognitive skills indicate that literacy is not some relatively simple addendum to or foundation for human cognitive competence. Rather, the conceptual framework and the empirical data suggest that, whereas one can be quite intelligent, well developed in oral language, and possessed of high aptitudes for learning, even though illiterate or poorly literate, it is highly unlikely that one can be highly literate and yet low in intelligence, poorly developed in language (vocabulary and syntax), and of low aptitude for learning.

The implication of this observation is that policymakers, such as those involved in the SCANS (U.S. Department of Labor, 1991), and educators involved in developing high levels of workforce readiness in students, must recognize that to develop all youths and adults to high levels of workforce readiness in the so-called "foundation skills" means, essentially, that *they must simultaneously be brought to high levels of intelligence, oral language, literacy and learning ability.* Fortunately, the significant correlations of all of these types of measures with years of education suggests that this might be accomplished through education. But it is likely to require a very extensive and expensive effort to reform the education system for in-school youth, and a similar effort to develop an adult workforce education system capable of accomplishing this task.

ACKNOWLEDGMENTS

Portions of this work was supported by a grant to the Applied Behavioral & Cognitive Sciences, Inc. from the National Center on Adult Literacy (NCAL) at the University of Pennsylvania. NCAL is part of the Educational

Research and Development Center Program of the Office of Educational Research and Improvement, U.S. Department of Education, in cooperation with the Departments of Labor and Health and Human Services. The positions and policies contained in this report are solely those of the author and do not necessarily reflect the positions or policies of the Applied Behavioral & Cognitive Sciences, Inc., the National Center on Adult Literacy, the Office of Educational Research and Improvement, or the U.S. Departments of Education, Labor, and Health and Human Services.

REFERENCES

Armstrong Laboratory. (1991, January). *Learning abilities measurement program (LAMP), phase 2, 5 year plan.* Brooks Air Force Base, TX: Human Resources Directorate, Det 5, Manpower & Personnel Division, Individual Attributes Branch.

Benton, L., & Noyelle, T. (1992). *Adult illiteracy and economic performance.* Paris: Organisation for Economic Co-operation and Development, Centre for Educational Research and Innovation.

Binkin, M. (1986). *Military technology and defense manpower.* Washington, DC: The Brookings Institution.

Bransford, J., Sherwood, R., Vige, N., & Rieser, J. (1986). Teaching thinking and problem solving. *American Psychologist, 41,* 1078–1089.

Brizius, J., & Foster, S. (1987). *Enhancing adult literacy: A policy guide.* Washington, DC: The Council of State Policy & Planning Agencies.

Carroll, J. B. (1987, February). The national assessment in reading: Are we misreading the findings? *Phi Delta Kappan, 68*(6), 424–430.

Chisman, F. P., & Woodworth, R. S. (1992). *The promise of JOBS: Policies, programs and possibilities.* Washington, DC: The Southport Institute for Policy Analysis.

Eitelberg, M. J. (1988). *Manpower for military occupations.* Alexandria, VA: Human Resources Research Organization.

Eitelberg, M. J., Laurence, J. H., Waters, B. K., & Perelman, L. S. (1984). *Screening for service: Aptitude and education criteria for military entry.* Alexandria, VA: Human Resources Research Organization.

Fremer, J. J. (1989). Testing companies, trends, and policy issues: A current view from the testing industry. In B. R. Gifford (Ed.), *Test policy and the politics of opportunity and allocation: The workplace and the law* (pp. 61–80). Boston: Kluwer Academic Press.

Gal, R., & Mangelsdorf, A. (Eds.). (1991). *Handbook of military psychology.* London: Wiley.

Hoffman, L. M. (1989, March). *Effectiveness of computer-based basic skills instruction: The Army's job skills education program.* Paper presented at the annual conference of the American Educational Research Association, San Francisco, CA. Washington, DC: U.S. Department of Education, Office of Educational Research and Improvement.

Kirsch, I. S., & Jungeblut, A. (1986). *Literacy: Profiles of America's young adults.* Princeton, NJ: Educational Testing Service.

Kyllonen, P. C., & Christal, R. E. (1990). Reasoning ability is (little more than) working-memory capacity?! *Intelligence, 14,* 389–433.

Maier, M. H., & Sims, W. H. (1986, July). *The ASVAB score scales: 1980 and World War II.* CNR 116. Alexandria, VA: Center for Naval Analysis.

Mikulecky, L. (1995). Workplace literacy programs: Organization and incentives. In D. Hirsch & D. Wagner (Eds.), *What makes workers learn: The role of incentives in workplace education and training* (pp. 129–149). Cresskill, NJ: Hampton Press.

Mikulecky, L., & Diel, W. (1980). *Job literacy.* Bloomington, IN: Reading Research Center, Indiana University.

National Alliance for Business. (1995, March). *Work America, 12,* Washington, DC: Author.

Office of the Assistant Secretary of Defense (Force Management & Personnel). (1987, August). *Population representation in the military services fiscal year 1986.* Washington, DC: OASD (FM&P).

Payne, D. L., Christal, R. E., & Kyllonen, P. C. (1986). Individual differences in learning abilities. In T. G. Sticht, F. R. Chang, & S. Wood (Eds.), *Cognitive science and human resources management. Advances in reading/language research. Vol. 4* (pp. 27–36). Greenwich, CT: JAI Press.

Payne, D. L., Schendel, J. D., Page, R. C., Cunningham, J. W., Mathews, J. J., Lamb, T. A., & Villanueva, A. (1992, May). *Fundamental skills needs assessment methods* (Final Technical Report AL-TP-1992-0011). Brooks Air Force Base, TX: Armstrong Laboratory.

Rouse, W., & Morris, N. (1985). *On looking into the black box: prospects and limits in the search for mental models* (Report No. 85-2). Atlanta, GA: Georgia Institute of Technology, Center for Man-Machine Systems Research.

Rumelhart, D. (1980). Schemata. In B. Bruce & W. Brewer (Eds.), *Theoretical issues in reading comprehension.* Hillsdale, NJ: Lawrence Erlbaum Associates.

Sternberg, R. (1977). *Intelligence, information processing and analogical reasoning: The componential analysis of human abilities.* Hillsdale, NJ: Lawrence Erlbaum Associates.

Sticht, T. G. (1975). *Reading for working: A functional literacy anthology.* Alexandria, VA: Human Resources Research Organization.

Sticht, T. G. (1977). Comprehending reading at work. In M. Just & P. Carpenter (Eds.), *Cognitive processes in comprehension* (pp. 221–246). Hillsdale, NJ: Lawrence Erlbaum Associates.

Sticht, T. G. (1995, March). *The military experience and workplace literacy: A review and synthesis for policy and practice* (NCAL Tech. Rep. TR94-01). Philadelphia, PA: National Center on Adult Literacy.

Sticht, T. G., & Armstrong, W. B. (1994, February). *Adult literacy in the United States: A compendium of quantitative data and interpretive comments.* Washington, DC: National Institute for Literacy.

Sticht, T. G., Beck, L., Hauke, R., Kleiman, G., & James, J. (1974). *Auding and reading: A developmental model.* Alexandria, VA: Human Resources Research Organization.

Sticht, T. G., Chang, F. R., & Wood, S. (Eds.). (1986). Cognitive science and human resources management. *Advances in Reading/Language Research* (Vol. 4). Greenwich, CT: JAI Press.

Sticht, T. G., Fox, L. C., Hauke, R. N., & Zapf, D. W. (1977, September). *Reading in the Navy* (Report Number FR-WD-CA-76-14). Alexandria, VA: Human Resources Research Organization.

Sticht, T., & Hickey, D. (1991). Functional context theory, literacy and electronics training. In R. Dillon & J. Pellegrino (Eds.), *Instruction: Theoretical and applied perspectives* (pp. –). New York: Praeger.

Sticht, T. G., Hofstetter, C. R., & Hofstetter, C. H. (1995, January). *Assessing adult literacy by telephone.* El Cajon, CA: Applied Behavioral & Cognitive Sciences, Inc.

U.S. Departments of Labor and Education. (1988). *The bottom line: Basic skills in the workplace.* Washington, DC: U.S. Government Printing Office.

U.S. Department of Labor. (1991, June). *What work requires of schools: A SCANS report for AMERICA 2000.* Washington, DC: U.S. Department of Labor.

Venezky, R. L., Kaestle, C. F., & Sum, A. M. (1987). *The subtle danger: Reflections on the literacy abilities of America's young adults.* Princeton, NJ: Educational Testing Service.

Waters, B. K., Barnes, J. D., Foley, P., Steinhaus, S. D., & Brown, D. C. (1988, October). *Estimating the reading skills of military applicants: Development of an ASVAB to RGL conversion table* (Final Report 88-22). Alexandria, VA: Human Resources Research Organization.

Wells, A. (1995). Workplace basic skills program in the United Kingdom: Why so few? In D. Hirsch & D. Wagner (Eds.), *What makes workers learn: The role of incentives in workplace education and training* (pp. 151–157). Cresskill, NJ: Hampton Press.

Wigdor, A. K., & Green, B. F. (1991). *Performance assessment for the workplace.* Washington, DC: National Academy Press.

Yerkes, R. M. (Ed.). (1921). Psychological examining in the United States Army. *Memoirs of the National Academy of Sciences* (Vol. 15). Washington, DC: National Academy of Sciences.

Zeidner, J. (1987, April). *The validity of selection and classification procedures for predicting job performance* (IDA Paper P-1977). Alexandria, VA: Institute for Defense Analysis.

Zeidner, J., & Drucker, A. (1983). *Behavioral science in the Army: A corporate history of the Army Research Institute.* Alexandria, VA: U.S. Army Research Institute for the Behavioral and Social Sciences.

Chapter 12

Assessing Employability Skills: The Work Keys™ System

Joyce R. McLarty
Timothy R. Vansickle
ACT
Iowa City, Iowa

In recent decades, concern has mounted that American workers—both current and future—lack the workplace skills necessary to meet the challenges of technological advances, organizational restructuring, and global economic competition (cf. National Commission on Excellence in Education, 1983; Secretary's Commission on Achieving Necessary Skills [SCANS], 1991). The Work Keys system from ACT (formerly American College Testing) is an innovative response to this problem. Work Keys is an integrated system that provides a continuous structure for documenting and improving individuals' generic workplace skills. Generic skills are those that are not specific to a particular job, but rather provide a foundation for learning and performing well on most jobs. Generic skills include various problem-solving, communications, and personal skills. By providing individuals with good, reliable information regarding both their existing skill levels and the skills that will be expected of them in the workplace, ACT Work Keys can help them make solid career decisions.

ACT consulted with employers, educators, and labor organizations to develop a list of generic workplace skills that are used in a wide range of jobs, are teachable in a reasonable period of time, and facilitate job analysis (ACT, 1992a). In 1992, initial Work Keys generic skills were selected on the basis of a review of the literature relating to employer-identified skill needs (Agency for Instructional Technology, 1989; ACT, 1987; Bailey, 1990; Carnevale, Gainer, & Meltzer, 1990; Center for Occupational Research and Development, 1990; Conover Company, 1991; Educational Testing Service,

1975; Electronic Selection Systems Corporation, 1992; Greenan, 1983; SCANS, 1990) and a survey of employers and educators who participated in the design of the Work Keys system. The latter were from seven states (Illinois, Iowa, Michigan, Ohio, Oregon, Tennessee, and Wisconsin) and a network of community colleges in California, all of which served as "charter" members of the Work Keys development effort. These charter members assisted in the design and review of plans and materials, and provided examinees for the prototype and field-test phases of assessment development. Twelve skills were selected for initial development based on those identified by educators and employers (McLarty, 1992). It was anticipated that this list would be modified over time in response to changing employer needs.

The similarities between the skills selected by ACT for the Work Keys system and the foundation skills identified by The Secretary's Commission on Achieving Necessary Skills (SCANS) are undoubtedly due to both having consulted similar resources to identify needed skills, inasmuch as the Work Keys skill set was first identified at about the same time as that for SCANS (SCANS, 1991). SCANS organized its foundation skills as Basic Skills, Thinking Skills and Personal Qualities; Work Keys groups its generic skills as Communication Skills, Problem Solving Skills, and Interpersonal Skills. The individual skills clustered within these two sets of categories, however, are quite similar. Recent research conducted as part of the National Job Analysis Study as a validation of the effort originally begun by SCANS (see Nash & Korte, chapter 4, this volume) will also inform future development of the Work Keys skill set.

The Work Keys system provides a metric which describes the skill requirements for individual jobs in terms of levels of proficiency. This metric provides the information that schools need in order to determine how to prepare students more completely for the workplace, and that businesses need in order to determine the qualifications of potential employees and to design job-training programs to help current employees better meet the demands of their jobs. By showing individuals a direct link between their education and training and their ability to qualify for jobs, and by providing them timely and accurate feedback on their progress in acquiring generic employability skills, Work Keys is designed to have a positive effect on learner persistence and achievement. By providing employers a skills metric directly linked to their generic job skill requirements, Work Keys is designed to help ensure that the applicants they hire have the necessary generic skills for competent job performance. By providing educators with information that relates generic skills to documented job requirements, Work Keys is designed to guide the development of well-focused and effective instruction.

Work Keys has been developed as a system with interactive components to address job analysis, assessment, and instructional support. These components are built around a central skill scale for each of the eight currently operational skill areas: Reading for Information, Applied Mathematics, Listening, Writing, Applied Technology, Locating Information, Teamwork, and Observation. Additional skill areas are in development.

This chapter describes key elements of the Work Keys system, focusing on those strategies that make this approach unique and on the issues brought to the fore by dealing directly and jointly with the assessment of people and jobs while applying common standards to each. The first section addresses the development of the Work Keys hierarchical skill scales. The second section focuses on building assessments of individuals that are linked to the skill scales using selected response and constructed response formats. Section three briefly describes the assessment of jobs using the Work Keys job profiling system. Section four describes joint use of job profiles and assessment results to provide meaningful information to individuals, educators, and employers.

DEVELOPING THE SKILL SCALES

Creating a Usable Metric

In assessment development, it is common for the assessment to be developed first and development of score scales to follow (Crocker & Algina, 1986; Nunally & Bernstein, 1994). This is necessary for any score scale that depends on examinee data for its construction (e.g., norm-referenced score scales) because in that case the skill scale and score scale are necessarily identical. The Work Keys *skill* scale can be conceptualized as an independent definition of the construct to be measured, a definition that is not based on the psychometric characteristics of the assessment. The *score* scale reflects the characteristics of the assessment and can be evaluated with respect to how well the assessment scores represent the designated skills and skill levels. More than one assessment approach can conceivably use the same skill scale. In the development of Work Keys, the need to link job analysis with the assessment of individuals argued for separation of the skill scale from the score scale so that both the assessment of jobs (job profiling) and the assessment of individuals could address the same skill scale. This circumstance required development of skill scales before development of the assessment and its score scale.

Several skill scale criteria were identified by Work Keys staff as critical for the operational system (McLarty, 1992). Work Keys skill levels would have to (a) be readily interpretable as a description of what the examinee can do, and the skills required by the job; (b) be appropriate for large-scale use and validation as part of a system for selecting qualified job applicants;

and (c) provide information useful for an examinee wishing to improve skills in order to meet job requirements, an educator or trainer wishing to assist examinees in improving their job-related skills, and an employer wishing to select well-qualified employees. A primary goal was that the skill scale metric should communicate the level of a generic skill the job requires and a person possesses to the test taker, the teacher, the trainer, the employer, the labor union leader, and other audiences. The Work Keys skill scale would have to communicate clearly and concisely to people making decisions based on the assessment results.

Since the scaling technique selected would serve as the basis for both the skill scales and the score scales, the choice of a suitable approach was critical. Two approaches to scaling seemed most promising. One was the scaling technique developed by Guttman in the 1940s (Stouffer et al., 1950). The other was a newer technique that was actually an outgrowth of the Guttman approach: Item Response Theory (IRT; Crocker & Algina, 1986). Because the Guttman technique was much more straightforward in its ability to communicate, it was initially decided to use that technique as the basis of the Work Keys system.

Guttman scaling is a deterministic model in which test items of increasing difficulty are administered so that the raw score of an examinee (the number of items responded to correctly) unambiguously identifies which items were answered correctly and which were not, thereby identifying the skill level at which the examinee is functioning. A Guttman scale is similar to a behaviorally anchored rating scale (Landy & Farr, 1983; Smith & Kendall, 1963) except that the score levels are defined by test items rather than by conceptual descriptions. If it could be achieved, the measurement of generic skills with a Guttman scale would yield an intuitively easy to understand method of showing what skill level a person possesses and what skill level a job requires in any given skill area.

In order for the Guttman approach to scaling to function properly, it was necessary to create skill scales that would behave hierarchically. Although there appeared to the Work Keys development team to be natural hierarchies in some skill areas, such as reading, it was less evident whether other skill areas would have this essential characteristic. The team defined a *natural* skill hierarchy as existing when it is intrinsically necessary for a person to have mastered one aspect of a skill before mastering the next above it in the hierarchy. Many aspects of reading (e.g., mastering words before sentences) and some of mathematics (e.g., mastering addition before multiplication) have this characteristic. Indeed, if the skill area was properly defined, it appeared that natural hierarchies could be observed in almost any skill area. Alternative definitions of the same skill area, however, might not result in a natural hierarchy. For example, language competency in English might be expected to form a natural hierarchy if the skill were defined broadly across all aspects of language and a wide range of skill levels. Some narrower definitions might result in multidimensionality (e.g., clusters of skills that are separate but do not form a hierarchy, such as speaking skill separated from reading skill). For development purposes, the team worked to identify strands of related expertise that appeared to be hierarchical within the skill area. In fact, development of content and cognitive strands that were broad enough to index job performance, narrow enough to be interpretable, and hierarchical enough to form the basis of a Guttman scale, was a key task of the early Work Keys development effort in each skill area.

Content strands and *cognitive strands* are terms the Work Keys development team uses to refer to constructs often used to organize test development efforts. Content strands reflect the knowledge base associated with particular skills and skill levels. For example, the Work Keys Applied Mathematics skill scale incorporates content strands associated with quantity, money, time, measurement, proportions and percentages, and averages (ACT, 1994a). The Work Keys Applied Technology skill scale incorporates content strands relating to mechanics, electricity, fluid dynamics, and thermodynamics (ACT, 1995a).

The use of cognitive strands was popularized in the *Taxonomy of Educational Objectives: Cognitive Domain* (Bloom, 1956). Work Keys uses a similar approach, but defines the cognitive levels for each skill scale specifically for that skill scale, incorporating such person-independent considerations as the number of steps necessary to reach the solution of a problem, the amount of extraneous information and other distractions in the situation, and the degree to which the content strands interact. For example, at the lower levels of Applied Technology, mechanics is a separate consideration from electricity, while at the higher levels, electromechanical interactions must be considered. This avoids potential problems associated with an item's tapping one cognitive level for a particular examinee and a different cognitive level for a different examinee, a possibility exacerbated by the very heterogeneous examinee population served by the Work Keys system. The Work Keys approach, then, is to merge the content and cognitive strands that are specific to each skill area, in order to define hierarchies that represent important characteristics of each skill area.

The Work Keys skill scales are, therefore, *intentional* scales, created to ensure that the resulting score scale will be both meaningful and hierarchical. Cognitive and content-related aspects of a skill are analyzed to identify and combine their component strands in order to generate hierarchical scales that will be meaningful to individuals, educators, and employers. Aspects of content and cognitive skills that do not contribute to the hierarchical scale are excluded from it. The quality of the resulting skill scales is then judged by the degree to which they serve as a common metric to link the job analysis to the assessments, and by their usefulness in identifying the skill levels required by jobs and possessed by individuals. Clustering approaches (identifying skills perceived as different but equally valuable) are intentionally excluded from the scale-building process, although such clusters could become the basis of separate skill scales.

Thus, the Work Keys skill levels are designed to be arbitrary but standardized, and particular to each skill. To explain, while a hat size of six is arbitrary but standardized, it is not expected to be comparable to a shoe size of six. A woman who wants to purchase a hat and shoes will need to measure both her head and her feet using scales that are appropriate for those parts, and no one would suggest that her feet are better than her head if she needs size seven shoes and a size six hat. Similarly, just as a person who needs size six shoes does not automatically need a size six hat, a Level 6 in Reading for Information is not the same as Level 6 in Locating Information. However, Level 6 in Reading for Information does mean the same skill level whether it is used to describe a job or a person, and Level 6 in Locating Information means the same thing when it is an individual's assessment score that it means when it is part of a job profile. The common metric, then, forges a link between assessment results and job analysis, but does not refer to any necessary relationship between skills.

Development Procedures

In the process of developing eight skill scales, the Work Keys development team has refined the following procedure for establishing hierarchical skill scales. Each Work Keys skill scale is developed initially by a panel of employers and educators. The panel first develops a broad definition (such as "workplace observation") of the skill area for which a scale is to be developed, identifies examples of tasks within this broadly defined skill domain, and narrows that domain to those examples that are important for job performance across a wide range of jobs (by excluding things like the observation of microscopic samples). Next, they organize the remaining tasks into strands (such as observation for the purpose of maintaining quality control). Within each strand, they order the tasks into a series of difficulty levels, with the lowest being the simplest and the highest being the most complex.

The number of levels is determined iteratively on the basis of the number of separate levels which appears to best fit the task groupings. The panel then abstracts the variables they believe cause a task to be more or less difficult. For example, less difficult observation for quality control involves (a) directed attention to one or a very small number of features which are (b) easy to differentiate from the standard and for which (c) unlimited time in which to make the determination is allowed. More difficult observation has (a) no specific direction as to what to attend to, (b) a large variety of features to be inspected simultaneously, (c) distractions, (d) a short time in which the determination must be made, and (e) very fine distinctions between the item inspected and the standard. This conceptual analysis process is repeated for each strand identified.

Finally, to facilitate combining the strands into a single test, the panel suggests which levels of each strand are at approximately the same levels of difficulty as given levels of other strands (the strands do not necessarily have the same number of levels). The exemplar tasks, the identification of elements of difficulty, and the suggested common levels across strands form the basis for creating a description of the skill area and its levels. This description is then reviewed by panel members and others and refined until it is as conceptually clear as possible.

For the eight skill scales developed to date, it appears that key elements in establishing the hierarchical scale have been communication proficiency (especially for those assessments focused on expressive or receptive language) and cognitive complexity related to reasoning, critical thinking, and problem solving. In general, content is assigned to the scale only after the cognitive dimension has been established. This is at variance with many curriculum outcome achievement tests in which the sampling of content forms the primary definitional basis.

Because parallel strands had to be merged to form the scales, the Work Keys development team elected to increase all elements of complexity simultaneously in defining the skill levels. In the case of the Work Keys skill Observation, for example, the bottom of the scale includes "direct cuing" of the element to be observed (e.g., "notice the correct placement of the silverware"), whereas only general directions about what the examinee needs to look for (e.g., "this is a properly set table") are provided at the top. Simultaneously, the scale includes no distractions at the lowest level but multiple distractions at the top level (e.g., several interruptions to the table-setting demonstration). In the real world, direct cuing may occur with, and general cuing without, the real world distractions. However, combining tasks of different complexity levels across strands might have introduced multidimensionality, compromising the hierarchical nature of the final scale and therefore its usefulness as a metric, so it was decided to intentionally covary the strands' complexity.

The same principle of combining parallel strands was employed within the assessments by covarying the complexity of stimuli and items. For Reading for Information, for example, the lowest level requires examinees to gain straightforward information from simple, direct text, while the highest level requires them to make complex inferences from dense and convoluted text. It is certainly possible to gain straightforward information from convoluted text, but including this combination could compromise the usefulness of the scale.

Work Keys scales define a series of hierarchical levels which serve as reference points for the assessment score scale and the job profiling scale. So far it has proved possible to define four to five levels clearly in each of eight skill areas. In the development process, the lowest level of each scale was pegged at about the lowest level an employer would care to assess. If less of that skill were required on the job, the employer would probably not formally assess that skill. The highest level was pegged at the highest level an employer would expect of an employee without specialized job-specific training. It is assumed that if an employer wants more of the skill than this, the employer will require or provide additional training. Once both "ends" of the scale have been established, as many levels as can be conceptually and psychometrically distinguished (jointly for skill scales, assessment score scales, and job profiling) are developed.

This approach appears to have worked well for most employers. The vast majority of jobs profiled have fallen within the range of skill levels established. However, a number of educators and a few employers have indicated the need for lower skill levels. The educators looking for lower skill levels are generally concerned with remedial needs at very basic literacy levels. The employers are addressing the need to select from pools of applicants with very limited skills. Neither group has suggested that these low levels would be considered adequate for most jobs. Should the demand be sufficient, however, lower levels of the Work Keys skill scales can be established.

DEVELOPING ASSESSMENTS

Once each skill scale had been defined, it was necessary to construct an assessment to measure individuals' skills relative to it. The Work Keys system assessments were designed to meet the following criteria.

1. The way in which the generic skill is assessed is generally congruent with the way the skill is used in the workplace.
2. The lowest level assessed is at approximately the level for which an employer would be interested in setting a standard.
3. The highest level assessed is at approximately the level beyond which specialized training would be required.
4. The steps between the lowest and highest levels are large enough to be distinguished and small enough to have practical value in determining workplace skills.
5. The assessments are sufficiently reliable for high-stakes decision making.

6. The assessments can be validated against empirical criteria.

7. The assessments are feasible with respect to administration time and complexity, as well as cost.

Most Work Keys assessments use a selected-response, multiple-choice format. Selected-response assessments have significant advantages with respect to reliability and feasibility for large-scale use. As a result, they were used wherever they could adequately model workplace use of the skill. For example, the examinee must demonstrate reading skills for Reading for Information, must calculate for Applied Mathematics, must find appropriate information for Locating Information, and must observe for Observation. It was clear that constructed response would be necessary for the Writing assessment, so Writing was developed in that format. For a variety of reasons described later, Listening was combined with Writing in the constructed-response format.

Applied technology skills might have been measured with a real-world benchwork performance assessment where examinees could demonstrate their skills by actually working on real equipment. However, that did not meet the feasibility criterion. In the selected-response format, Applied Technology can indicate whether the examinee would know what to do in a given situation, but not whether the individual can perform the skill. Presumably, however, the individual who does know what to do is more likely to do it than the one who does not. The situation is similar for Teamwork. (See chapters 15 and 16 in this volume by Grummon and by O'Neil, Chung, and Brown for alternative approaches.) A video stimulus and multiple-choice format were chosen for the Teamwork assessment, with the understanding that this assessment could indicate whether examinees would know what should be done, but not under what conditions they could or would do it themselves. Teamwork does not assess the motivation to behave in a particular way. Rather, Teamwork, like the other Work Keys assessments, is an achievement test. The examinee is asked, "If your goal is to support the team and accomplish the work task, what should you do?" The actual goal examinees would have in the situation is not considered, as that goal would be guided by their personal values and the individual company's policies and culture.

The Selected-Response Model

The different response formats for the assessments require somewhat different development approaches. For the selected-response approach, assessments are built directly on the skill scale definitions provided by the panel. Work Keys development team members first identify the number of distinct levels to be created. A draft "test blueprint" is developed from the panel's skill level descriptions. This blueprint is unlike many such documents used in education in that it is not a list of the content topics or objectives to be covered and the number of test items to be assigned to each. Rather, the blueprint is more similar to a scoring rubric used for the holistic scoring of constructed-response assessments (White, 1994). Each level (score point) is defined in terms of its characteristics, and exemplar test items are created to illustrate it. While it was sometimes appropriate to assign content to a unique level, in most cases it was the complexity of the stimulus and question that determined the level to which a particular problem was assigned. Also, it was generally the case that multiple content strands merged, especially toward the top of the scale. For example, electrical and mechanical principles appear separately at the lowest levels of the Applied Technology skill scale and assessment, while electromechanical systems appear at higher levels. Similarly, the alternatives for a single multiple-choice question may include multiple content classifications, modeling a well-integrated curriculum, yet making the typical approach to test blueprints (which assumes that each item measures one and only one objective) unusable.

The process used for the development of Work Keys multiple-choice test items is similar to that used for many standardized assessment programs, including other programs developed by ACT (Anastasi, 1982; Crocker & Algina, 1986). Both stimuli and response alternatives meet basic requirements associated with high quality achievement tests. Once the general test blueprint has been developed, ACT test specialists write sufficient numbers of prototype test items to create one full-length test form. This is administered to at least two groups of high school students and two groups of employees; typically one group of students and one of employees will be in the same city with the process replicated in another state, varying the situation (e.g., if the first site is a suburban setting, the second may be an inner city). Numbers of examinees have varied according to the test format, using more for multiple-choice than for constructed-response tests. Typically the target is at least 200 students and 30 employees divided across the two sites for each prototype test form. Examinees are asked to provide comments and suggestions about the prototype form, while educators and employers are invited to review the prototype and comment on it. Based on the information from the prototype testing, the specifications for assessment development are adjusted, and a written guide for item writers is prepared.

Work Keys test items are developed by individuals familiar with various work situations as well as with the specific skill being assessed. Item writers external to ACT are selected for their insight into the use of a particular skill in different employment settings. This approach was chosen because both content and contextual accuracy are critically important for Work Keys. A test question that contains inaccurate content may be distracting even if the specific content does not affect the examinee's ability to respond correctly to the skills portion of the question. Inaccurate facts and improbable circumstances or consequences from a series of procedures or actions are not acceptable. An examinee who is knowledgeable of the particular workplace should not identify any of the assessment content, circumstances, procedures, or keyed responses as unlikely, inappropriate, or otherwise inaccurate.

Given the wide range of employability skills assessed, verifying content accuracy for Work Keys is challenging. Assessments such as Teamwork require extraordinary efforts to verify keyed responses because different workplaces may endorse different responses to the same situation. To help Work Keys staff to detect any possible problems, the item writers write a justification for the correct response and for each distractor for each test item. Both the items and the justifications are checked and, if necessary, modifications are made.

Since Work Keys assessments are designed to be workplace relevant, depictions of the workplace in them must be realistic in all relevant respects, even when item content is not affected. Stimulus materials must be relevant and appropriate to the workplace described, and questions asked must be ones likely to be encountered in it. The assessment must avoid giving the impression that it has been created by assigning a set of content questions at random to a set of workplace settings. Each item or passage set should have an integrity of its own, blending genuine workplace needs and issues into the assessment format.

Creating realistic stimuli is a central task for Work Keys item writers because of the wide variety of workplaces represented in the assessments and the need to avoid creating an excessive reading burden (e.g., by including lengthy and detailed descriptions not essential to responding to the question). This may require narrowing the context to include only those workplace features required to answer the question and to create a relevant context for it. The item writer is required to ensure that details are realistic and that the amount of detail provided is neither so limited that the question becomes academic nor so plentiful that the examinee is overburdened. One special concern is to avoid the "academic" question, defined as one in which an individual in the situation depicted would have no reason to care about the answer. This could be because the answer to the question would not help that individual achieve his or her goals, or because an individual in that situation would already know the answer. Another challenge to test development arises in the use of video stimuli. It is critical to avoid "carry-over" from one item set to another, and for that reason ACT has used a different setting and different actors for each stimulus. Wherever possible, filming is done on location to enhance content and contextual accuracy.

Once the test questions and stimuli have been created and edited, all are reviewed by external consultants for content and contextual accuracy and for fairness to racial, ethnic, and gender groups (ACT, 1995c). Any adjustments to the test items are made prior to pretesting them. In order to provide the data required for both classical and IRT-based statistics, each item is administered to a sample of about 2,000 examinees. For practical reasons, most of these examinees have been students, although smaller samples of employees have also been assessed for each pretest. The resulting item analyses are used to create equivalent test forms in which both the overall characteristics of the test and the within-level content, complexity, and psychometric characteristics are made as similar as possible. At least two forms have been created for each Work Keys assessment. A separate step using spiraled forms administered to randomly equivalent groups is used for equating the forms (Kolen & Brennan, 1995).

Guttman-Based Scoring

Work Keys scoring and scaling, like the skill scales, is designed to be congruent with the ideas of Louis Guttman (Guttman, 1944, 1950; Stouffer, et al., 1950). The purpose of a Guttman scale is to reduce all of the information in the assessment to a simple, joint rank ordering of examinees and assessment items. From this ordering, one can predict the correct and incorrect responses of a particular examinee to the assessment items. Examinees are expected to answer correctly items below their level on the scale, and to answer incorrectly items above their level. For Work Keys, this means that an examinee is expected to be a master of all levels of skill up to and including his or her estimated level of skill, and a nonmaster of higher levels.

In Work Keys multiple-choice assessments, skill levels are represented not by a single assessment item, but by six, eight, or nine items per test form, depending on the skill area. The nature of these test items is that correct and incorrect responses to these items can never be predicted with absolute certainty based on the examinee's mastery of the skill. Some allowance must be made for this uncertainty. One such allowance was to define mastery in terms of the percentage of items an examinee should be expected to answer correctly. The criterion, recommended by a joint committee of educators and employers, was 80% of the items correct. That is, an examinee answering 80% of the items for a given skill level should be considered to have mastered that level of skill. This acknowledges the possibility that an examinee who is a master of a given skill level might, due to carelessness or some other unaccountable random error, incorrectly answer one or two items representing that level of skill on a test.

The Guttman-style inferences about a particular examinee's assigned level of skill are drawn from the pattern of within-skill-level percentage correct scores of the examinee. An examinee is expected to answer at least 80% of the assessment items correctly at each level of skill, up to and including his or her assigned level, and to answer fewer than 80% of the items correctly at each higher level of skill. Recognizing the impossibility of using an exact 80% criterion with the available numbers of items at each level, the criteria were set at five of six, six of eight, and seven of nine, of the items answered correctly at each level for the different assessments.

Guttman scaling carries a great deal of meaning and is easily interpretable by users. Several indices of scale quality, based on examinees and/or items not fitting expected patterns in the simultaneous ordering, can be computed. This "misfit" is essentially a type of error estimate reliability. Work Keys used two critical indices from the Guttman-scaling procedure: the coefficient of reproducibility and the coefficient of scalability. These indices provide an estimate of the fit of the data to the Guttman model. The coefficient of reproducibility, which has an accepted critical value of .90, indicates how well the scores are able to reproduce the item response matrix. The coefficient of scalability is used to calculate reproducibility as a percentage of improvement with respect to the maximum reproducibility possible, given item p- values (Dunn-Rankin, 1983). The threshold for the coefficient of scalability was set at .60.

While these indices describe the scale's quality at the individual item level, the focus of the Work Keys assessments is on a set of items at a particular level, rather than on an individual item. To accomplish the Guttman scaling used in the Work Keys system, it was necessary to make a series of mastery/nonmastery inferences, one for each level of skill represented on the assessment, on the basis of the examinee's within-level performance. If an examinee answered fewer than 80% of the items for a given level correctly, for example, the mastery/nonmastery score for that level would be 0; otherwise the score would be 1. An examinee's pattern of mastery on all skill levels could thus be represented by a four or five character string, depending on the number of skill levels included in the test, where each character was a 0 or a 1. For example, the string, 11000 would mean the examinee mastered the first (easiest) two skill levels, but did not master the last three (hardest) skill levels. Such a string is consistent with the Guttman inferences that the Work Keys system is designed to support.

Because responses to test items have an error (uncertainty) component, mastery/nonmastery patterns such as 01000 occurred. These patterns present a dilemma for scoring and interpretation because they are inconsistent with strict Guttman assumptions. In a perfect Guttman scale, no misfitting patterns would occur. Because of the possibility that lower levels on the skill scale could be constructed at a later date, levels on Work Keys multiple choice tests have been named so that the lowest level is Level 3, therefore five-level scales go from 3 to 7 and four-level scales go from 3 to 6. A misfitting pattern results if, for example, an examinee responds correctly to the criterion number of items at Levels 3, 4, 5, and 7, but not at Level 6 (a pattern of 11101). In theory, the examinee would not be able to meet the criterion at Level 7 without meeting it at Level 6 as well. One explanation for this occurrence is that the examinee has mastered Level 6 skills, but failed to demonstrate this fact on this particular test or occasion of measurement due to carelessness, accident, or weakness on one or more of the particular items representing Level 6 on this test form. Inferences of mastery based on six to nine selected response items are less reliable than those based on larger numbers of items.

In Work Keys, using the Guttman based scoring method, the highest contiguous level achieved is reported as the level of skill attained by an examinee. For example, an examinee who obtains a 0–1 level matrix of 11000 for Applied Mathematics would be given a "score" of Level 4. An examinee obtaining an atypical 0–1 level matrix, such as 10100 would receive a Work Keys score report indicating a score of Level 3 with a flag indicating that he or she might be capable of scoring at a higher level.

The percentage of atypical or misfitting patterns like this one is a relevant statistic by which to judge the adequacy of the scale as it is being used. It is also possible to subject the 0–1 level matrix to a Guttman-style analysis in which the levels are treated as if they were individual items. This 0–1 level matrix yields Guttman indices of nearly 1.00 in most cases. Table 12.1 contains both the item- and level-based coefficients of reproducibility and

TABLE 12.1
Guttman Indices for Item and Level 0-1 Matrices and Percentage of Atypical Profiles for Multiple-Choice Assessments

Assessment	Number of Levels	Item 0-1 Matrix[a]		Level 0-1 Matrix[b]		% of Atypical Profiles
		CR	CS	CR	CS	
Applied Mathematics	5	.90	.61	.99	.91	6.0
Applied Technology	4	.87	.57	.95	.89	4.0
Locating Information	4	.90	.58	.99	.93	3.0
Reading for Information	5	.91	.59	.98	.90	9.0
Teamwork	4	.88	.56	.96	.89	11.0

Note. CR = Coefficient of Reproducibility (criterion = .90). CS = Coefficient of Scalability (criterion = .60).
[a]The item matrix, although not used by ACT for Work Keys assessment analysis, represents traditional Guttman scaling (Guttman, 1944, 1950; Stouffer et al., 1950).
[b]The level matrix treats level mastery as if each level was an item. This is the approach used by ACT for Work Keys assessment information analysis.

of scalability based on the responses of a sample of more than 14,000 eleventh- and twelfth-grade students for each of the five multiple-choice, dichotomously scored Work Keys assessments. Table 12.1 also presents the percentages of atypical profiles for the five Work Keys selected-response assessments.

The Guttman-based approach resulted in skill scales that had very desirable characteristics with respect to interpretability. Scoring on the level basis, as was done operationally, resulted in very high coefficients for both scalability and reproducibility. The lower coefficients for item-level scaling were expected because items within each level would be sufficiently close in difficulty not to scale well. However, the score reliability for the selected-response tests based on the 80% within-level scoring rules was lower than desired (see Table 12.3 on p. 311).

IRT-Based Scoring

After careful study, it was determined that an Item Response Theory (IRT) strategy (Crocker & Algina, 1986; Hambleton, Swaminathan, & Rogers, 1991; Nunnally & Bernstein, 1994) could be used to (a) retain the content-defined scale levels necessary to provide the critical links between the skill scales, the assessments, and the job profiling while (b) changing scoring for the selected-response assessments to a probabilistic approach focused on applying the 80% rule to estimated performance on the within-level item pools. For additional information regarding this change, see McLarty and Schulz (1996). IRT accommodates the uncertainty inherent in using small numbers of test items in predicting examinees' percentage correct scores for the level as a whole, because IRT gives the probability of a correct response to each test item. If the model fits the data, these item-response based probabilities are an accurate and reliable basis for making inferences about an examinee's mastery of a skill level. An expected percentage correct score over a collection of items can be obtained by taking 100 times the average probability of a correct response to the items.

This provides a theoretical and empirical basis for using all of the items on the test, not just those representing a particular level of skill, to estimate an examinee's mastery of a skill level. That is, all of the items, not just the Level 3 items, can be used to infer an examinee's mastery at Level 3 of a skill area. The theoretical basis for this is the unidimensionality assumption that is central to both Guttman and IRT scaling. Unidimensionality means that all of the items on the test work together to measure the same thing (Crocker & Algina, 1986). If an examinee takes only Level 4 items and performs very well on them, one can be reasonably sure that the examinee is a master of the same skill at Level 3 as well. If the examinee answers slightly fewer than 80% of the Level 3 items correctly, this discrepancy might be attributable to random error, due to the specific representation and small number of Level 3 items on the test. The unidimensionality assumption is used in interpreting an examinee's skill level: As with Guttman scaling, an examinee is presumed to be a master of all levels of skill up to and including the reported level, and a nonmaster of higher levels. This assumption can be applied to the total score to make more reliable inferences about an examinee's mastery than can be made by considering only performance on items representing a particular skill level.

To do this, the Work Keys development team used IRT (Hambleton et al., 1991) to estimate a "latent ability" value, called theta (θ), that corresponds to the total score an examinee earns on the test. At the same time, the statistical properties of the individual test items in the pool of items for all test forms were estimated. Since alternate forms of each test were administered to randomly equivalent groups, the statistical properties of test items from different test forms were expressed on a common IRT scale. The probability of a correct answer to each and every item representing each skill level, given the items' statistics and a theta corresponding to the total score earned by an examinee on a particular test form, was calculated. These probabilities were averaged to give the expected percentage of items an individual with a given total score would correctly respond to at each skill level.

Table 12.2 shows the relationships between total raw scores on Form 10AA of Reading for Information (ACT, 1992b) and the expected percentage correct score on the within-level item pools for that skill area. Total raw scores below 7 are not considered because this level of performance is below that expected by chance, (i.e., by guessing). Examinees scoring below "guessing" would be assigned to skill level "Below 3." An examinee who earns a total score of 14 is expected to get 83% of Level 3 items correct, 63.1% of Level 4 items correct, and 42.1% of Level 5 items correct, and so on. This pattern of percentage correct scores corresponds to a Guttman pattern of 10000, so the level of skill to be reported for such an examinee would be Level 3. A total score of 14 is the lowest total score for which Level 3 mastery is expected, based on the 80% correct criterion, so 14 can be considered the cutoff score for classifying an examinee as Level 3 or higher. The lowest total raw score for which mastery of Level 4 is expected, based on the 80% correct criterion, is 18, so 18 becomes the cutoff score for Level 4.

Cutoff scores for other skill levels are determined in a similar fashion. By using this approach, one can evaluate an examinee's performance (expected percentage correct) with respect to the entire pool of items at each level, rather than with respect to only the sample of items on a given form of the test.

The expected percentage correct corresponding to the cutoff score for a given skill level was selected to be *closer* to 80% than that of any other total score. The expected percentage is slightly less than 80% about half the time so that, on *average*, the operational criterion equals the targeted

TABLE 12.2
Item Pool Percent Correct by Level and Total Number Correct

Work Keys Assessment
Reading For Information Form 10AA

Item Pool Level

Score Level	Total Items Correct	3	4	5	6	7
Below 3	7	34.2	29.9	24.5	18.9	19.9
	8	43.6	35.1	26.8	19.5	20.0
	9	52.1	39.8	29.0	20.2	20.2
	10	59.8	4.3	31.3	20.8	20.4
	11	66.7	48.8	33.7	21.5	20.5
	12	72.9	53.4	36.3	22.3	20.8
	13	78.3	58.2	39.2	23.3	21.0
3	14	83.0	63.1	42.4	24.4	21.3
	15	86.9	68.1	46.1	25.9	21.6
	16	90.2	73.1	50.3	27.7	22.1
	17	92.8	78.0	54.9	30.2	22.7
4	18	94.9	82.6	60.1	33.5	23.5
	19	96.5	86.8	65.6	38.1	24.5
	20	97.7	90.4	71.4	44.1	26.1
	21	98.5	93.3	77.0	51.7	28.2
5	22	99.1	95.4	82.1	60.0	32.1
	23	99.4	96.9	86.5	68.1	34.9
	24	99.7	98.9	90.1	75.3	40.0
6	25	99.8	98.7	93.1	81.5	46.7
	26	99.9	99.2	95.4	86.6	55.8
	27	99.9	99.6	97.1	90.8	67.4
	28	100	99.8	98,4	93.9	79.7
7	29	100	99.9	99.2	96.0	90.7
	30	100	100	99.6	97.0	96.9

criterion of 80%. Another responsibility in establishing cutoff scores is to make sure the cutoff score represents equal difficulty across forms. This means that the cutoff scores occasionally, collectively represent a slightly higher or lower standard than 80% for a given test form in order to ensure comparability across test forms.

Table 12.2 reveals two very important trends in the percentage correct scores. First, within any given level of skill, the expected percentage correct score increases with the total score. Second, for any given total score, the expected percentage correct score decreases as the skill level increases. The first trend shows that mastery of a given skill level is positively related to the total score. It is essential to establish this before using the total score as the basis for assigning examinees to a level of skill. The second trend means that all of the expected patterns of mastery/nonmastery across skill levels will be consistent with Guttman-style interpretations. This approach was successful because of the unidimentionality and hierarchical characteristics built into the Work Keys tests during the development process. Thus, these data support the interpretation that an examinee is a master of skill levels up to and including his or her reported level, and a nonmaster of higher levels.

The chief advantage of the IRT-based number correct scoring strategy is its higher reliability. The reliability of IRT-based number correct scoring and highest contiguous mastery scoring were estimated using the IRT model. Because both methods are essentially based on number correct scores (either summed over all the items on the test or summed within levels), IRT can be used to estimate the reliability of both methods, providing the model has an acceptable fit to the data. For both methods of scoring, the IRT model was used to estimate the error variance and total variance of level scores, and the reliability coefficient was estimated as 1 minus the ratio of error to total variance (Crocker & Algina, 1986, p. 352).

Table 12.3 shows the estimated reliability of the IRT-based number correct method and the highest contiguous mastery rule (Guttman-pattern scoring) for assigning examinees to skill levels. The IRT-based number correct method is substantially more reliable. On the 0 to 1 scale for reliability, differences of 0.1 as seen for Reading for Information, Form 10AA, are statistically and practically significant. Notice, however, that the reliability of level scores, even those obtained by the number correct method, is substantially lower than the reliability of number correct scores. This is due to the reduced number of score points in the level scale.

Another advantage of the IRT-based number correct method of level scoring is that with it, the 80% correct standard can be implemented more precisely than it can be with the within-level scoring method. With 6, 8, or 9 items per level on Work Keys assessments, the closest one can come to implementing an 80% standard is 83% (5 correct out of 6), 75% (6 correct out of 8) or 78% (7 correct out of 9). The expected percentage correct corresponding to total scores generally comes closer to 80% because of the larger number of possible total scores on the test, compared to the small number of within-level total scores.

The most important, practical indicator of the fit of the IRT model to the Work Keys data is that examinees' actual within-level performance was

TABLE 12.3

Internal Consistencies Obtained From the Guttman Scaling and IRT-Based Number Correct Scoring and the Score Distributions for Five Work Keys Selected-Response Assessments

	Applied Mathematics	Applied Technology	Locating Information	Reading for Information	Teamwork
Number of Items	30	32	32	30	36
Level score reliability: pattern scoring[a]	.75	.63	.69	.62	.45
Level score reliability: IRT approach	.80	.72	.73	.72	.60
Total score reliability: IRT approach[a]	.85	.77	.82	.79	.70
	Percentage of Examinees Across Forms by Skill Level (pattern/IRT)				
< Level 3	7/8	51/70	14/20	5/6	10/13
Level 3	20/21	29/21	20/25	10/7	26/36
Level 4	29/32	14/7	46/46	29/38	35/32
Level 5	31/27	5/1	19/10	35/30	26/19
Level 6	10/10	0/0	1/0	18/17	3/1
Level 7	2/3			2/2	
Mean level score	4.22/4.19	2.73/2.41	3.72/3.45	4.55/4.51	3.85/3.59

[a]Based on the currently operational form.

statistically consistent with their expected percentage correct score. If the expected percentage correct score was above 80, examinees usually answered at least 5 out of 6, 6 out of 8, or 7 out of 9 of the items representing that skill level, correctly. Discrepancies did occur, but no more often than one would expect, given the response probabilities through the IRT model. When discrepancies exist, inferences based on the total number correct score are equally valid in theory, but more reliable than those based on within-level scoring.

Another practical indication that the IRT model fits the data is the match between empirical and predicted frequency distributions. An *empirical* frequency distribution is the proportion of the pool of examinees scoring at each skill level when number correct or highest contiguous mastery scoring is applied to a set of data. A *predicted* frequency distribution is the proportion of examinees scoring at each skill level that is predicted when the behavior of the scoring rule (number-correct or highest contiguous mastery) is modeled by use of the IRT item statistics and examinee ability (theta) distribution. The IRT analysis does not dictate a match between these distributions, so a match between them is one indication that the IRT statistics can be relied upon to predict other practical outcomes, such as the hypothetical test/retest reliability coefficients in Table 12.3.

For the many reasons described, scoring for the Work Keys selected response assessments has been changed from the deterministic model (item-based pattern scoring) to the probabilistic model (total score-based IRT approach). Since scores continue to be reported as Work Keys levels, this change is not obvious to examinees and others using the score reports. To ensure the continuity of trend data, the comparability of scores for individuals in the same applicant pool, and the appropriateness of pre-post comparisons, however, ACT has recalculated old scores using the new procedures.

The Constructed-Response Model: Work Keys Listening and Writing

To make efficient use of examinee time while modeling a workplace situation in which employees receive and record information in order to transmit it to others, the Work Keys development team chose to combine the assessments of Listening (receptive language) and Writing (expressive language) in a single administration. The Work Keys Listening and Writing assessment functions as a single assessment but generates two separate scores. For this assessment, examinees listen to six prompts on audiotape and are encouraged to take notes while each prompt is played twice. Then they are instructed to write out the message, as if for another person, using proper business English. The resulting response can be scored for Listening (accuracy of information), Writing (adequacy of prose), or both (ACT, 1994b, 1995b).

Creating a hierarchical scale for constructed-response assessments, such as the Work Keys Listening and Writing assessment, turned out to be challenging because the elements of complexity (how hard is the question?) and competency (how well can the examinee respond?) are both explicit. In contrast, in a selected-response assessment, the complexity of the test question is explicit, and is comprised of the complexity of the stimulus and question as well as the resemblance among the alternative choices. The competency of the examinee's response is inferred from the number of correct responses or, for IRT, from the probability of a correct response conditional on the complexity (difficulty) of the problems posed. For many selected-response testing programs, the effects of the competency of the examinee and the complexity of the problems presented remain both blended and transparent in the reported score. This makes it easy to ignore the complexity of the problems (test items) as a separate factor, unless one wishes to create a hierarchical scale, as Work Keys has done.

In a constructed-response situation, both complexity and competency factors are explicit and require investigation. The complexity of the problem is determined in test construction by selection of the stimulus and the question posed, and the competency of the response is judged by directly applying scoring criteria to the examinee's response. The Work Keys development team chose to create prompts at different levels of complexity for the Listening and Writing assessment. Four levels of prompts were created by increasing the length (number of words) and the number of pieces of information included at each level. The scoring rubrics were focused on the competency of the examinee's response: how much of the information was captured accurately for Listening, and the adequacy of the prose (grammar, syntax, spelling, and the like) for Writing.

Pretesting of large numbers of prompts showed a modal correlation between Listening and Writing scores of .52, suggesting that there was enough difference between the scores to infer two distinct, if probably correlated, skills. The pretest study also examined the relative impact of both dimensions (prompt complexity and examinee competence) on examinee scores for each skill area, in order to determine the best way to combine these dimensions when creating the hierarchical level scales. Analysis indicated that very little score variance for either skill was associated with the complexity of the prompts (Vansickle, 1992). Examinees performed similarly across prompts ranging in complexity in the Listening and Writing assessment for both Listening and Writing scores. As a result, the decision was made to score both skills based on overall scores, without differentiating based on the prompt complexity level. This is analogous to scoring a total test comprised of items of heterogeneous difficulty based on one point per item rather than weighting the more difficult items differentially. The Listening and Writing level scores were thereby linked solely to the examinee competence dimension.

Work Keys multiple-choice assessments were given arbitrary level numbers beginning at Level 3, to acknowledge that the lowest score was well above "none" of the skill, and to allow for later construction of lower levels without recourse to negative numbers should those be needed. That strategy was not appropriate for the constructed-response assessments. Based on the scoring rubric, which had five scale points, the score scale was assigned scores of Levels 1 to 5, with 0 indicating no response or an unscorable paper (off topic, profane, not in English, or the like). The scale level assigned to an examinee was the highest contiguous level at which the examinee received at least 9 of the 12 possible scores (two scorers for each of six prompts) in order to approximate the 80% criterion used for the other tests. Because of the scoring method, atypical score patterns could not occur for the constructed response assessment (Vansickle, 1992).

Use of the combined Listening and Writing assessment raised questions about the effects of each skill on the assessment of the other. Because individuals who received very low Listening scores would not have retained enough of the prompt to receive "on topic" scores for Writing, a criterion minimum Listening score of Level 1 was established to help ensure a valid score in Writing. Conversely, although not strictly parallel, it was determined that an individual with a Writing score below Level 2 would not be able to write well enough to provide the scorer with sufficient information on which to base a valid Listening score. Therefore a criterion score of Level 2 in Writing was established to help ensure valid Listening scores. In addition, ACT recommended that the Listening and Writing assessment be used for selecting employees only where the related job required both skills, since requiring either skill for the assessment but not for the job would be inappropriate (Cascio, 1982; Dunnette & Hough, 1990).

Generalizability Analyses

As soon as operational and research data were available, a series of generalizability studies was conducted for the Listening and Writing scores. Generalizability studies partition the effects of different components (facets) of the assessment process on the variance of examinee scores. Ideally, the skill of the *Person* is the predominant effect, with little or no effect due to the specific test *Form* the person took, the *Tasks* within that form, the *Raters* who scored the written responses, or any of the interactions among these (Brennan, 1992a, 1992b; Shavelson & Webb, 1991). Results of three of the generalizability studies are reported in Table 12.4. Addi-

TABLE 12.4
Generalizability Results for Listening and Writing Scores

Study 1
(D-Study, N = 7097, Raw Score 0-5 Scale)

Effect	Listening		Writing	
Persons	.207	20%	.441	45%
Tasks	.202	20%	.070	7%
Raters	.000	0%	.000	0%
Persons by tasks	.499	49%	.289	30%
Persons by raters	.006	1/2%	.029	3%
Tasks by raters	.000	0%	.000	0%
Persons by tasks by raters	.108	10 1/2%	.146	15%

Study 2
(Bootstrapped Analysis Using Level Scores; Same Data as for Study 1)

Effect	Listening		Writing	
Persons	.314	62%	.543	81%
Tasks	.031	6%	.010	1%
Persons by tasks	.160	32%	.118	18%
Reliability				
G[a]	.663		.822	
PHI[b]	.622		.809	

Study 3
(Forms Effects Using Level Scores; N = 167)

Effect	Listening		Writing	
Persons	.263	50%	.406	71%
Forms	.088	17%	.000	0%
Persons by forms	.174	33%	.166	29%
Reliability				
G[a]	.602		.710	
PHI[b]	.502		.710	

[a]G is a reliability-like coefficient appropriate for relative (e.g., norm referenced or ranking) decisions (Brennan, 1992a, 1992b; Shavelson & Webb, 1991).
[b]PHI is a reliability-like coefficient appropriate for absolute (e.g., criterion referenced) decisions, such as assigning examinees to Work Keys levels.

tional studies are reported by McLarty, Gao, and Stientjes (1996). Studies 1 and 2 used operational data from 7,097 high school juniors and seniors in one state who all took the same test form. The correlation between Listening and Writing scores for this sample was .52. Study 1 was a D-study (decision study) that estimated the effects associated with using two raters to score each of six tasks. Results are shown using the raw score scale. The Writing assessment had 45% of the effects associated with Persons, with an additional 30% associated with the interaction of Persons with Tasks, indicating that some Persons performed differentially better on some tasks than on others. This is a common effect reported in the literature on performance assessment of writing skills (cf. Brennan, Gao, & Colton, 1995). In Study 1, the Listening assessment showed greater Task effects (20%) as well as a much larger Person by Tasks (49%) effect than appeared for the Writing assessment, resulting in a smaller percentage of the effect for Persons (20%). Recall that since the same tasks and the same responses were being scored for both skills, these results also support the notion that two distinct skills are being assessed.

Study 2 used the same data as Study 1. It investigated the effect of using examinees' average scores across prompts and raters, rounded to the nearest integer, instead of the pattern-based scoring described above. Since each examinee received only one score per skill area, a bootstrapping method was used to estimate the reliability of the reported scores (Efron, 1982). Results indicated higher reliability for the Writing scores than for the Listening scores using this method. The effects associated with these reliabilities are similar to those for Study 1, with much larger Person by Tasks effects for Listening than for Writing scores.

Study 3 used results from a separate research study in which each examinee took two of the three test forms, in counterbalanced orders (Gao, 1996). This study was designed to evaluate forms effects, in order to determine whether equating would be required in order to report results from the multiple forms on a common scale. Forms effects were noted for Listening and, based on these results, equating will be used to ensure that the scores reported from the different forms are comparable. Person by Forms effects were noted for both Listening and Writing scores, with associated reductions in the reliability coefficients.

It is hypothesized that there is probably some aspect of Listening that is not fully circumscribed by the test specifications. Three possibilities are (a) effects of the complexity levels of the prompts could have reappeared now that scoring has been improved and rater effects are nearly nonexistent; (b) something inherent to Listening skills, and possibly analogous to modes of discourse in Writing, might be affecting individual examinees differentially; and (c) because Work Keys uses workplace contexts from across a variety of settings, differential familiarity with particular workplace settings might be affecting individuals' responses. These possibilities remain to be investigated.

Based on Study 2 and on studies that were carried out on the operational data, it was found that the reliability of the scoring could be improved by using an average score rather than the pattern-based approach originally selected. Even where the average score was rounded to an integer to provide the level score, this improvement was significant. For Listening, reliability was estimated at .41 for the pattern score, .56 for unrounded averages, and .50 for rounded averages. For Writing, reliability was .68 for the pattern score, .82 for unrounded and .71 for rounded average scores. Accordingly, Work Keys has modified the score conversion process for the Listening and Writing assessments. However, rather than using rounded average scores, the reporting scale is based on a total score summed over six prompts and two raters with a cutoff for each level one standard error of measurement below the total score that would be associated with full mastery of that level. For example, full mastery at Level 3 would be associated with a total score of 36 (12 scores of 3). The cutoff score for Level 3 is therefore 33 for Listening (based on a standard error of 3) and 34 for Writing (based on a standard error of 2).

JOB PROFILING

It is common to conduct a job analysis to see what tasks are required for a particular job and to then use the results to build a content valid test for that job (Cascio, 1982; Dunnette & Hough, 1990). This approach works best where a test will be built specifically for each job. This is not the case with the Work Keys system, where the same test is designed to assess generic skills and skill levels associated with many different jobs. Thus, while general information about the types of generic skills and skill levels needed in the workplace was used in developing the skill scales, the primary uses of job analysis by Work Keys are to (a) establish the content relatedness between a specific job and the existing Work Keys skill area and, if the job and skill are found to be content related, to (b) establish the Work Keys skill level required by the job. This latter task in effect establishes a skill standard for that job.

As with other judgment-based standard-setting methods, job profiling (job analysis relative to Work Keys skill scales) must address the key issues of who should make the judgments, on what basis, and by what method. Because those setting the standards must be intimately familiar with the job, the decision was made that the subject matter experts should, wherever possible, be job incumbents. Although some companies choose to include supervisors and trainers as subject matter experts, ACT recommends that the standard-setting group be comprised predominately of individuals who are doing or have recently done the job that is being profiled (ACT, 1995d; Anastasi, 1982).

The subject matter experts make their judgments based on their knowledge of the job under consideration and on a knowledge of the Work Keys skills and skill levels. The latter is information which is provided by Work Keys job profilers during job analysis. Because of the depth of understanding that must be gained by the subject matter experts before they can accurately align the skills they perform in their jobs with the Work Keys skill scales, the job profiling process is carried out in small group settings, much like focus groups. Each group of about eight subject matter experts is guided by an authorized Work Keys job profiler. Potential job profilers are selected based on relevant previous training and experience and are authorized by ACT after completing a week of intensive training in the job profiling process and the Work Keys skills and skill levels.

Job profiling is conducted in two phases: task analysis and skills analysis. In task analysis, the subject matter experts identify the tasks which are accomplished on the job and then rate them for both importance and relative frequency. These two ratings are multiplied together to create a criticality index that is used to order the tasks. The subject matter experts review the ordered task list, make any changes they feel are necessary to appropriately represent the job, and then narrow the list to about 30 of the most critical tasks. The discussion of these tasks provides a shared understanding of the job that is critical for the second phase of the analysis, skills analysis.

In skills analysis, the subject matter experts first link the Work Keys skills (such as Teamwork) to the task list, identifying each task that requires the particular skill. Once the set of tasks requiring that skill has been identified, the subject matter experts, using successive approximation, determine which level of the skill is required for that set of tasks. The subject matter experts begin with a description of a skill level that the job profiler believes is just below the level needed on the job. They determine whether their job requires skills which are above, below, or about the same as the level described. If they feel that the skills they must have are higher, they are shown the next higher level; if they feel the skills they must have are lower, they are shown the next lower level. If they feel that the skills they must have are about the same, they are shown both the next higher and the next lower levels. No decision is reached until the subject matter experts have considered at least three levels of the skill: the one they have chosen and at least one above and one below it (unless they have chosen the highest or lowest level available). Subject matter experts also have the option of determining that the level required is below or above the levels measured by Work Keys.

Job profiling requires about half a day to orient the subject matter experts to the process and to develop the task list. The efficiency of this process results from the job profiler already having observed the job, reviewed a job description, and completed a preliminary task list using ACT's SkillPro software (ACT, 1995e). Profiling the job relative to individual skills requires varying amounts of time, with the first always requiring additional time as the subject matter experts learn the skills analysis process. Four to five Work Keys skills can be analyzed in about 4 hours, with all eight requiring about 7 to 8 hours. Thus, it is possible to complete a full job analysis using all eight skill areas in less than a total of 100 hours of subject matter expert time. If the resulting profile is to be used for high stakes purposes such as selecting individuals to fill positions, ACT strongly recommends that the process be replicated with at least one additional, independent group of subject matter experts to ensure that the resulting profile is both accurate and reliable. Larger and more diverse groups of positions within a job (i.e., where individuals with the same job title perform somewhat different functions for the company) may require additional profiling sessions to adequately model job skill requirements. This job profiling process addresses the content validation standards of the Equal Employment Opportunity Commission (1978).

Based on the work of several hundred authorized profilers, ACT is rapidly developing an extensive database of job profiles. Based on 446 jobs analyzed, 91.7% had profiles within the Work Keys Applied Technology skill scale. For the remaining skill areas, with 444 to 613 jobs profiled, the percent of profiles within the skill scale ranged from 96.5% for Applied Mathematics to 99.2% for Reading for Information.

While each profile is useful for selection purposes only to the company that conducted the profiling, summaries of profiles are helpful to educators as they guide individuals preparing for jobs. To support this use, ACT has published *Work Keys Occupational Profiles* (ACT, 1997).

Additional analyses of profiles may also be helpful. For example, Table 12.5 provides an expectancy table analysis for the job of Maintenance Mechanic based on 29 job profiles in the database. A compilation like this one can be informative to educators planning applied academics curricula to prepare students for specific occupations. Importantly, the expectancy table makes the critical point that while there is some consistency in the requirements of particular jobs, there is also considerable heterogeneity. A Maintenance Mechanic, for example, is likely to need a high level of skill in Applied Technology, but the Teamwork requirements may vary with the specific job setting.

While the jobs profiled to date do not represent a systematically collected sample of jobs nationally or for any specific region, it would be possible to select and weight them so as to comprise such a sample. As additional

TABLE 12.5
Job Skill Requirements Expectancy Analysis for Maintenance Mechanic

DOT Code: 638.281-014

World-of-Work Family L: Industrial Equipment Operations and Repair

The table below is based on an analysis of the 29 Work Keys profiles for that job (28 job profiles and 1 occupational profile) which were in the Work Keys database on September 15, 1995. These profiles come from eight states.

How to read this table. Percentages are based on the jobs profiled in that skill area. For Reading for Information (RI) there were 29 jobs profiled, none of which required reading skills below the lowest level for that skill, which is Level 3. In fact, none of them required skills at Level 3, but 13% of the jobs required Level 4 skills, 60% required Level 5 skills, 17% required Level 6 skills, and 10% required Level 7 skills (the highest measured by the Work Keys scale). RI = Reading for Information; AM = Applied Mathematics; L = Listening; W = Writing; LI = Locating Information; TW = Teamwork; AT = Applied Technology.

Skill Area	RI	AM	L	W	LI	TW	AT
N of Profiles[a]	29	29	25	25	26	23	25
Level	Percent of Profiles by Skill Level[b]						
Below lowest or not needed	0	0	0	0	0	0	0
1			0	4			
2			0	22			
3	0	6	8	30	4	4	4
4	13	16	40	33	41	8	15
5	60	31	52	11	41	8	15
6	17	34			15	63	59
7	10	9					
Above highest	0	3	0	0	0	0	7

[a]Some skills were not profiled for some jobs/occupations.
[b]Total percentages may not add to 100% because of rounding.

profiles are completed, it should also be possible to detect regional or industry specific variations that might have an impact on training needs, and to provide an overview of the skills needed.

USING ASSESSMENTS AND PROFILES TOGETHER

Because Work Keys job profiling and the Work Keys assessments address the same skill scales, it is possible to compare assessment results with job requirements in a direct and intuitively comprehensible manner. This comparison is designed to support sound selection practices by an employer who profiles the job and then uses assessment results as part of the employment decision process to establish a pool of individuals who meet the minimum skill levels identified by the profile. Such comparisons relate the profile of one job to the skills and skill levels of each individual applicant.

It is also possible to aggregate skills profiles over both jobs and individuals. Work Keys job-skill comparisons compare the percentage of jobs profiled at each level with the percentage of examinees whose assessment results were at that level. Table 12.6 is a job-skills comparison for seven Work Keys skills. These data should be interpreted with great caution because neither the examinees nor the jobs included represent a systematically collected sample of any particular population. Most of the examinees are high school students, and a disproportionate number of jobs are in manufacturing and related occupations.

For Reading for Information, the percentages of examinees and jobs by level are relatively symmetrical, with a larger proportion of examinees than jobs at the very top level. In each of the remaining skill areas, job requirements exceed examinee skill levels across the scale. These differences are smallest for Applied Mathematics and Teamwork and largest for Listening and Applied Technology. Recall that given the nature of the Work Keys skill scales, these high skill levels typically do not require additional content coverage. Rather, increased skill is indicated by the ability to use the content in more complex applications. The highest levels typically require separating relevant from irrelevant information; dealing with unfamiliar terms, procedures, or formats; and troubleshooting procedures that have been done incorrectly. These are skill levels which the examinees in the sample, including college-bound juniors and seniors in high school, appear not to have mastered.

CONCLUSION

The Work Keys system is unique in that it provides external referents, empirical job analysis, and assessment results. These can be used to guide individuals as they prepare for the workforce, educators and trainers as they assist in this preparation, and employers as they seek qualified applicants. The focus on generic skills is consistent with the growing need for individuals to continue to learn on the job. Job specific skills, properly aligned to the job, meet day-one performance requirements. Generic skills provide a base for company-specific learning, additional generalization of job specific skills to address new technology and procedures, and the development of new skills to support occupational growth and change. These are critical to establishing and maintaining healthy businesses and satisfying careers.

TABLE 12.6
Number and Percentage of Jobs Profiled at Each Level for the Eight Operational Assessments

	Reading for Information		Applied Mathematics		Listening		Writing		Locating Information		Teamwork		Applied Technology		Observation	
	No.	%	No.	%	No.	%	No.	%	No.	%	No.	%	No.	%	No.	%
Level 7	24	3.95	47	8.06												
Level 6	74	12.17	103	17.67												
Level 5	226	37.17	121	20.75	229	44.90	55	10.91	56	10.65	142	32.49	108	26.41	3	30.00
Level 4	189	31.09	162	27.79	180	35.29	148	29.37	199	37.83	82	18.76	77	18.83	3	30.00
Level 3	95	15.63	150	25.73	88	17.25	203	40.28	202	38.40	112	25.63	122	29.83	3	30.00
Level 2					8	1.57	82	16.27	69	13.12	101	23.11	102	24.94	1	10.00
Level 1					5	0.98	16	3.17								
Level 0					0	0	0.00	0	0.00							
Total	608		583		510		504		526		437		409		10	

Number and Percentage of Examinees Scoring at Each Level for the Eight Operational Assessments

	Reading for Information		Applied Mathematics		Listening		Writing		Locating Information		Teamwork		Applied Technology		Observation	
	No.	%	No.	%	No.	%	No.	%	No.	%	No.	%	No.	%	No.	%
Level 7	7,817	9.30	1,918	1.53												
Level 6	7,261	8.63	11,574	9.26												
Level 5	32,644	38.82	28,846	23.07	8	0.03	9	0.03	81	0.09	4,071	14.88	138	0.19	0	0.00
Level 4	23,074	27.44	37,480	29.97	1,111	4.19	4,313	15.78	9,264	9.95	8,480	30.99	2,105	2.89	0	0.00
Level 3	11,732	13.94	29,942	23.94	15,210	57.32	13,088	47.88	44,414	47.70	8,480	30.99	6,223	8.53	0	0.00
Level 2	1,573	1.87	15,285	12.22	8,968	33.80	7,972	29.16	22,341	23.99	5,958	21.77	17,982	24.66	0	0.00
Level 1					868	3.27	1,512	5.53	17,010	18.27	376	1.37	46,468	63.73	0	0.00
Level 0					368	1.39	443	1.62								
Total	84,092		125,045		26,533		27,337		93,110		27,365		72,916		0	

The Work Keys system links individuals, educators, and employers in a variety of mutually supportive arrangements. In developing assessments and job profiling to meet the high-stakes requirements of employee selection, Work Keys has met or exceeded the requirements of a wide variety of other applications such as establishing standards for educational or training programs, evaluating such programs, certifying skill mastery, and identifying the training needs of individuals. By distributing these programs and services through a national network of Work Keys Service Centers, ACT has made them available locally as well as through ACT's centralized services.

Additional work remains to be done. "On the drawing boards" are several additional generic skill areas. This and other Work Keys development will be informed by the ongoing results of the National Job Analysis Study as they become available (see Nash & Korte, this volume). Additional efforts in progress address further coordinated use of occupational and job-specific job profiling and skills assessments, the use of computer-administered and computer adaptive assessments, links to the ACT World-of-Work Map (Vansickle & Prediger, 1990) to support career counseling and to other ACT programs, and a variety of enhancements to the current system. Also in progress are additional empirical, criterion-related validity studies and a variety of other research efforts. The results of these studies will guide further development of the Work Keys system and its appropriate use by individuals, educators, and employers.

ACKNOWLEDGMENTS

The authors would like to thank Xiaohong Gao and Matthew Schulz for the analyses and insights they provided with regard to the generalizability of Listening and Writing scores and to the conversion from Guttman to IRT-based scoring, respectively.

REFERENCES

Agency for Instructional Technology. (1989). *Workplace readiness: Education for employment. Personal behavior, group effectiveness and problem solving for a changing workplace.* Unpublished report, Bloomington IN: Author.

American College Testing. (1987). *Study power* [Series]. Iowa City, IA: Author.

American College Testing (1992a). *A strategic plan for the development and implementation of the Work Keys system.* Unpublished report, Iowa City, IA: Author.

American College Testing. (1992b). Work Keys: Reading for information, form 10AA [Assessment]. Iowa City, IA: Author.

American College Testing. (1994a). *Applied mathematics* (Work Keys: Targets for instruction series). Iowa City, IA: Author.

American College Testing. (1994b). *Writing* (Work Keys: Targets for instruction series). Iowa City, IA: Author.

American College Testing. (1995a). *Applied technology* (Work Keys: Targets for instruction series). Iowa City, IA: Author.

American College Testing. (1995b). *Listening* (Work Keys: Targets for instruction series). Iowa City, IA: Author.

American College Testing. (1995c). *Work Keys guide for reviewers.* Iowa City, IA: Author.

American College Testing. (1995d). *Work Keys SkillPro* [Computer software]. Iowa City, IA: Author.

American College Testing. (1997). *Work Keys occupational profiles.* Iowa City, IA: Author.

Anastasi, A. (1982). *Psychological testing* (5th ed.). New York: Macmillan.

Bailey, L. J. (1990). *Working: Skills for a new age.* Albany, NY: Delmar.

Bloom, B. S. (Ed.). (1956). *Taxonomy of educational objectives: The classification of educational goals. Handbook 1. Cognitive domain.* New York: McKay.

Brennan, R. L. (1992a). *Elements of generalizability theory* (rev. ed.). Iowa City, IA: American College Testing.

Brennan, R. L. (1992b). Generalizability theory. *Educational Measurement: Issues and Practice, 11,* 27–34.

Brennan, R. L., Gao, X., & Colton, D. A. (1995). Generalizability analyses of Work Keys Listening and Writing Tests. *Educational and Psychological Measurement 55*(2), 157–176.

Carnevale, A. P., Gainer, L. J., & Meltzer A. S. (1990). *Workplace basics: The essential skills employers want.* San Francisco: Jossey-Bass.

Cascio, W. F. (1982). *Applied psychology in personnel management* (2nd ed.). Reston, VA: Reston.

Center for Occupational Research and Development. (1990). *Applied mathematics* [Curriculum]. Waco, TX: Author.

Conover Company. (1991). *Workplace literacy system* [Brochure]. Omro, WI: Author.

Crocker, L. M., & Algina, J. (1986). *Introduction to classical and modern test theory* (pp. 68–83). New York: Holt, Rinehart, and Winston.

Dunnette, M. D., & Hough, L. M. (Eds.). (1990). *Handbook of industrial and organizational psychology* (2nd ed., Vol. 1). Palo Alto, CA: Consulting Psychologists Press.

Dunn-Rankin, P. (1983). *Scaling methods.* Hillsdale, NJ: Lawrence Erlbaum Associates.

Educational Testing Service. (1975). *Cooperative assessment of experiential learning. Interpersonal learning in an academic setting: Theory and practice* (CAEL Institutional Report No. 2). Princeton, NJ: Author.

Efron, B. (1982). *The jackknife, the bootstrap, and other resampling plans.* Philadelphia: Society for Industrial and Applied Mathematics.

Electronic Selection Systems Corporation. (1992). *AccuVision systems for personnel selection and development* [Assessment]. Maitland, FL: Author.

Equal Employment Opportunity Commission. (1978). Uniform guidelines on employee selection procedures, *Federal Register, 43,* 38290–38315.

Gao, X. (1996). *Sampling variability and generalizability of Work Keys Listening and Writing scores* (ACT Research Report). Iowa City, IA: American College Testing.

Greenan, J. P. (1983). *Identification of generalizable skills in secondary vocational programs* [Executive summary]. Urbana, IL: Illinois State Board of Education.

Guttman, L. (1944). A basis for scaling qualitative data. *American Sociological Review, 9,* 139–150.

Guttman, L. (1950). The problem of attitude and opinion measurement. In S. A. Stouffer, L. Guttman, E. A. Suchman, P. F. Lazarsfeld, S. A. Star, & J. A. Clausen (Eds.), *Measurement and prediction.* New York: Wiley.

Hambleton, R. K., Swaminathan, H., & Rogers, H. J. (1991). *Fundamentals of item response theory.* In R. M. Jaeger (Series Ed.), *Measurement methods for the social sciences series: Vol. 2.* Newbury Park, CA: Sage.

Kolen, M. J., & Brennan, R. L. (1995). *Test equating: Methods and practices*. New York: Springer-Verlag.

Landy, F. J., & Farr, J. L. (1983). *The measurement of work performance: Methods, theory, and application*. San Diego, CA: Academic Press.

McLarty, J. R. (1992, August). Work Keys: Developing the assessments. In J. D. West (Chair), *Work Keys supporting the transition from school to work*. A symposium presented at the one hundredth annual meeting of the American Psychological Association, Washington, DC.

McLarty, J. R., Gao, X., & Stientjes, M. (1996, April). *Listening and writing: Two skills for the price of one?* Paper presented at the annual meeting of the National Council on Measurement in Education, New York.

McLarty, J. R., & Schulz, M. E. (1996, April). *Supporting Guttman inferences through an IRT model*. Paper presented at the annual meeting of the American Educational Research Association, New York.

National Commission on Excellence in Education. (1983). *A nation at risk: The imperative for educational reform. A report to the nation and the Secretary of Education*. Washington, DC: U.S. Government Printing Office.

Nunnally, J. C., & Bernstein, I. H., (1994). *Psychometric theory* (3rd ed.). New York: McGraw Hill.

Secretary's Commission on Achieving Necessary Skills (SCANS). (1990). *Identifying and describing the skills required by work*. Washington, DC: U.S. Government Printing Office.

Secretary's Commission on Achieving Necessary Skills (SCANS). (1991). *America 2000: An education strategy*. Washington, DC: U.S. Government Printing Office.

Shavelson, R. J., & Webb, N. M. (1991). *Generalizability theory: A primer*. In R. M. Jaeger (Series Ed.), *Measurement methods for the social sciences series: Vol. 1*. Newbury Park, CA: Sage.

Smith, P. C., & Kendall, L. M. (1963). Retranslation of expectations: An approach to the construction of unambiguous anchors for rating scales. *Journal of Applied Psychology, 47*, 149–155.

Stouffer, S. A., Guttman, L., Suchman, E. A., Lazarsfeld, P. F., Star, S. A., & Clausen, J. A. (1950). *Measurement and prediction*. New York: Wiley.

Vansickle, T. R. (1992, August). Work Keys: Developing a usable scale for multi-level, criterion-referenced assessments. In J. D. West (Chair), *Work Keys supporting the transition from school to work*. A symposium presented to the annual meeting of the American Psychological Association, Washington, DC.

Vansickle, T. R., & Prediger, D. J. (1990). *Mapping occupations: A longitudinal study based on vocational interests* (ACT Research Report Series No. 11). Iowa City, IA: American College Testing.

White, E. M. (1994). *Teaching and assessing writing: Recent advances in understanding, evaluating, and improving student performance*. San Francisco: Jossey-Bass.

Chapter 13

A Comprehensive
Performance-Based System
to Address Work Readiness

Stanley N. Rabinowitz
WestEd/Far West Laboratory for Educational Research and Development

The need to aggressively prepare students for the workplace and upgrade current workers' skills has taken a central place among the concerns of practitioners and policy makers alike. At the secondary education level, proponents of reform call for an abandonment of the traditional two-tiered approach characterized by abstract academics for the college bound and general or vocational training for the rest. In its stead are models that link school and work with varying degrees of formality, allowing secondary students (and their teachers) to explore career opportunities and obtain hands-on experience with the skills and behaviors required for success on the job. All students are expected to benefit from this more integrated, experience-based curriculum. The more popular of these models include Cooperative Education, Tech-Prep, Career Majors, and Youth Apprenticeships (Mendel, 1994).

Several assumptions underlie this career-preparation reform movement. Most basic is the view that American economic competitiveness is dependent on a highly skilled workforce capable of adapting to changing conditions and technologies. Successful worker training and retraining programs emphasize the need for both flexibility and high levels of skill attainment.

Policy makers and practitioners are seeking new models to guide their different but complementary roles. Increasingly, greater freedom and flexibility at the local level is granted in exchange for higher degrees of accountability for program goals, particularly students' achievement of high level academic and career-related standards. Policy makers, concerned with the apparent meager results obtained from large expenditures of public funds through the public school system or worker training programs, are looking for hard data, typically test results, to justify continued funding of programs. Local educators seek new assessment techniques to replace the limited multiple-choice options traditionally available (Baker, O'Neill, & Linn, 1993; Wiggins, 1993). This search is leading toward the promise of performance-based methods. For all constituencies, the ability to determine comparability of results across initiatives would enhance the credibility of training efforts and allow sophisticated cost-benefit models to inform future allocations of diminishing pools of money (Monk, 1992).

Developing a uniform assessment tool measuring both individual mastery of important skills and adaptability to new learning is a daunting task. When the myriad of models, programs, settings, content areas, and resources available across the nation is factored in, the difficulties confronting an assessment grow enormously. This chapter describes a workplace readiness assessment *system*—The Career-Technical Assessment Program (C-TAP)—designed to be both comprehensive and flexible. Use of C-TAP is shown to satisfy the needs of both the practitioner and the policy maker. Users are afforded a great deal of local control in its implementation, yet results show great promise of generalizing across content areas and program levels. C-TAP is a system of assessment components driven by career-related standards and their academic underpinnings.

In the next section, a description of components comprising C-TAP is presented. This is followed by a discussion of C-TAP's philosophical underpinning and principles guiding its development. Next we describe the model used for field testing and revision. We conclude with a discussion of scoring methods and related research efforts underway to increase the generalizability of the system and its applications.

C-TAP COMPONENTS

The C-TAP system consists of several cumulative and on-demand components designed to assess if a student or entry-level worker has mastered important standards needed for success on the job or at the next level of career training. The specific components of the C-TAP system are described next.

Cumulative Assessments

Portfolio. The C-TAP portfolio is the centerpiece of the assessment system. The portfolio serves as a vehicle for organizing and presenting a collection of the student's work for purposes of assessment as well as for presentation to prospective employers or advanced training institutions. The portfolio includes:

- *letter of introduction*, including a description of the contents of the portfolio and career aspirations, as well as an analysis of the work included within the portfolio;
- *career development package*, consisting of a resume, job or college application, and a letter of recommendation;
- *work samples* linked to key program content standards, demonstrating hands-on ability to apply knowledge and skills to job-related tasks;
- *writing sample*;
- *Supervised Practical Experience evaluation form.*

Supervised Practical Experience. A growing number of secondary programs provide students with a work-based learning experience in a setting related to their career interests. An evaluation form developed for this Supervised Practical Experience allows the supervisor to rate important generic career skills embodied in the following career performance standards: personal skills, interpersonal skills, thinking and problem solving skills, communication skills, occupational safety, and technology literacy. Additionally, students will be rated on skills specific to their career-preparation program that are required "on the job." A copy of the C-TAP Supervised Practical Experience evaluation form is presented in Appendix A.

Project. All students are required to complete and present an approved assessment project during the course of their program. Projects involve either the planning and development of a tangible end product or a written description and analysis of a significant process-oriented experience based on a "real world" application from the student's field. Project ideas are developed jointly by the student and teacher according to specific guidelines. A committee of educators and business and industry representatives oversee the development of students' work.

On-Demand Assessments

Project Presentation. Upon completion of his or her project, the student delivers an oral presentation describing the details of the project to a review panel and responds to questions from the panel. The panel may consist of teachers, parents, students, and industry representatives. The student's presentation is evaluated according to specified criteria including oral-presentation skills and ability to reflect on the project experience.

Written Scenario. Written scenarios present complex and realistic problems from the career or program major. Assessees have 45 minutes to respond in writing to the written scenario prompt; they are judged on their ability to apply content knowledge and problem solving skills to address the problem presented in the scenario.

PHILOSOPHICAL FOUNDATION

C-TAP has been designed to support and enhance a range of educational and worker training reform efforts. It was developed originally to support the innovations encouraged by the Carl D. Perkins Vocational and Applied Technology Act of 1990 (Perkins II), particularly curriculum integration and student and program accountability. C-TAP's features expanded over time to accommodate more sophisticated models of systemic worker-training reform, relying on career technical teachers and their academic counterparts as well as the active participation of industry (business and labor) partners represented by the School-to-Work Opportunities Act of 1994. In this section we describe these features in the context of the innovations they attempt to help implement.

Flexibility

At its core, C-TAP remains a highly flexible system of assessment. Paradoxically, this is accomplished, in part, due to a set of prescribed assessment components and scoring rules. By requiring individuals to complete uniform components, in this case the *portfolio, project,* and *written scenario,* individual student interests and unique program characteristics can drive the specific implementation of these pieces. The prescribed components allow the comparability across students and programs and also ensure that standards of mastery are equivalent across settings. Individual teachers and students retain the flexibility to develop specific applications that take into account local learning conditions, available resources, and individual interests. For example, in completing the project, students are encouraged to develop and implement their own ideas based on their interests and available resources. Students have great leeway to develop products which combine research and hands-on applications.

Similar flexibility is found throughout the system. C-TAP has been able to accommodate legitimate differences both within and across content areas. For example, Agriculture Animal Science programs vary across California, with different regions focusing on large or small animals. C-TAP has had to develop assessment tasks that take into account such regional differences. Formatting requirements for reports and other correspondence differ greatly between Business Education programs and their counterparts in other subject areas. C-TAP scoring rubrics are flexible enough to allow such differences while permitting comparability of scores across content areas.

The C-TAP system has had to adapt over the course of its development to reflect local conditions and program differences. The student project was derived from an on-demand assessment piece which was found to be too burdensome for local educators and unadaptable to local program emphases and resources. Based on feedback from employers and field-test teachers, the writing sample included in the portfolio evolved from a more prescriptive research paper. Such changes reflect an important driving point—as long as the student work is guided by high-level program standards and presented within the general system components (to allow comparability across diverse settings), a myriad of specific applications is acceptable. The important role standards play in this equation is discussed next.

Standards Driven

A key feature of the new wave of workplace training models is the development of and unifying role played by important knowledge and skills necessary for workplace success. Standards-driven reform allows current and future workers to benefit from full knowledge of the keys to success. Local, state, and national standards-development efforts across academic, generic workplace readiness, and specific industry applications, have given curriculum developers much to choose from in developing challenging, hands-on learning tools (Ananda, Rabinowitz, Carlos, & Yamashiro, 1995).

Assessments must be linked to clear and concise standards, reflecting major knowledge and skill areas. Most importantly, standards convey expectations of what workers and students should know and do (content standards) and what proficient performance should look like (performance standards; Kendall & Marzano, 1994). An example of a C-TAP content standard containing both a cognitive and behavioral component is one related to safety on the job. Students are expected to "understand occupational safety issues" and demonstrate proper handling of equipment and hazardous materials (C-TAP, 1995b; see Appendix B for a listing of several content standards assessed by C-TAP). The performance standard reflecting proficient achievement of this content standard is embedded in successful completion of a variety of C-TAP assessment components, as defined by educators and industry representatives. (See Ananda and Rabinowitz, 1995 for a comprehensive description of various models and definitions of systems of standards.)

While several levels and models of standards are available, C-TAP assesses, in an integrated fashion, standards representing:

- **program specific** including industry core, occupational family standards, and in some new applications, occupation specific standards;
- **generic workplace readiness** (SCANS, 1991; Council of Chief State School Officers Workplace Readiness Consortium, 1995);
- **academic foundations** covering core academic areas required for success in the program-specific area (e.g., technical reading for computer science programs; trigonometry for manufacturing programs).

An important early step in C-TAP's development was the revision of the Model Curriculum Standards (California Department of Education, 1994) that define the career-technical programs across California. The primary goals of the revision were to ensure that the new standards reflect the expectations for success in the workplace or in job training programs, and to develop solid academic foundations for lifelong learning. The Model Curriculum Standards were revised and validated by teams of educators and business representatives, the key clients for the use of the standards. This collaborative effort ensured that all important skills were included in the standards, that they were applicable in both training and work settings, and that all constituencies assumed responsibility for their dissemination and implementation. This model has been followed for subsequent C-TAP applications.

Performance Based

The methodologies collectively referred to in the literature as performance-based assessments have gained momentum as a tool for educational and workplace reform. This is largely because performance-based assessment tasks attempt to reflect "real life" situations facing workers and so are perceived as being more "authentic" than traditional paper-and-pencil tasks. Performance-based assessments are also engaging tasks, requiring the assessee to produce a quality product or performance (Wiggins, 1993). As compared to more traditional performance assessments that were used for decades in vocational education and industry, the newer performance-based methodologies such as C-TAP make greater demands on academic foundations and problem-solving in addition to actual task performance. Another distinction is that the newer methodologies tend to focus on holistic rather than fragmented samples of performance (Wirt, 1995).

Although both traditional paper-and-pencil tasks (i.e., multiple-choice) and performance-based tasks (e.g., portfolios, simulations, projects) can play important roles in specific applications, the decision was made for C-TAP to be primarily performance-based. This is because paper-and-pencil tests often lack credibility and are vulnerable to inappropriate test-taking behaviors on the parts of test takers and proctors, such as improperly teaching to the test. Performance-based assessment tasks represent more direct ways of assessing competency and skill, allowing for stronger links among assessments, standards, and instructional/training goals. A major challenge facing systems such as C-TAP, compared to traditional paper-and-pencil tests, is the substantial resource investment (human and monetary) required for development, administration, scoring, and training.

Careful scrutiny of the technical adequacy of such systems is also crucial (Linn, 1994). Brennan and Johnson (1995) argue that generalizability of results from performance assessments may suffer with their use unless a large number of tasks are included and student-task variance is controlled. Similar concerns are expressed in terms of equating (Green, 1995) and setting performance levels (Jaeger, 1995) of performance-based assessment approaches. Gearhart, Herman, Baker, and Whittaker (1993) raise important questions about the use of portfolios in high stakes environments which has consequences for all student work not completed in the typical on-demand assessment setting.

Technical adequacy is an overriding concern for C-TAP. Significant resources have been allocated to design and implement studies focused on setting performance levels that generalize across content areas and students, and measuring classroom-level consequences related to its use. Such an approach is consistent with evolving views of validity (Messick, 1995; Moss, 1995) and gets to the heart of what C-TAP is attempting to accomplish, that is, support true reform in the preparation of entry-level workers. Jaeger (1995) reports success at "setting performance standards on complex, multidimensional performance assessments" (p. 20). C-TAP's success in initial portfolio benchmarking studies supports this finding (C-TAP, 1995a).

Increasingly, new approaches and criteria are under pressure to classify and evaluate the results and consequences of performance based assessment efforts (Ananda & Rabinowitz, 1996). Expanded views of traditional concepts of validity and reliability hold promise for the widespread implementation of C-TAP and related performance-based systems of assessment (Delandshere & Petrosky, 1994; Moss, 1994). Nor should we exaggerate the technical adequacy of more established techniques such as multiple-choice assessments. Careful scrutiny of such measures, even in high-stakes applications such as high school graduation exit examinations, show rare examples of carefully developed and implemented generalizability and validity studies (Ananda & Rabinowitz, 1996).

Integrated

As mentioned, the C-TAP assessment system measures program specific, generic workplace readiness and academic foundations standards in an integrated fashion. Specifically, this means that each assessment component measures the knowledge and skills that comprise each of these classes of standards. For example, the C-TAP portfolio has several prescribed tasks that, taken together, allow conclusions to be drawn about the work readiness of the candidate. To successfully complete each of the tasks, the assessee must demonstrate mastery of the three levels of standards measured by the system. This is not to say that each task measures each type of standard with equal thoroughness. Integration does not require this. Rather, the system takes advantage of the strong features of each component. For example, the portfolio's *Career-Development Package* focuses principally on generic workplace readiness skills, whereas *Work Samples* are used essentially as important measures of specific occupational content.

Several advantages accrue from this integrated approach. First, it mirrors the way assignments are typically experienced in the workplace. Workers do not separately use their skills; instead, all knowledge and experience comes into play as the worker performs job-related functions. Consequently, C-TAP results are not differentiated by content area. Successful completion of a task presumes sufficient mastery of requisite skills. Second, the use of integrated classroom activities enhance levels of learning by allowing complex ideas to be presented in real-world contexts (Grubb, 1991, 1995). C-TAP training and implementation experiences are filled with numerous instances of collaboration between academic and career-technical teachers, raising the skill levels of each and providing students with more and better support than would be available from the teachers separately (C-TAP, 1995a). Finally, integrated assessment forces schools to examine practices and curricula that reinforce traditional, ineffective (isolated) models of educating students, particularly those at risk of failure. C-TAP's value as a tool to support systemic reform has been recognized in numerous instances (Rabinowitz, 1995a, 1995b).

System Driven

The C-TAP system is comprised of a variety of assessment components and tasks. The approach is referred to as "triangulation," or the accumulation of multiple forms of evidence to reach a comprehensive decision (Mathison, 1988). A triangulated approach allows for assessment of skills through multiple and complementary means, thus increasing the reliability of the resultant score and any decisions made from that score. For example, the Career Preparation Standards are assessed in C-TAP most directly by the Supervised Practical Experience and Career Development Package. Each of these components approaches the measurement of these standards from a different perspective. Taken together, this complementary assessment ensures greater generalizability of decisions concerning students' mastery of the Career Preparation Standards.

As described, C-TAP includes both *on-demand* and *cumulative* components. On-demand tasks require responses "on the spot" in a discrete period of time. Examples of on-demand components include actual job performance over real and simulated conditions, written scenarios, and writing prompts. Cumulative components are completed over a substantial period of time. Portfolios, projects, and on-the-job practicums are examples of cumulative assessment components.

There are several reasons for advocating a system that includes on-demand and cumulative components. First, most jobs involve a combination of activities, some of which can be completed over time, others with immediate timelines. Approaches to assessment must address both types of skills if they are to be valid and acceptable to the industry and education communities. Next, reform requires significant changes in the way students and workers are trained. Cumulative tasks help support a problem-solving based, "thinking" curriculum. Third, inclusion of on-demand components helps ensure more comprehensive coverage of standards; this may not occur in a cumulative-only system where more individual choice is typically permitted. Fourth, the combination of on-demand and cumulative tasks provides teachers and trainers sufficient time to complete all components with limited burden. Perhaps most important, including various types of assessment components allows examinees to demonstrate competency in different ways, capitalizing on each individual's areas of strengths, and satisfying equity and other legal concerns described later.

Beyond the assessment components themselves, C-TAP can also be seen as part of an ideal system of workplace preparation (Ananda & Rabinowitz, 1995). As Fig. 13.1 indicates, different types and combinations of standards come to the fore over the course of training for a student and retraining of a worker. Comparability and generalizability of evaluation results at both the student and program level are greatly benefited by the existence of a comprehensive system flexible enough to be used at each level and purpose. C-TAP has been designed to accommodate such varying conditions, goals, and purposes, ranging from informal feedback between students

TYPE OF STANDARD

Program Level	Academic	Generic Workplace	Industry Core	Occupational Family	Occupational Specific	Assessment Partners**	Certification Level
Occupational Specific • Post-Secondary Training Institution • Industry Setting	X	X	X	X	X	*Industry* / Education	♦ Job Entry ♦ Career Specialization
Grade 12	X	X	X	X	X	*Education* / Industry	Occupational Family
Grade 10	X	X	X			*Education* / Industry	CIM
Middle School	X	X				*Education* / Industry	None

FIG. 13.1. Ideal Industry Skills Training Model. Adapted from Ananda and Rabinowitz (1995). Italics indicate lead partners.

and teachers up through the more rigorous requirements of a formal certification system. Obviously, the technical requirements of such a system differ across these purposes. Goals and stakes play a significant role in the potential consequences of assessment results and thus in the behaviors of teachers and students prior to and during the administration period.

C-TAP has been developed with certification of secondary career-cluster program completers as its primary purpose. Thus the Model Curriculum Standards and Career Preparation Standards on which it is based have received extensive review and validation by industry and education representatives familiar with the requirements of postsecondary training programs and entry-level workers within and across industries. The minimum level of success has been set at the "proficient" level of C-TAP performance. Teams of cross-industry and education representatives are also working to ensure that the proficient label is applied comparably across content areas. Preliminary results suggest agreement levels exceeding 80% when C-TAP portfolios from a variety of occupational clusters are reviewed by mixed teams representing heterogeneous program areas and industries.

C-TAP DEVELOPMENT PRINCIPLES

In order for the resultant assessment system to be consistent with its philosophical foundations, the development of the C-TAP system and assessment tasks within each component needed to be highly systematic and follow a set of principles described next.

Nonredundant

C-TAP development recognized the existence of other pieces of the assessment puzzle. No attempt was made to assess knowledge or skills best measured by other instruments or other more efficient formats (e.g., multiple choice). For example, numerous standardized tests exist to assess basic reading ability or general mathematics proficiency. Consequently, C-TAP measures in applied contexts only those reading and mathematics skills specifically related to success in the job or program in question. Educators or employers are encouraged to examine C-TAP results in conjunction with evidence from other assessment sources to gain a full picture of a student's or job candidate's skills and potential success.

Consistency With Complementary Reform Initiatives

Successful implementation of C-TAP requires a strong commitment from teachers, students, and administrators. The extent to which the requirements and philosophy of C-TAP was consistent with and helped support the greater reform agenda already underway in schools enhanced teacher's abilities to incorporate performance-based methodologies with the support of administrators and business partners. Thus, C-TAP was designed to complement major secondary reform initiatives such as tech-prep or the program-major driven *Second to None: A Vision of the New California High School* (California Department of Education, 1992), the overall state testing program, and initiatives based on the issuance of certificates of initial and advanced mastery (Oregon Progress Board, 1991). At the national level, C-TAP supports the philosophy and implementation of both Perkins II and the School-to-Work Opportunities Act, as well as the movement toward greater flexibility at the local level in exchange for adherence to higher standards for all students.

Joint Development by Industry Representatives and Educators

C-TAP assessment tasks have been developed collaboratively by educators and industry representatives, helping to ensure fidelity to the requirements of the work place and feasibility in educational and training settings. Industry representation included line-level workers and supervisors, as they have the most current hands-on experience with existing job requirements and tasks.

Several specific benefits have come out of this collaboration. In more than one instance, industry representatives have been able to inform their education counterparts on recent changes in the scope of practice on the job. Not only were the educators able to implement these changes in the classroom, but the assessment tasks being developed were spared an early obsolescence, their fate had they been implemented as originally designed by the educators. Joint meetings have also raised the level of respect and overall awareness across both camps of the important role each plays in the training of current and future workers.

Time Considerations

A serious challenge facing developers of performance-based assessment systems is to temper the immediate needs for such instruments with the reality that reliable and valid assessments take a great deal of time to field test and implement. Examination of political and technical problems facing several current "state-of-the art" systems points to serious underestimations of the requisite development time on the part of policy makers and contractors (Ananda & Rabinowitz, 1996). C-TAP components have enjoyed the benefit of multiple opportunities to field test and revise, resulting in tasks that are accessible to a wide range of students with varying strengths and skill levels and that are able to be embedded into secondary reform curricula with minimal disruption.

Equity and Legal Defensibility

Equity concerns have been addressed throughout the C-TAP development process. Many of the content areas C-TAP has assessed suffer significantly from gender stereotyping (e.g., health careers and industrial technology programs). Differences in the performance of males versus females, or among ethnic and racial groups, are carefully reviewed and analyzed to ensure they are not the result of biased assessment methods (Baker & O'Neill, 1994). The needs of special populations are also under consideration; accommodations are dictated by legal requirements and the purposes for which the assessments are used.

Legal issues related to an assessment-based certification system or other potential uses are also a serious concern. Demonstration of job relevance, technical adequacy, and absence of bias are key to the legal defensibility of a certification system (Mehrens & Popham, 1992; Nettles & Nettles, 1995). As per the expectations of the "Standards for Educational and Psychological Testing" (AERA/APA/NCME, 1985), all C-TAP tasks undergo review by bias and other committees before they are implemented for high-stakes applications.

C-TAP FIELD-TEST PROCESS

As previously discussed, development of performance-based assessment systems, particularly one as comprehensive and flexible as C-TAP, is a time consuming process. Each C-TAP component and associated sample tasks were developed using a three-step process, as follows:

1. *Task Development.* Groups of teachers and industry representatives participated in development sessions. Following training by the WestEd/Far West Laboratory assessment staff on the Model Curriculum Standards (California Department of Education, 1994) to be covered and key features of performance-based assessment items, task developers worked individually and in groups to produce a large number of potential prototype tasks for further review and pilot testing.

2. *Pilot Testing.* The most promising of the prototype tasks developed in Step 1 were administered to a sample of students in a variety of career-preparation classes. The purpose of pilot testing was to determine how well the tasks fit into learning situations in the pilot classrooms and if students were able to respond to the sample tasks as expected. An important

feature of C-TAP is the teacher's ability to use the assessment process as part of the everyday activities in the classroom. Fewer than 50% of piloted assessment tasks advanced to the more rigorous field test phase; such decisions are primarily based on qualitative judgments by classroom teachers on the tasks effectiveness in conjunction with a global review of student work.

3. *Field Testing.* Those tasks which survived the pilot test process were subjected to a formal field test administered to a select sample of students in a variety of career-technical programs (e.g., vocational classes, Career Academies, Tech-Prep sites) and settings (urban, suburban, and rural). An important purpose for field testing was to determine logistical support needed for full implementation and technical adequacy (e.g., reliability, generalizability, and validity) of the system for anticipated purposes. Different models of scoring were attempted, to increase the reliability and decrease the burden on the classroom teacher and other probable scorers. Care was taken to ensure that students from special populations (e.g., English-language learners, special education) were included in the field test.

More than 20,000 students in over 200 high schools participated in development and field test activities. The pilot and field test process greatly informed the logistical support required for full implementation of C-TAP, as well as strategies for scoring of student work. These two issues are discussed next.

LOGISTICAL CONSIDERATIONS

Implementation of C-TAP at the program and classroom level requires a commitment to change on the part of teachers, administrators, and students. Teachers who stick with the system through a second implementation year report significant improvement in their ability to incorporate major aspects of the program into their instructional and assessment activities. In the first year, most teachers treat the components as add-ons to their traditional classroom activities. By the second year, the components become more naturally integrated into everyday learning events.

To speed up the assimilation process, training for new teachers relies heavily on the use of C-TAP "veterans." Workshops focus on the lessons of their experiences and feature a variety of classroom-tested solutions to implementation challenges. Experienced teachers also serve as C-TAP mentors, providing ongoing assistance via telephone, e-mail, and in some cases, site visits.

Support from administrators significantly affects overall implementation rates. Only the strongest teachers can be successful in isolation; their limited success has shown little generalization beyond that specific teacher's classroom. Administrators, both schoolwide and in the various departments, have assisted implementation in a variety of ways including:

- scheduling, particularly joint planning time for teachers within and across departments. Other scheduling considerations include longer class time blocks, which assists in the completion of portfolio and project entries.
- sanctioning and supporting the participation of academic teachers, especially English teachers, in the career-technical classroom (and vice versa). English teachers play a key role (often taking the lead) in instructing and scoring many of the written components of the portfolio. Joint credit for assignments is often granted in the career-technical and academic class.
- resources made available such as equipment (e.g., word processors), newsletters to parents and board members, and substitute teachers and travel support for teachers to attend workshops and review other implementation sites.
- the attitude among top district and school administrators that reform is valued in general, C-TAP is an avenue to achieve important changes, and results can make the difference between a successful implementation and just another initiative to be waited out.

Given these logistical conditions for success, C-TAP implementation has been found to be most successful at sites already involved in school reform activities. Districts involved in programs such as Tech-Prep (Grubb, 1995), career academies (Stern, Raby, & Dayton, 1992), and site-based decision making (Wohlstetter & Odden, 1992), to name just a few, are more likely to provide both the tangible and intangible support needed for success.

Teacher Training

Staff at WestEd/Far West Laboratory have developed and instituted a number of activities and support documents to train teachers in the implementation of C-TAP. This chapter will focus on two of these services—teacher/student guidebooks and training workshops—which are designed to work together to prepare teachers and students to the challenges they face.

Guidebooks. Separate guidebooks have been designed for teachers and students to lead them through important implementation issues and to provide logistical support (C-TAP, 1995b, 1995c). All C-TAP teachers and students receive individual copies respectively; ownership of "personal" copies has been found to be crucial for student buy-in.

The C-TAP Teacher Guidebook (C-TAP, 1995b) consists of four major sections and a series of appendixes. Section I introduces the teacher to the goals and components of the system, provides field-tested suggestions for logistical support and ways to motivate students, and concludes with models and activities to track student progress and ensure content coverage across the components. Sections then follow on specific implementation of the portfolio, project, and written scenario, including summaries of all major requirements and activities, tips for introducing major ideas, and forms for tracking student work.

The C-TAP Student Guidebook (C-TAP, 1995c) focuses on instructions for students to participate in the assessment, including sample completed work for all entries to serve as models. Written in a style and format conducive to student interest, the Student Guidebook has been effective in motivating students and assisting them directly in the completion of activities and components.

Training Workshops. Ideally, the use of guidebooks occurs as part of a comprehensive professional development plan. Initially, groups of teachers are introduced to the goals and requirements of C-TAP and led through the use of the guidebooks via a series of carefully crafted discovery activities. Teams consisting of career-technical teachers across content areas, academic colleagues, and administrators are encouraged to attend training, to provide support and permit a team-approach to teaching and scoring. Networks of teachers consisting of veteran mentors and novices are developed and begin interacting during the initial training day. Following the first training meeting, telephone contact times between trainers and teachers, to review progress and answer questions, are scheduled. Site visits are also arranged as needed.

Additional workshops are scheduled throughout the year, coinciding with the introduction of major assessment components to students such as the project or written scenario. Field-tested techniques to assist and motivate students are stressed throughout (C-TAP, 1995b). Teachers have an opportunity to bring with them students' work in progress, and to review their results and problems with colleagues across the region. Throughout all workshops, exemplars of student work across all levels of achievement form much of the basis of training. Examination of carefully selected model student work is an invaluable activity for teachers and students alike, to grasp the requirements of the system and guide their own efforts. The development of these exemplar tasks are described in the section on scoring.

TABLE 13.1
1994-1995 C-TAP Teacher Survey--
Selected Responses

Question	% High	% Medium	% Low	% Not Sure
Knowledge of revised Model Curriculum Standards for your career-technical area:	52	36	8	3
Knowledge of and experience with new approaches to new student assessment prior to this year:	26	50	23	2

Teacher Logistics Survey

In the spring of 1995, a sample of 126 participating C-TAP teachers responded to a series of questions relating to their experience with C-TAP implementation and support. Summaries of questions addressing logistical and administrative support and the value of various support services, as well as teacher knowledge of key curriculum and assessment concepts underlying the C-TAP standards-driven system, are included next.

As shown in Table 13.1, respondents report substantial knowledge of the Model Curriculum Standards driving their career-technical program, with 52% reporting a "high" level of knowledge and 36% reporting a "medium" level. Respondents also report a fair amount of prior knowledge and experience with new approaches to student assessment (high = 26%; medium = 50%; and low = 23%).

When asked about the effectiveness of various sources of information for implementing C-TAP (see Table 13.2), Far West Laboratory workshops were

TABLE 13.2
1994-1995 C-TAP Teacher Survey--
Effectiveness of Sources of Information for Implementing C-TAP

	% Highly Effective	% Effective	% Ineffective	% Missing
Workshops	52	35	6	7
Teacher Guidebooks	46	48	2	4
Student Guidebooks	51	41	2	6
Sample Student Work	37	42	9	12

rated positively by 87% of teachers, with 52% of respondents rating them "highly effective" and 35% as "effective." Teacher and student guidebooks were also rated positively, with 98% rating them as "highly effective" or "effective." Samples of student work were rated "highly effective" by 37% and "effective" by 42% of respondents for an overall approval rating of 79%. This is particularly noteworthy because samples of student work were limited in number and not as widely disseminated in 1994–95 as they will be in future years.

Table 13.3 shows how respondents rated the level of support they received from different potential sources. As shown, the three sources with the highest ratings were Far West Laboratory staff, California Department of Education staff, and department chairs. Specifically, 75% of respondents reported receiving "significant support" from Far West Laboratory staff, while 45% and 43% reported receiving "significant support" from California Department of Education staff and department chairs, respectively. This finding is consistent with the training model used, with Far West Laboratory taking the lead role.

Finally, as shown in Table 13.4, teachers were asked to identify the most important factors essential for C-TAP implementation in the classroom. The three factors most frequently checked as essential were more class time for C-TAP (69%); integrated curriculum (52%); and more professional development related to C-TAP (44%). In contrast, only 7% of respondents identified more parent understanding and different professional development as essential factors.

TABLE 13.3
1994-1995 C-TAP Teacher Survey
Level of Support for Implementation Received From:

	% Significant Support	% Little Support	% No Support	% Missing
School principal	36	30	23	11
District administrator	26	28	32	15
Department chair	43	12	23	22
Other career-technical teachers	28	34	27	11
Academic teachers	16	37	34	12
Local business and industry representatives	16	33	34	17
Parents and local community members	14	21	48	17
Far West Laboratory staff	75	17	3	5
California Department of Education staff	45	23	18	14

TABLE 13.4
1994-1995 C-TAP Teacher Survey
Check the Five Most Important Factors Essential for Successful C-TAP Implementation in
Your Classroom*

	% Checking this Factor
More class time for C-TAP	69
Integrated curriculum	52
More professional development related to C-TAP	44
More school-wide support of portfolios	39
Significant numbers of students in common with other teachers using C-TAP	38
More experience with portfolios	35
Previous experience with performance-based assessment	33
More school-wide support of C-TAP	27
More experience with Model Curriculum Standards (MSC)	22
Career academies	20
Previous experience with interdisciplinary assessment	19
Block scheduling	19
Team teaching	19
Tech prep	19
Cooperative learning	14
Site-based management	9
Different professional development related to C-TAP than we received	7
More parent understanding of portfolios	7
Different C-TAP guidelines	0

Note. *Represents proportion identifying response as one of 5 most important factors.

SCORING

Scoring presents a substantial challenge for C-TAP. Several recent efforts at large-scale portfolio assessments, for example, have not been sufficiently reliable to satisfy traditional psychometric standards (Kahl, 1995). For C-TAP, our scoring approach has focused on the following concerns: How can we achieve acceptable levels of agreement among judges? How consistent are estimates of student performance across different content areas?

The C-TAP scoring process is geared toward achieving disciplined judgments of scorers, and builds off the lessons learned from large-scale portfolio scoring efforts, such as the Kentucky Instructional Results Information Systems and the California Learning Assessment System (Ananda & Rabinowitz, 1996). Thus, much effort is spent on training of scorers, including the identification of formal examples of student work at critical score points in order to better "anchor" individual scoring judgments. To avoid duplication, only the specifics of the scoring rubric and process are described next with respect to the C-TAP portfolio. A similar process is used for both the C-TAP project and written scenario components.

Scoring Rubrics

Consistency in scoring depends on a shared sense of content requirements, dimensions around which to group student work, and standards of performance. The C-TAP teacher and student guidebooks specify such requirements and standards, helping set the stage for more consistent judgments when actual scoring takes place. Such scoring takes place in the summer, upon submission of student work from the preceding school year.

C-TAP assessment components are evaluated holistically as *Basic, Proficient,* or *Advanced,* based on examination of the overall quality of the portfolio relative to how well the work demonstrates mastery of the program's Model Curriculum Standards. *Basic* means the performance is unsatisfactory and does not meet requirements. *Proficient* means the performance is very good, meeting requirements and demonstrating important skills and abilities. *Advanced* means the performance is outstanding, going beyond what is required and demonstrating important skills and abilities.

The holistic rating as Basic, Proficient, or Advanced is based on broad evaluation dimensions that cut across the required student work. For the portfolio, four such dimensions have been identified: *content, career preparation, analysis,* and *communication.* These dimensions, as follows, represent the critical features by which students demonstrate competence and mastery of standards in their career-technical area.

- Content: breadth, depth, and application of knowledge and skills related to the career-technical Model Curriculum Standards.
- Career Preparation: Understanding of career preparation and personal employability attributes.
- Analysis: Ability to apply analytical skills to the gathering of information and evaluation of own work.
- Communication: Effective use of communication skills.

A rubric, or set of scoring guidelines, has been developed based on these four dimensions. The rubric facilitates evaluation and helps ensure consistency of overall ratings by defining specific features of the dimensions that distinguish performance levels. The rubric, shown in Fig. 13.2, distinguishes among Basic, Proficient, and Advanced levels of performance for the four evaluation dimensions. Analogous rubrics have been developed for the project and written scenario.

Scoring Process

The portfolio scoring process involves participating C-TAP teachers, and others with relevant content expertise (e.g., academic teachers, administrators, industry representatives). Participants are brought together for

Table 10.2: C-TAP Portfolio Rating Guide

	Basic	Proficient	Advanced
CONTENT • Knowledge of major ideas and concepts in career–technical standards • Knowledge of how skills in career–technical standards are applied	• Shows gaps in knowledge; misunderstands major ideas and concepts • "Hands–on" work demonstrates minimal knowledge or skill	• Shows knowledge of major ideas and concepts; covers the content of important career–technical standards • "Hands–on" work demonstrates a variety of skills	• Shows clear understanding of major ideas and concepts; explains how ideas and concepts relate to each other • "Hands–on" work demonstrates superior skill
CAREER PREPARATION • Career planning • Personal qualities needed for employment	• Shows little or no evidence of planning for a career • Does not identify own personal qualities needed to be successfully employed	• Shows evidence of planning and developing a career • Identifies own personal qualities needed to be successfully employed	• Shows excellent understanding of career planning; describes a realistic plan that leads to clear career goals • Highlights own personal qualities needed to be successfully employed throughout the portfolio
ANALYSIS • Evaluation of own skills and work • Investigation and information gathering	• Gives incomplete or sketchy evaluation of own work • Does not gather information from several sources	• Gives accurate evaluation of own work • Shows ability to find and use information from several sources	• Shows understanding and insight in evaluating own work. • Demonstrates superior ability to gather and combine information from various sources
COMMUNICATION • Attention to audience • Using own ideas • Organization and clarity • Accuracy, neatness, and completeness • Language mechanics, sentence structure, and vocabulary	• Shows little or no awareness of the audience • Writing is not original; copies the ideas of others • Ideas are presented in a disorganized way • Work lacks accuracy and completeness; appearance interferes with communication of ideas • Writing contains errors in language use that make ideas difficult to understand	• Effectively presents self and ideas to outside reviewer • Writing is original • Writing is clear and organized • Work is accurate, neat, and complete • Writing contains few language errors; ideas are not difficult to understand	• Self and ideas "come alive" to outside reviewer • Writing is original and may be creative • Writing is clear and well organized throughout portfolio • Work is accurate and complete; appearance helps the communication of ideas • Writing is almost free of language errors and is easy to understand

FIG. 13.2. Basic, Proficient, and Advanced levels of performance for the four evaluation dimensions.

multiday scoring activities during the summer. The first activity is identification of formal examples, or exemplars, of student work at critical score points (e.g., Basic, Proficient, Advanced). Training toward this end begins with a review of portfolio entry requirements and an orientation to the portfolio rubric. Participants, working individually, review student work and nominate the complete portfolios they feel best exemplify Basic, Proficient, and Advanced student work. Following selection of complete portfolios, individual entries (e.g., letters of introduction, career development packages, work samples, writing samples) are reviewed as possible examples. The selection of individual entries allows representation of a broad range of acceptable student experiences and approaches.

Each individual presents his or her nomination to the group. The relative strengths and weaknesses of each nominated portfolio or portfolio entry are discussed. On the basis of this discussion, the final examples are selected. Where necessary, changes are noted and made to ensure that the examples fully meet the portfolio requirements and exemplify Basic-, Proficient-, or Advanced-level quality of student work.

Following these workshops, C-TAP project staff work to further refine the student portfolios selected as exemplars across the different content areas. This step includes review of exemplars from different content areas to help ensure consistency across these areas. The refined exemplars are then submitted to content and policy experts for review and approval. The end products of these activities are benchmarks, or expert-determined examples of student work at each performance level for use in training scorers.

The next level of scoring activity involves actual scoring of student work by selected raters (again, C-TAP teachers, academic teachers, industry representatives with appropriate content knowledge). As with the selection of exemplars, scorer training begins with a review of the portfolio requirements and the scoring rubric. However, in this instance, the scoring rubric is presented along with the exemplars of student work. This allows scorers to internalize the rubric, aided by concrete examples at critical score points. Furthermore, scorers within content areas are also oriented to exemplars from other content areas, to help ensure comparability of scoring across the different contents areas. Finally, the work of the individual scorers is carefully monitored to ensure that they adhere to the specified scoring rubric.

Investigating a New Scoring Approach

Some recent development suggest that holistic scoring entails a considerable loss of information as well as lower reliability in comparison to analytic scoring (Kahl, 1995). To address this issue, the project staff is investigating a new scoring approach, built directly from the existing one. This new approach will call for scorers to assign a rating (e.g., Basic, Proficient,

Advanced) to each dimension separately before making an overall holistic rating. Scorers will then be asked how they used the individual dimension ratings to inform their overall holistic judgment. The holistic rating will be compared to an overall rating based on combined dimension scores.

NEXT STEPS

Much research and development work has gone into the building and testing of the C-TAP system to date. Payoff is already being felt at the student and program level. Many students have been able to use their completed work—in particular, their portfolios—to obtain employment or entrance into postsecondary training programs. Teachers report important differences in the way they teach and in the motivation and achievement of students, including those from special populations such as English language learners or special education (C-TAP, 1995a).

Additional work is required for schools to take full advantage of C-TAP. In this section we describe some ongoing efforts to increase the value and accessibility of C-TAP to broader populations and purposes.

"How-to" Manual

C-TAP development in specific content areas has proceeded according to the guidelines and principles described throughout this paper. There is a growing interest, due to growing interest in expanding the system to additional content areas and fiscal restraints, to speed up the development process. Our preliminary indications are that this is possible for relatively less formal uses, such as program evaluation without high stakes student consequences, since lower stakes applications require less formal technical evidence of adequacy.

To promote this development into new content domains, C-TAP staff have developed a "how-to" manual to guide local users to expand the C-TAP system into areas not yet formally developed (C-TAP, 1996). Evidence from the 1994–95 field test indicates that the development cycle can be shortened by a full year if usage is limited to lower stake applications with their related lower technical requirements. The "how-to" manual, used in conjunction with the guidebooks and training model, is seen as an efficient means of expanding C-TAP in a time of rapid work place change and limited available resources for development projects.

Applications Beyond Career Technical Programs

Most C-TAP implementation has occurred in the context of career-technical reform. However, an increasing number of "traditional" academic programs are finding value in the integrated portfolio/project/written scenario system supported by the guidebooks and teacher workshops. C-TAP portfolios have been developed in English classes, both with and without a career focus. C-TAP staff are developing exemplar student work in science, mathematics, and other academic areas with career and general applications.

Ideally, the career-technical and academic content are integrated in the training of secondary and postsecondary students. However, a growing number of sites are reporting that C-TAP, begun in a less integrated fashion than desirable, helps speed up the rate of integration. This can be attributed to the development of overlapping vocabulary and sets of practices derived from the use of C-TAP by participating teachers (career technical and academic) and the natural integration that performance-based teaching and assessment practices foster.

Career Preparation Assessment

Many schools and training sites are not far enough along in reform activities, particularly in the full integration of academic and career-development curriculum, to smoothly implement the full C-TAP system and requirements. Many such sites have found it valuable to phase in pieces such as the portfolio. Others have turned to a more generic, portfolio driven, C-TAP derivative system known as the Career Preparation Assessment. The Career Preparation Assessment portfolio (Career Preparation Assessment, 1996) includes the same basic components and terminology as its C-TAP counterpart. However, the primary standards it assesses are the generic Career Preparation Standards, listed in Appendix B.

Similar to C-TAP, the Career Preparation Assessment has been developed jointly by educators and business representatives. Focus groups consisting of both constituencies have advised WestEd staff throughout the development and pilot test phase. Ongoing work with field test sites indicates that the Career Preparation Assessment holds promise as a tool to introduce workplace concepts and performance-based assessment methodologies to all secondary students and teachers. Additionally, the Career Preparation Assessment appears to be a significant tool to bring together career-technical and academic educators along with business and community representatives and parents to support overall secondary and postsecondary reform.

CONCLUSION

The C-TAP system of assessment has shown great potential at supporting systemic change at the secondary and postsecondary levels of education and worker training. This is accomplished primarily by changing the roles of teacher and students. Students become increasingly responsible for planning, implementing, evaluating, and revising their work; teachers shift to a more mentoring role. This active learning mode is expected to benefit students as they exit school or training and enter the workforce.

As with any assessment system, technical and feasibility considerations will play a large role in C-TAP's degree of implementation and the confidence in its results. We have argued that models of technical adequacy appropriate for performance-based assessments need to evolve along with their use. Such research is ongoing with the C-TAP system, particularly around the key issues of generalizability of results and consequences for teachers and students around various applications.

Full implementation of C-TAP also requires strong commitment at all levels of the education system, from the classroom teacher through the administration, out into the community. Significantly, C-TAP and its derivatives such as the Career Preparation Assessment provide a forum and context for this collaboration to occur with demonstrated benefits to employers, teachers, and students.

REFERENCES

American Educational Research Association, American Psychological Association, and National Council on Measurement in Education. (1985). *Standards for educational and psychological testing.* Washington, DC: American Psychological Association.

Ananda, S., & Rabinowitz, S. (1995). *Developing a comprehensive occupational certification system* (U.S. Department of Labor Contract # F-4323-3-00-90-33). Washington, DC: Institute for Educational Leadership.

Ananda, S., & Rabinowitz, S. (1996). *Lessons learned from the implementation of large-scale performance assessment systems.* Manuscript submitted for publication.

Ananda, S., Rabinowitz, S., Carlos, L., & Yamashiro, K. (1995). *Skills for tomorrow's workforce.* San Francisco, CA: Far West Laboratory for Educational Research and Development.

Baker, E., & O'Neil, H. (1994). Performance assessment and equity: A view from the USA. *Assessment in Education, 1,* 11–26.

Baker, E., O'Neil, H., & Linn, R. (1993). Policy and validity prospects for performance-based assessment. *American Psychologist, 48,* 1210–1218.

Brennan, R., & Johnson, E. (1995). Generalizability of performance assessments. *Educational Measurement: Issues and Practice, 14,* 9–12, 27.

California Department of Education. (1992). *Second to none: The report of the California high school task force.* Sacramento, CA: Author.

California Department of Education. (1994). *Career-technical education framework and standards.* Sacramento, CA: Author.

Career-Technical Assessment Program. (1995a). *Career Technical Assessment Program 1994–95 final report.* San Francisco, CA: Far West Laboratory for Educational Research and Development.

Career-Technical Assessment Program. (1995b). *Career Technical Assessment Program 1994–95 teacher guidebook.* San Francisco, CA: Far West Laboratory for Educational Research and Development.

Career-Technical Assessment Program. (1995c). *Career Technical Assessment Program 1994–95 student guidebook.* San Francisco, CA: Far West Laboratory for Educational Research and Development.

Career-Technical Assessment Program. (1006). *"How to" guidebook for the development of C-TAP assessment tasks.* San Francisco, CA: Far West Laboratory for Educational Research and Development.

Career Preparation Assessment. (1996). *Career Preparation Assessment portfolio guidelines.* San Francisco, CA: Far West Laboratory for Educational Research and Development.

Carl D. Perkins Vocational and Applied Technology Act of 1990, 10 U.S.C. 2321. (1990).

Council of Chief State School Officers Workplace Readiness Consortium. (1995). *Consensus framework for workplace readiness assessment.* Washington, DC: Council of Chief State School Officers.

Delandshere, G., & Petrosky, A. (1994). Capturing teachers' knowledge: Performance assessment a) and post-structuralist epistemology, b) from post-structuralist perspective, c) and post structuralism, d) none of the above. *Educational Researcher, 23*(5), 11–18.

Gearhart, M., Herman, J., Baker, E., & Whittaker, A. (1993). *Whose work is it? A question of large-scale portfolio assessment* (CSE Tech. Rep. 363). Los Angeles, CA: National Center for Research on Evaluation, Standards, and Student Testing.

Green, B. F. Comparability of scores from performance assessments. *Educational Measurement: Issues and Practices, 14,* 13–15, 24.

Grubb, W. (1991). *The cunning hand, the cultured mind: Models for integrating academic and vocational education.* Berkeley, CA: National Center for Research in Vocational Education.

Grubb, W. (1995). *Education through occupations.* New York: Teachers College Press.

Jaeger, R. (1995). Setting standards for complex performances: An iterative, judgmental policy-capturing strategy. *Educational Measurement: Issues and Practice, 14,* 16–20.

Kahl, S. (1995, April). *Scoring issues in selected statewide assessment programs using non-multiple-choice formats.* Paper presented at the annual meeting of the American Educational Research Association, San Francisco, CA.

Kendall, J., & Marzano, R. (1994). *The systematic identification and articulation of content standards and benchmarks.* Aurora, CO: Mid-continent Regional Educational Laboratory.

Linn, R. (1994). Performance assessment: Policy promises and technical measurement standards. *Educational Researcher, 23,* 4–14.

Mathison, S. (1988). Why triangulate? *Educational Researcher, 17*(2), 13–17.

Mehrens, W., & Popham, J. (1992). How to evaluate the legal defensibility of high-stakes tests. *Applied Measurement in Education, 5*(3), 265–283.

Mendel, R. (1994). *The school-to-career movement: A background paper for policymakers and foundation officers.* Indianapolis, IN: The Lilly Endowment.

Messick, S. (1995). Standards of validity and the validity of standards in performance assessment. *Educational Measurement: Issues and Practice, 14,* 4–8.

Monk, D. (1992). Education productivity research: An update and assessment of its role in education finance reform. *Educational Evaluation and Policy Analysis, 14,* 307–322.

Moss, P. (1994). Can there be validity with reliability? *Educational Researcher, 62,* 229–258.

Moss, P. (1995). Themes and variations in validity theory. *Educational Measurement: Issues and Practice, 14,* 5–13.

Nettles, M., & Nettles, T. (Eds.). (1995). *Equity and excellence in educational testing and assessment.* Boston, MA: Kluwer Academic Press.

Oregon Progress Board. (1991). *Oregon benchmarks: Setting measurable standards for progress, report to 1991 legislature.* Salem, OR: Author.

Rabinowitz, S. (1995a). Beyond testing: A vision for an ideal school-to-work assessment system. *Vocational Education Journal, March,* 27–29, 52.

Rabinowitz, S. (1995b, April). The Career-Technical Assessment Program: A tool for comprehensive integrated reform. *Second to None, 3,* 1–2.

School-to-Work Opportunities Act of 1994, 20 U.S.C. 6101. (1994).

Secretary's Commission on Achieving Necessary Skills. (1991). *What work requires of schools: A SCANS report for America 2000.* Washington, DC: U.S. Department of Labor.

Stern, D., Raby, M., & Dayton, C. (1992). *Career academics: Partnership for reconstructing American high schools.* San Francisco, CA: Jossey-Bass.

Wiggins, G. P. (1993). *Assessing student performance.* San Francisco, CA: Jossey-Bass.

Wirt, J. (1995). *Performance assessment systems: Implications for a national system of skill standards, volume II* (Tech. Rep.). Washington, DC: National Governors' Association.

Wohlstetter, P., & Oden, A. (1992). Site based management. *Educational Administration Quarterly, 28,* 529–549.

APPENDIX A

Complete this section to evaluate the student's performance on additional Model Curriculum Standards, areas of training, and/ or occupational competencies.

Area _____   Cluster _____

Standards, Areas of Training, Competencies	Advanced	Proficient	Needs Improvement	N/A
1. _____	_____	_____	_____	_____
2. _____	_____	_____	_____	_____
3. _____	_____	_____	_____	_____
4. _____	_____	_____	_____	_____
5. _____	_____	_____	_____	_____
6. _____	_____	_____	_____	_____
7. _____	_____	_____	_____	_____
8. _____	_____	_____	_____	_____
9. _____	_____	_____	_____	_____
10. _____	_____	_____	_____	_____

Comments on student's performance _____

_____

_____

Evaluator's signature _____   Date _____

Title _____

Student's comments (optional) _____

_____

_____

Student's signature _____   Date _____

Instructor's signature _____   Date _____

(Attach training plan, if available. Store completed evaluation form in student's portfolio.)

APPENDIX A

Supervised Practical Experience Evaluation Form

Directions: Evaluate and rate the performance of this student. Complete the form (front and back) using the Ratings described below. Add comments about the student's performance and sign the form. Discuss the evaluation with the student.

Student _____   School/Center _____

Job title _____   Training period _____

RATINGS:	
Advanced	– outstanding performance, demonstrates superior skill
Proficient	– very good performance, demonstrates satisfactory skill
Needs Improvement	– inadequate performance, skill needs improvement
Not applicable (N/A)	– performance was not observed on this skill

	Advanced	Proficient	Needs Improvement	N/A
I. Personal Skills				
Appearance	_____	_____	_____	_____
Attitude/Responsibility	_____	_____	_____	_____
Time management	_____	_____	_____	_____
Continued learning	_____	_____	_____	_____
Ethical behavior	_____	_____	_____	_____
II. Interpersonal Skills				
Interaction	_____	_____	_____	_____
Shared responsibility	_____	_____	_____	_____
Leadership	_____	_____	_____	_____
Conflict resolution	_____	_____	_____	_____
III. Thinking & Problem Solving Skills				
Creative thinking	_____	_____	_____	_____
Critical problem solving	_____	_____	_____	_____
Information gathering	_____	_____	_____	_____
Mathematical reasoning	_____	_____	_____	_____
IV. Communication Skills				
Speaking	_____	_____	_____	_____
Listening	_____	_____	_____	_____
Following directions	_____	_____	_____	_____
Writing	_____	_____	_____	_____
Telephone skills	_____	_____	_____	_____
V. Occupational Safety				
Application	_____	_____	_____	_____
Materials and equipment	_____	_____	_____	_____
VI. Technology Literacy				
Application	_____	_____	_____	_____
Overall Rating	_____	_____	_____	_____

(over)

APPENDIX B: CAREER PREPARATION STANDARDS

Career Preparation Standard 1: *Personal Skills.* Students will understand how personal skill development affects their employability. They will exhibit positive attitudes, self-confidence, honesty, perseverance, self-discipline, and personal hygiene. They will manage time and balance priorities as well as demonstrate a capacity for lifelong learning.

Career Preparation Standard 2: *Interpersonal Skills.* Students will understand key concepts in group dynamics, conflict resolution, and negotiation. They will work cooperatively, share responsibilities, accept supervision, and assume leadership roles. They will demonstrate cooperative working relationships across gender and cultural groups.

Career Preparation Standard 3: *Thinking and Problem Solving Skills.* Students will exhibit critical and creative thinking skills, logical reasoning, and problem solving. They will apply numerical estimation, measurement, and calculation, as appropriate. They will recognize problem situations; identify, locate, and organize needed information or data; and propose, evaluate, and select from alternative solutions.

Career Preparation Standard 4: *Communication Skills.* Students will understand principles of effective communication. They will communicate both orally and in writing. They will listen attentively and follow instructions, requesting clarification or additional information as needed.

Career Preparation Standard 5: *Occupational Safety.* Students will understand occupational safety issues including the avoidance of physical hazards in the work environment. They will operate equipment safely so as not to endanger themselves or others. They will demonstrate proper handling of hazardous materials.

Career Preparation Standard 6: *Employment Literacy.* Students will understand career paths and strategies for obtaining employment within their chosen fields. They will assume responsibility for professional growth. They will understand and promote the role of their field within a productive society, including the purpose of professional organizations.

Career Preparation Standard 7: *Technology Literacy.* Students will understand and adapt to changing technology by identifying, learning, and applying new skills to improve job performance. They will effectively employ technologies relevant to their fields.

Chapter 14

Workplace Readiness Portfolios

Jonathan Troper
CRESST/University of California, Los Angeles

Catherine Smith
Michigan Department of Education, Lansing, MI

Kids are different today. Until recently, students produced schoolwork and promptly forgot about it. At most, parents held onto assignments that received special praise or had sentimental value and squirreled them away to be admired at some unknown point in the future. Today, students can be seen maintaining files of their previous work, sometimes complete with autobiographies and analyses of the competencies they have demonstrated in classroom assignments and work accomplished out of school. They call these files *portfolios*. Until they are evaluated, however, they are simply folders containing students' documents.

Across the country, teachers are training their students to keep portfolios (e.g., Baker, 1993; Kentucky Department of Education, 1994; Koretz, Stecher, Klein, & McCaffrey, 1994; LeMahieu, Gitomer, & Eresh, 1995; Smith, 1993; and see Arter & Spandel, 1992, p. 44, for projects and clearinghouses). Moreover, numerous districts and states have developed systems to train teachers in portfolio assessment (e.g., Herman, Gearhart, & Aschbacher, 1996; Kentucky Department of Education, 1994; Koretz et al., 1994; LeMahieu et al., 1995; Smith, 1993). Why? Portfolio assessment promises to do what standardized tests cannot: to capture authentic evidence of complex skills (such as communication, responsibility, or leadership) in a way that supports good instruction, encourages self-assessment, includes extra-curricular accomplishments, and does not require day upon day of performance testing (Herman et al., 1996; Paulson & Paulson, 1994; Smith, 1993; Wolf, 1989).

Prompted by employers (e.g., Michigan Department of Education, 1996), schools, districts, and states are using portfolios as a means to focus students on developing general workforce competencies, such as those described in the SCANS reports (U.S. Department of Labor, 1991), and to assess students' proficiency in those competencies to determine their readiness to succeed as members of the workforce (e.g., California: see the chapter in this book by Stanley Rabinowitz; Utah: Richard Keene, personal communication October 13, 1995; Michigan: Michigan State Board of Education, 1993; Smith, 1993). A workforce readiness portfolio (WRP) system motivates or mandates teachers to have each student build a file of documents that demonstrate the skills and habits he or she has that make him or her ready for the world of work.

Because most readers of this chapter will not have had portfolios when they were in school, the first half of the chapter describes Michigan's workforce readiness portfolios and two methods that have been used to score them. We weave this description into three focal WRP implementation issues: the purposes of WRP assessment, the content of WRPs, and the scoring systems used to evaluate the content. In the second half of the chapter, we describe the range of choices people must make when they implement WRPs. We use the Michigan experience to highlight five of those choices: who mandates portfolio assessment, the resources required for development and implementation of WRP assessment, and the purposes, scoring, and technical quality of WRP assessment. We also present evidence that several factors beyond the skills intended for measurement by the assessment influence the content of students' portfolios. Such factors include the purposes of portfolio assessment, the scoring system, a school's portfolio culture (i.e., instructional practices and attitudes that differ between schools), and the wording of educational law regarding portfolios. We will highlight low-stakes uses for portfolio assessment (i.e., uses without significant reward or punishment for the student or the school based on assessment results) because some research findings have suggested that portfolios are not technically ready for high-stakes purposes (Herman et al., 1966; Herman, Gearhart, & Baker, 1993; Koretz et al., 1994; Linn, Baker, & Dunbar, 1991; Troper, Baker, & Smith, 1994b; Valencia & Calfee, 1991). Finally, we make some recommendations for implementation of WRP assessment and suggest directions for future research.

THE MICHIGAN WORKFORCE READINESS
PORTFOLIO SYSTEM: PURPOSE, CONTENT,
AND SCORING

Assessments are developed to meet purposes on several levels. The assessment system usually supports overarching educational goals, in this case improving key workforce skills in students (Michigan State Board of Education, 1993). At a more specific level, assessments of students provide information either at the level of measuring individual students' knowledge or skills, or at some more aggregated level, such as by sampling students within institutions, such as schools or districts. Portfolio content is mandated by schools or chosen by students to meet the purposes of the assessment. The method of scoring portfolios is designed to meet those purposes, and shapes the content teachers and students choose to put in portfolios. This section describes the overarching instructional goals and specific purposes, typical content, and scoring system for workplace readiness portfolios (WRPs) in Michigan.

Overarching Purpose: Improving Key Workforce Skills

Modern educational systems are designed to prepare all students to be ready for a number of careers. A WRP assessment system should fit this overarching goal (e.g., Michigan State Board of Education, 1993). Several studies designed to identify the skills necessary for success across careers have converged on a common set of skills. Assessment of students' readiness for the workforce should focus on such a common set (O'Neil, Allred, & Baker, 1992; chapter 1, this volume).

O'Neil et al. (1992) analyzed these studies and found four categories of skills. First are basic skills, which include the familiar communication skills of reading, writing, speaking, and listening, as well as arithmetic. Second are higher order thinking skills, which include problem solving, thinking creatively, and reasoning. Third are interpersonal and teamwork skills, which include participation on teams, negotiation, leadership, and working with diversity. Fourth are personal characteristics and attitudes, which include responsibility, self-management, career planning and development, and self-esteem. In addition, using technology was cited as important in at least two of the studies.

While mastering each of these skills can be viewed as worthwhile instructional objectives for all students, most of these skills do not have a home in any single academic discipline. In fact, some of them are more likely to be learned outside of school. Nothing unites them except the possibility of application in a common setting—the workplace. No single test is likely to produce information on all skills. Portfolios are an attractive alternative to testing because they can display authentic items (i.e., from real-life situations) generated at multiple locations while measuring a diverse array of instructional objectives.

Specific Purpose: For Students and Institutional Improvement

Typically, portfolios are vehicles for ongoing, in-class assessment by students and teachers and one-time, large-scale assessment by an outside agency. The former allows for complex description of the state of a student's skills (including ability to present material in a portfolio) and is used typically to help the student, parents, and teachers decide what skills and knowledge the student should focus on learning. The latter usually supports program evaluation or high-stakes decisions about funding (e.g., Kentucky Department of Education, 1994; Koretz et al., 1994; LeMahieu et al., 1995). External assessment may also support high-stakes decisions about graduation or college admission for individual students. External uses usually require boiling down assessment results into a single score or set of scores.

The Michigan portfolio system is designed to support both ongoing, classroom assessment "to enable students to discover, document, and develop their employability skills" and external assessment "for program evaluation and school improvement" (Stemmer, Brown, & Smith, 1992, pp. 32, 34). As we discuss later, there is evidence that these purposes may conflict if not managed well. In order to give students the freedom to discover, each student is given the power to choose the specific content of his or her portfolio. In order to guide students and teachers to cover the general workplace skills previously listed, the state Department of Education has produced a list of key skills and a scoring system of "benchmarks" to provide guidelines for what counts as evidence for each skill.

Content of a Workforce Readiness Skills Portfolio

A student's workforce readiness portfolio provides evidence for the student's attainment of the general skills listed earlier. A weak portfolio may be a jumble of assignments, awards, and test scores, "simply a container of student work or assessments" (Paulson & Paulson, 1994, p. 4). A persuasive portfolio is deliberately organized by the student to display items that tell any portfolio reader a story about what and how the portfolio author has learned in some area (Arter & Spandel, 1992; Paulson & Paulson, 1994).

Excellent workforce readiness portfolios include exhibits such as letters of recommendation, awards, photographs of student work or of students at a place of employment, and actual work samples. These are woven together with explanations and analyses as to how each exhibit demonstrates specific skills (Smith, 1993; Troper, Baker, & Smith, 1994a). A student might, for example, use a photograph of the student training others to handle customers, a letter from the employer commending the student for excellence in consistent attendance and interpersonal skills, and an explanation of context from which the photo and letter came to demonstrate the skills of responsibility, leadership, and oral communication. Another student might demonstrate the same three skills using a local newspaper article describing the student's career as captain of the basketball team and a paragraph explaining how the article illustrates those skills.

Scoring Workplace Readiness Portfolios: Two Methods

Evaluating or scoring portfolios is carried out differently depending on assessment purposes (Herman et al., 1996). High-stakes purposes, such as allocating funding to schools or determining a student's placement in an ability track or admission into a program or college, usually require the production of a single score per portfolio so that students or schools can be judged against each other or against some standard. Low-stakes purposes, such as in-class diagnosis, may benefit from a scoring system that produces a score for each instructional objective. For classroom uses, it can work like the grading of any assignment. The teacher judges portfolios unconstrained by a scoring system or, in some cases, develops a classroom scoring rubric in concert with students or other teachers. For external assessment uses, a few people are selected and trained to use a standardized scoring system. In a hybrid system, such as Kentucky's or Vermont's, a standardized system is taught to large numbers of teachers acting as external scorers so that they can also use the scoring system in the classroom (Kentucky Department of Education, 1994; Koretz et al., 1994).

Scoring systems influence portfolio contents. Teachers and students work to make portfolios match the criteria by which the scoring system will evaluate portfolio exhibits. To encourage excellent portfolios and help readers interpret portfolio exhibits, the Michigan scoring system considers an exhibit adequate proof[1] that the student has demonstrated that skill in at least one situation only if the student has also included a written analysis reflecting on how the exhibit demonstrates the skill (Michigan State Board of Education, 1993; Stemmer & Smith, 1993). Without a student's written analysis for a specific skill, a portfolio is considered unscorable for that skill by the Michigan scoring system because the student has not shown the intent to demonstrate that skill. However, in the early stages of portfolio development, most students are not expected to create excellent portfolios (Michigan State Board of Education, 1993), and statistics from the first statewide scoring show that the vast majority (almost 99%) of portfolio exhibits did not have any students' written analyses attached (Michigan Department of Education, 1996). Because so few portfolios were expected to register on the Michigan scale, Michigan is also using a modified version of a scoring system developed at UCLA through the Centers for Research in Evaluation, Standards, and Student Testing (CRESST) that does not require analyses and does not intend to measure evidence of a student's actual skill level, but rather detects whether a portfolio contains credible exhibits on each skill.

The UCLA/CRESST scoring system was derived from those skills determined to be important in workplace readiness skills studies (e.g., Mehrens,

---

[1]Proof in a legal sense of showing the intention to demonstrate the skill plus action that is credible.

1989; O'Neil, Allred, & Baker, 1992; Stemmer & Smith, 1993; U.S. Department of Labor, 1991) in conjunction with Michigan teachers and administrators involved in piloting WRPs. The skills were described at a broad level. That way, they could be easily comprehended by students, teachers, and employers.

Table 14.1 shows the UCLA/CRESST scoring system as it was used in a study at UCLA scoring a sample of 43 WRPs from nine Michigan schools piloting portfolios (Troper et al., 1994a, 1994b). In this study, five raters were trained to distinguish between WRP exhibits that were and were not credible examples for each skill (Troper, Baker, & Smith, in press). When the raters had reached 80% agreement with prescored exhibits, they were considered ready to score 43 actual portfolios. The training design was based on concept learning theory. Whereas rater training often uses only examples to provide raters with prototypes for each concept or skill level being scored, this training included nonexamples to help raters make clear distinctions and avoid overgeneralization (Tennyson & Cocchiarella, 1986).

A more complex scoring system was developed by the Michigan Department of Education and implemented by National Computer Systems. It measures whether or not portfolios meet "benchmarks" (see Table 14.2)

TABLE 14.1
Items in the CRESST Employability Skills Portfolio Scoring Rubric

---

A. General impression (holistic measures: 1 = very poor, 7 = superb
    1. General impression of skills
    2. Quality of portfolio presentation regardless of content
B. Communication skills[a]
    3. Writing
    4. Reading
    5. Speaking
C. Personal management skills[a]
    6. Career development
    7. Flexibility and initiative
    8. Responsibility
D. Interpersonal and team skills[a]
    9. Leadership
    10. Responsiveness to others
    11. Contributing to a team
E. Thinking/problem solving and technical skills[a]
    12. Carrying out a planned project
    13. Solving problems or proposing solutions other than multistage projects
    14. Understanding systems
    15. Science and technology
    16. Mathematics

Note. [a]Analytic measures: each measure scored "Yes" if the portfolio contained a credible exhibit showing that the student used that skill at a level that seemed adequate for generic entry-level work and "No" if not. From Troper, Baker, and Smith (in press). Reprinted with permission of the authors and UCLA/CRESST.

TABLE 14.2
Benchmark Items From the Michigan State Department of Education Employability Skills
Portfolio Scoring Rubric

Academic Skills	Personal Management Skills	Teamwork Skills

1. Communication
  Writing
  Speaking/listening
  Reading
2. Mathematics
3. Science and technology
4. Problem solving
	5. Responsibility
	6. Organization
	7. Flexibility and initiative
	8. Career development
		9. Team communicating
		10. Responsiveness to others
		11. Contributing to a team
		12. Team membership
		  Leadership
		  Following

Note. From Stemmer and Smith (1993).

for each of twelve workforce readiness skills (Michigan Department of Education, 1996; Michigan State Board of Education, 1993; Stemmer & Smith, 1993). Unlike a traditional test or even a performance assessment, the Michigan system does not purport to measure multiple levels of student skill. Instead, it uses a system of subskills (called "diamonds" after the diamond-shaped graphics used to differentiate them from the main benchmark skills—see Table 14.3 for some examples) to help students and teachers locate more specifically students' strengths and weaknesses.

The UCLA/CRESST and Michigan benchmark scoring systems are similar in three ways. First, the skills measured by the UCLA/CRESST scoring system are defined at the same level of generality as the Michigan benchmarks and are almost identical. Second, both scoring systems use the same rules for evaluating the credibility of exhibits. Third, neither system measures students' actual skill level. However, the two scoring systems also differ in two key aspects. First, the Michigan scoring system does not consider an exhibit as evidence of a skill unless it is accompanied by an analysis written by the student, explaining how the exhibit demonstrates that the student used a skill. Second, whereas the UCLA/CRESST system looks at exhibits as credible or not credible indicators that a student has used a skill at the level of generality of the Michigan benchmarks, the Michigan system determines whether an exhibit plus its analysis provide evidence that the student has used a diamond subskill. Only if a student has provided evidence for all the diamond subskills under a benchmark is credit given for the benchmark.

The skills in Table 14.2 come from the employability skills profile developed by the Michigan Governor's Employability Skills Task Force and validated in a survey of 8,000 Michigan employers (Mehrens, 1989), and they fit well with the common skills identified in O'Neil et al.'s (1992) comparison of national studies. Table 14.3 shows assessment benchmarks developed by employer committees in 1992 and tried out in 12 Michigan pilot schools in 1993.

How Raters Evaluate Exhibits or Evidence

Since portfolio scoring systems rely on human raters to evaluate evidence of a student's achievement, the rules by which raters make decisions about that evidence are central. Here, we describe briefly what raters are trained to look for in each WRP scoring system. Later, we will discuss some potentially controversial issues in scoring.

In the UCLA/CRESST scoring system (Troper et al., 1994a, 1994b), portfolios receive a *yes* score for a skill when the portfolio contains a credible exhibit showing that the student has used that skill at a level that seems (to trained raters) adequate for a generic entry level job, and a *no* if it does not. Raters are trained to rule out exhibits that are not credible, such as unsupported student descriptions of their own skills. Exhibits are categorized as being a direct demonstration of a skill, a third party statement that the student has shown the skill, or the student's own statement that they have used the skill. In this case, a student's own statement is an unsupported claim as opposed to an analysis of an exhibit that itself is credible. An exhibit counts as credible when it comes from a credible third party, such as in a letter of recommendation from an employer or a newspaper article about the student, or is demonstrated directly by the artifact in the portfolio, such as an essay exploring the match between a student's skills and the requirements of a career. For most skills, a student's own statement is not counted as credible. However, for certain skills, a student's own statement is accepted as credible if it is supplemented with corroboration from a reliable third party or direct demonstration or gives enough detail to be believable (see Troper et al., in press, for further details).

One drawback to this system is that the level of accomplishment students must demonstrate through their exhibits is rather low so as not to burden raters with the difficult and time-consuming task of judging the quality of students' work. For instance, any test counts as a credible exhibit as long as it received a grade of B or higher for a classroom test or a score of 50th percentile or higher for a standardized test—a lenient scoring criterion. Thus, for skills where credible exhibits are produced in abundance in most schools (e.g., writing and math), most portfolios could easily receive *yes* scores for those skills even though we may suspect that many or most students are weak in that area. On the other hand, credible exhibits are not easy to come by in classrooms for most skills. For instance, the exhibits used for teamwork and leadership generally come from outside classrooms, such as sports news photos and articles from newspapers or letters of recommendation.

TABLE 14.3
Evidence Required for Three of the Twelve Michigan Skills Benchmarks

---

1. Communication

    Writing

    - writing a report by organizing important information and expressing different points of view
    - writing legibly, using correct spelling and grammar
    - filling out forms; writing orders

    Speaking and listening

    - speaking clearly to convey idea
    - using business or technical terms as needed
    - listening actively by asking appropriate questions
    - giving and taking instructions
    - presenting a report and answering questions

    Reading

    - reading simple and complex instructions such as manuals
    - evaluating the quality of written materials you produce or use
    - combining and using a variety of information sources: manuals, maps, charts, graphs

5. Responsibility

    - having a good school/work attendance record
    - demonstrating self-control where minimum directions and supervision are given
    - planning for a decision that significantly impacts your life plans (e.g., choosing a college/career path)
    - meeting school/work deadlines

11. Contributing (to a team)
    - understanding and contributing to the mission of the team
    - representing the team to others that assist in the mission
    - demonstrating loyalty to the team and showing commitment to the team's growth and improvement
    - helping the team identify goals, and contributing to achieving them
    - making and following a set of rules and procedures that will contribute to the mission
    - helping the team develop to meet needs in the future

Note. In order for a portfolio to meet one of these three benchmarks, the portfolio must contain credible evidence (exhibits plus written analysis) to demonstrate use of all related subskills marked with a bullet. A credible analysis plus exhibits covering less than all of the subskills counts as partial evidence for the benchmark. Reproduced with permission of Michigan State Board of Education (1993).

The data provide a more complex picture. It is instructive to compare the results from the UCLA study of portfolios from pilot schools (Troper et al., 1994b) with those from recent scoring of a statewide sample of 1,050 11th-grade WRPs from 42 randomly sampled Michigan schools that began implementing them after the legislative mandate (Michigan Department of Education, 1996). As seen in Table 14.4, which reports scores using the UCLA/CRESST scoring system, while credible writing exhibits were found in the vast majority of portfolios (88% in the pilot school sample and 76% in the statewide sample), the lenient-scoring problem was not so pronounced in the statewide sample for other academic skills where typical exhibits were simply tests, (e.g., 57% in math, 55% in science), perhaps because many of the tests in portfolios in the statewide sample had low grades (less than B or lower than 50th percentile). In fact, in the statewide sample, credible exhibits for several skills were found in very few portfolios (e.g., 7% for problem solving, 6% for responsiveness to others, and 15% for speaking). Because the law mandated portfolio exhibits on career exploration, the vast majority of portfolios (76%) included credible exhibits for that skill.

The pilot schools' portfolios contained far more credible exhibits for teamwork and personal management skills than did portfolios from schools in the statewide sample. For instance, credible exhibits for responsiveness to others were found in 35% of pilot portfolios compared to 6% of statewide portfolios, and for flexibility and initiative in 44% of the pilot sample compared to 16% of those statewide. Also, within each sample, clear differences between schools registered on the UCLA/CRESST scale. In the pilot sample, one school's portfolios showed credible exhibits for more skills than the others, and raters noted that the portfolios from this school were noticeably better (Troper et al., in press). In the statewide sample, the vast majority of students' analysis statements came from two schools (Michigan Department of Education, 1996). On the other hand, similar proportions in both sets of portfolios contained credible exhibits for academic skills. We interpret these findings to suggest that portfolios' contents reflect aspects of school curriculum, including the extent to which the portfolio culture at a school or in a district promotes student reflection on each workforce readiness skill. By portfolio culture, we mean, for instance, how enthusiastic teachers and administrators are about portfolios, how often students are given time to work on portfolios, and what instructions students are given about how to choose and reflect on portfolio contents.

As noted, portfolios are considered by the Michigan scoring system to be incomplete for a diamond subskill and thus not scoreable for that subskill if they contain no student analysis for that skill. There are three reasons for this. First, this decision makes a student's development of a WRP over 5 years a sounder educational experience, allowing him or her to

TABLE 14.4
Percentages of Portfolios Showing Credible Exhibits of Each Employability Skill When
Scored Using the UCLA/CRESST Scoring System

	1993 Pilot Schools			1995 Statewide Sample[a]	
Workplace Readiness Skill	None[b]	One or More[b]	Disagree[c]	None	One or More
Communication skills					
Writing	2	88	9	24	76
Reading	9	53	37	45	55
Speaking[d]	63	21	16	85	15
Career					
Career exploration[d]	47	44	9	23	76
Personal management skills					
Flexibility	35	44	21	84	16
Responsibility	40	51	9	66	34
Teamwork skills[d]					
Leadership[d]	51	23	26	86	14
Responsiveness to others	47	35	19	94	6
Contributing to a team	26	53	21	69	31
Team communicating[d]	--	--	--	95	5
Problem solving skills					
Organization and planning	67	9	23	68	32
Problem solving	79	2	19	93	7
Technical skills					
Understanding systems[d]	91	0	9	--	--
Science and technology	35	44	21	45	55
Math	33	60	7	43	57

Note. N = 43 portfolios from 9 schools for the pilot sample. N = 1050 portfolios from 42 school for the statewide sample. Reproduced with permission of Troper et al. (in press) and Michigan Department of Education (1996).
[a]With the statewide sample, pairs of raters scored each exhibit by consensus. Reliability estimates from individually scored test exhibits are not yet available.
[b]Percentages of portfolios where all or all but one of five raters agreed that the portfolio contained a credible exhibit for the skill. Raters scored the pilot sample individually.
[c]Percentages of portfolios where more than one rater disagreed that the portfolio contained a credible exhibit for the skill.
[d]Definitions for these skills differed between two versions of the UCLA/CRESST scoring system used for the pilot sample and the statewide sample. In the case of Teamwork, a benchmark called Team Communicating did not have an exact analog in the UCLA/CRESST scoring system. For the statewide sample, the method of rating was the same, but the actual skills rated corresponded to the Michigan benchmarks, which differed somewhat from the original UCLA/CRESST skills. For further details, see Troper et al. (1996) and Michigan State Board of Education (1993).

show a growing understanding and application of skill by explaining why specific exhibits show skills. Second, writing analyses was intended to help students see the parallels between adolescent experience and adult workforce situations, to make transfer from one to the other easier. Finally, that the Michigan rules require an analysis for an exhibit to be considered evidence of a skill helps provide contextual information that makes up for reviewer ignorance and that could reduce reviewer bias. For instance, urban reviewers at UCLA scoring exhibits without analyses could not see the workplace skills inherent in some Michigan farm exhibits.

In summary, the UCLA/CRESST and Michigan scoring systems are complementary. The UCLA/CRESST scoring system registers whether or not a portfolio has a credible exhibit for each of fourteen broad workplace readiness skills, while making no judgment as to the quality or depth of skill. The Michigan scoring system registers whether or not a portfolio has credible evidence that the student has demonstrated each of many subskills for each of twelve benchmarks. Credible evidence for a subskill requires one or more credible exhibits plus the student's written analysis explaining how the exhibit(s) demonstrates the use of the subskill. Groups of subskills add up to a benchmark, comparable to the broad skills found in the UCLA/CRESST scoring system. If a portfolio shows credible evidence for all subskills within a benchmark, the student is credited with having met the benchmark. If not, then the missing diamond subskills are considered diagnostic information pointing to areas of potential improvement. The UCLA/CRESST scoring system detects variations between portfolios that do not contain students' analyses. Such portfolios do not even register on the Michigan scoring system. While most portfolios currently do not include analyses, we expect that in mature WRP cultures, the UCLA/CRESST scoring system will be swamped by excellent portfolios and the Michigan scoring system will be better able to detect variations.

IMPLICATIONS OF CHOICES IN DESIGNING PORTFOLIO ASSESSMENT

Policy makers designing portfolio assessments build their systems based on choices along several dimensions. Portfolios are too new for research to have delved into the implications of every dimension along which portfolio programs may differ. However, researchers have identified the important dimensions (see Table 14.5) and provide some guidance for making choices (see Arter & Spandel, 1992 for an excellent tutorial; Herman et al., 1996; Michigan State Board of Education, 1993; Moss et al., 1992). Future research efforts can untangle the effects of the various choices made by educators implementing portfolios. We discuss just a few critical dimensions, mostly in light of the Michigan experience.

The Michigan experience with WRPs can shed light on some of the dimensions in Table 14.5. We focus on five of particular interest: who mandates using portfolio assessment, resources provided, the dual purpose of portfolio assessment, technical review, and scoring. Other dimensions, such as maturity, are woven into the discussion. In some cases, WRPs differ from portfolios intended to measure other knowledge or skills.

Who Mandates

While many of the benefits claimed for portfolios occur in the classroom, portfolios are often implemented in response to administrative or legislative mandates (e.g., Kentucky Department of Education, 1994; Michigan Department of Education, 1996). For instance, in Michigan, the Employability Skills Task Force recommended voluntary WRP assessment for Grades 8 through 12. Then the state legislature *required* that schools provide portfolios for these students, but with few mandates as to the content of the portfolios (standardized test scores, grade transcripts, evidence of career exploration, and anything else the student wanted to put in to show achievement, including materials from outside of school). Many schools have chosen to tailor the content of those portfolios to the workforce readiness benchmarks. When schools are legally required to carry out initiatives that require significant amounts of work, their first response is often, "What is the minimum we can do to meet the requirements of the law?" This response prevents the possibility of the development of a mature portfolio culture, with portfolios an instructional practice well integrated into the curriculum.

From the Michigan experience, it is our impression that voluntary participation (combined with extra resources—see next section) resulted in higher quality portfolios. In Michigan, 22 school districts participated voluntarily during the pilot phase, applying for and receiving very small grants (Stemmer & Smith, 1993). Scoring sessions on WRPs from those schools in 1994 found many portfolios meeting the Michigan benchmark standards for at least one skill. Pilot schools that produced the best portfolios all had at least one very enthusiastic teacher or administrator who championed the process. Portfolios from those schools often contained evidence that demonstrated diamond skills and sometimes met entire benchmarks (Michigan Department of Education, 1996). One district produced a video tape and kit describing how to produce portfolios that became a hot sales item when the rest of the schools in the state were mandated to begin implementation. In contrast, recently scored WRPs from schools that waited to produce portfolios until the mandate went into effect included not a single portfolio meeting a single skill benchmark (i.e., having credible exhibits for each of the diamond subskills for the main skill and students' analyses explaining why their exhibits were proof that they had displayed

TABLE 14.5
Dimensions Along Which Portfolios and Portfolio Assessments Differ

Dimension	Typical Choices
Who mandates portfolios	State, district or local education agency; school administration; individual teacher.
Resources provided	Time, staff development, consulting, and logistical support such as materials.
Who determines portfolio contents	Student; state, district, or local education agency; school administration; individual teacher.
Required contents	None versus suggested guidelines versus requirements for specific items, such as the student's analysis explaining how each exhibit demonstrates skills to be measured by the assessment, or specified exhibits, such as, resume or letter of recommendation.
Who assesses	Student with help of teacher or peers; peers; teacher; panel of teachers and local employers; outside assessor, such as state education agency.
Teacher role	Professional informed by assessment versus implementer of bureaucratic accountability measures.
Purpose of assessment/uses of resulting scores	Ongoing student development versus summative score; analysis of individuals versus institutions; low-stakes (e.g., personal development, communication with parents and employers, or school improvement) versus high-stakes (job, graduation, admission to college, or determination of school funding).
Timing	What grade? How often? During the year (allows possibility of diagnostic feedback) or at the end?
Stakeholders	Involvement and goals: To what extent are teachers, students, parents, and employers involved in the design and implementation? Do stakeholders agree with assessment purpose (especially high vs. low-stakes)?
Criteria	Reviewer's own judgment versus standardized scoring rubric. Clarity. Teacher and student comprehension of criteria.
Criteria development	Criteria developed by experts and handed down to users versus evolved over time through discussions among educators.
Scoring	Holistic (one score across whole portfolio) versus analytic (e.g., one score per skill).

(Continued)

TABLE 14.5
(Continued)

Dimension	Typical Choices
Technical quality	Assessment system analyzed versus unanalyzed for scoring reliability and validity for assessment purposes, including alignment of assessment consequences with overarching instructional goals.
Reporting format	Written narrative versus categorical or numerical score(s); single score versus profile of many scores.
Level of aggregation and analysis	Individual student score or profile versus school or district score or profile.
Staff development	Minimal versus extensive; portfolios integrated into, versus peripheral to, staff development.
Integration	Integration of portfolios into instruction, for example, portfolios act as feedback for planning of instruction or instructional modules designed to produce portfolio exhibits.
Maturity	Maturity of portfolio culture: portfolio-wiseness of teachers and students, stage of diffusion of best practices.

that diamond subskill). Rather, most portfolios contained the absolute minimum required by the law (Michigan Department of Education, 1996).

On the other hand, implementing a system as complex and different from traditional educational methods as portfolio assessment may work best when educators at all levels are involved in the development of the portfolio assessment system. That way, educators develop a common comprehension about the nature and purpose of the portfolio assessment system (Gitomer, 1993; LeMahieu et al., 1995). This takes significant resources that only a large institution, such as a district or state, can provide. Furthermore, we have evidence that the wording of the Michigan mandate has had an influence on the content of instruction. Career Development is specifically mentioned in the Michigan law. Statistics in Table 14.4 show that students in the statewide random sample provided more credible exhibits on Career Development than they did for any other skill, despite the fact that teaching math and writing is a universal practice, while teaching career development is rare.

Whether schools are motivated by voluntary participation or a mandate, successful implementation of portfolio assessment may require sincere buy-in from participants. If so, whoever mandates or calls for voluntary participation must provide inspirational leadership that motivates such buy-in and the resources to sustain the program over time.

Resources Provided

Portfolio implementation and scoring can be expensive (Koretz et al., 1994; LeMahieu et al., 1995). The greatest cost is teachers' time. It takes time for teachers or other raters to score portfolios. After several hours of training for recent WRP scoring, raters required between 5 minutes and 1½ hours to rate each portfolio, depending on its complexity. More importantly, it takes time to learn how to teach portfolios and to work on them regularly with students. It takes teachers time to figure out even the most basic logistics. Issues such as where to keep portfolios and what materials to use to hold documents do not have simple answers agreed on by most schools (Stemmer & Smith, 1993). Then it takes years to become proficient at training students to produce complete and persuasive portfolios. Stemmer and Smith estimated in 1993 that 10 years would be required for implementation of a fully developed WRP assessment system. Two years later, this still sounds realistic. Preliminary statistics from the recent statewide scoring show that fewer than 1% of all portfolio exhibits included written analyses, despite the emphasis on their importance in the materials produced by Michigan state administrators (Michigan Department of Education, 1996). Only 6% of all portfolios in the statewide sample contained an analysis for any exhibit. This compares to results from 15 pilot schools that bid for and won small grants in 1994 to try out the scoring process. They put in far more staff development on employability skills and on how to provide proof of skills than did almost all of the schools in the statewide sample. Most of the students in these 15 schools produced at least one credible analysis to accompany an exhibit, and many of their portfolios showed credible analyses for a dozen or more exhibits.

Dual Purposes of Portfolio Assessment

Research has suggested a potential conflict between the classroom purpose of portfolios—ongoing assessment to support instruction and empower students to self-assess—and external, accountability uses of assessment results, especially if external agents wish to see the assessment produce valid individual student scores or if the stakes are high. One vision of a mature portfolio culture sees teachers helping develop scoring criteria and students acting like professional authors, revising their work with the assistance of their peers, parents, and teachers. This, however, conflicts with the need for portfolio scores in a large-scale assessment to reflect individual student achievement (Gearhart & Herman, 1995) and for portfolios to be comparable enough for technically sound scoring (Herman et al., 1966). Gearhart and Herman found evidence that different teachers gave different amounts of assistance to their students in writing pieces entered into student portfolios, blurring the validity of portfolio scores as measures of individual achievement.

Furthermore, students might be reluctant to take risks in their portfolios when the stakes are high. Feeling free to take risks is important for learning (Dweck & Leggett, 1988). But controversial exhibits in Michigan WRPs, such as persuasive essays on abortion or drug use, have been found to lower the scores given by portfolio raters (Troper, Baker, & Smith, 1994a). Also, if portfolios are used to grant students admission to college or employment, portfolios could be excellent sales jobs hiding weak skills. We feel that the best use of WRPs is in the classroom and in the students' personal use of their own portfolio, such as making better career and college decisions through greater awareness of their skills. Low-stakes assessments may support instruction better than high-stakes. It is also possible that the development of two portfolios—one for ongoing classroom assessment and a presentation portfolio for display to outside reviewers—may help protect the messy processes involved in classroom uses from the more stylized requirements of external assessment.

Some Issues in Standardized Scoring of Portfolios

Scoring portfolios according to a standardized scheme involves setting criteria by which individual portfolio exhibits and/or the portfolio as a whole will be judged and making those criteria public, defining measurement scales for those criteria, and setting standards for levels of performance (i.e., describing what about an exhibit or a portfolio determines its position on each measurement scale; Arter & Spandel, 1992). This is not a simple exercise. Many of the decisions about these three steps can be quite controversial.

Criteria for assessment represent the constructs being measured by the assessment. They should flow from the intended uses of assessment scores and the theoretical structure of the content being assessed (Messick, 1995). For instance, the qualities or skills assessed in workplace readiness portfolios should be those that make a difference in the success of real workers. In addition, criteria should be clearly communicated and meaningful to all involved in portfolio implementation, including students, teachers, and parents (Herman et al., 1996; LeMahieu et al., 1995). They must also be fair to all, not measuring extraneous variables unequally distributed among students (Herman et al., 1966; Messick, 1995). Culling criteria from studies of the skills used in a wide variety of workplaces, such as the studies mentioned early in this chapter, is intended to provide a fair and theoretically sound foundation. Furthermore, grouping the subskills described in those studies so that they fit everyday language (e.g., teamwork, responsibility) makes it possible to communicate them to students, teachers, and parents so that all parties find them meaningful.

Turning criteria into scales can be done in different ways (Herman et al., 1996). Holistic scoring measures a whole portfolio on a single dimension, such as overall employability (Herman et al., 1996). Analytic scores each measure a specific aspect of a portfolio or one exhibit or related group of exhibits at a time (Herman et al., 1996). For example, the fourteen UCLA/CRESST skill scores and all the Michigan benchmark scores are analytic. Holistic portfolio scoring may be cheaper because it requires less time for training and for scoring each portfolio. However, analytic scores may be more likely to provide specific enough feedback to improve learning (Herman et al., 1996). In either case, scales must contain intervals large enough so that raters can distinguish one level from another. Because we believed that raters could not discern students' actual skill levels from a portfolio, we designed both the Michigan and UCLA/CRESST scoring systems without requiring raters to distinguish between adequate and excellent skill. Instead, we instructed raters simply to judge for each skill whether or not each portfolio contained an exhibit that met the credibility requirement, that is, that the exhibit was a reasonably believable indication that the student had demonstrated the skill at a level adequate for a generic entry-level job.

Fitting exhibits to points on measurement scales requires rules of evidence to guide raters' judgment. Determining what standard of proof to require is not easy. Can a student's own statement count as evidence of a skill? In almost all cases, we decided not. Should students provide evidence of the use of a skill in multiple contexts? This could allow us to believe that the student's skill was generalizable. However, requiring a portfolio to show multiple uses of a skill would result in huge, costly portfolios. We decided that an exhibit that was credible and showed that a student had used the skill was enough.

Four other issues provided some thorny problems. First, some exhibits could be phony. Restricting exhibits to in-class experiences would allow teachers to ferret out phony exhibits, but would make the evidence gathering unfair to students who could only provide evidence of workplace skills from experiences outside of school. Second, assessing exhibits produced by a group requires portfolio reviewers to rely on explanations found in students' analyses. If a student claims to have done most of the work, the reviewer must accept the claim.

Third, some portfolios contain very similar exhibits, but the student authors write divergent analyses. On this issue, the Michigan and UCLA/CRESST scoring systems take different positions. The Michigan scoring system evaluates the exhibit only for skills for which the student makes a claim in a written analysis. In contrast, the UCLA/CRESST scoring system counts an exhibit as credible if it meets certain criteria (see section "How Raters Evaluate Exhibits or Evidence," earlier in this chapter) regardless of which skills a student declares or even whether the student has written an analysis at all. Both positions are taken for fairness. The Michigan system refrains from assessing a student against a standard the student is not trying to meet. In the case of an exhibit that could have been used to demonstrate a skill for which a student has not made a claim, the reviewer can write suggestions to the student to declare additional skills in his or her analysis. The UCLA/CRESST system refrains from penalizing a student who does not realize that an exhibit is credible for a certain skill.

Fourth, we found suggestive evidence that superficial aspects of portfolios might bias raters. In the recent scoring in Michigan, it was noted that portfolios from each school tended to look alike, and look different from those of other schools. It is possible that reviewers may prejudge portfolios once they have seen several from the same school that are all memorably weak or strong.

Technical Quality

Conceptions of validity are expanding as researchers reexamine assessments as tools used in institutional contexts for purposes broader than measurement. Herman and Winters (1994) identified three questions involved in validity and reliability issues. "What do scores mean?" "Do inferences from the scores lead to accurate decisions about students? programs? schools?" and "Are the scores consistent or stable?" (p. 50). In addition, rising importance is being attributed to the alignment between the actual consequences of an assessment and the overarching instructional goals of the educational institutions in which the assessment is being implemented—now called "consequential validity" (see Linn et al., 1991; Messick, 1995; Moss, 1995).

What Do Workforce Readiness Portfolio Scores Mean? The validity of the UCLA/CRESST scoring system was tested by examining the magnitude of correlation between UCLA/CRESST ratings and employers' judgments of the same portfolios; the results were encouraging (see Table 14.6). Employers were asked to imagine that they had just had identical, adequate interviews with identical people, but one had left the portfolio behind. They then rated their willingness to hire the candidate with the portfolio based entirely on the portfolio's contents.

Employers' willingness to hire correlated moderately to strongly with many of the UCLA/CRESST scales,[2] especially those measuring skills strongly demanded by employers (e.g., Mehrens, 1989; U.S. Department

---

[2]Further description of the analyses done on this data can be found in Troper, Baker, and Smith (in press).

of Labor, 1991)—teamwork ($r = .88$) and personal management ($r = .85$) and the holistic scales measuring their general impression of a portfolio author's skills ($r = .83$) and quality of portfolio presentation ($r = .88$). Note that, as would be expected, low correlations were found for scales with restricted ranges, that is, where almost all portfolios either had credible exhibits (writing) or had no credible exhibits (understanding systems, problem solving, and organization and planning). In the case of career exploration, the lack of correlation may have been due to credible exhibits appearing in a few otherwise weak portfolios from a special education school or due to an inappropriate criterion, that is, employers do not decide on whom to hire based on the candidate's having explored careers.

That the quality of portfolio presentation also correlated strongly with willingness to hire was a reminder that this study measured portfolios and did not claim to test students' actual employability skills. To measure to what extent WRP scores reflect students' actual workforce readiness skills might require longitudinal studies of students' success on actual jobs.

Were the Scores Consistent? Interrater agreement was relatively good for many scales (see Table 14.6). Interrater agreement was above 70% for almost all UCLA/CRESST scales. For the seven-point holistic scales, Cronbach's alpha was found to be above .75. Such levels suggest that the scales are not ready for high-stakes assessment, but can be used where individual portfolio scores are aggregated. In the case of reading, low interrater agreement is likely to have been due to the difficulty of coming to agreement on what counts as credible evidence for the skill (Troper et al., 1994b).

Because each student typically had only one exhibit for each skill, it was impossible to measure interexhibit consistency. Furthermore, stability of scores over time is not a meaningful issue when portfolios are assembled and revised over a long period. It has been argued that raters may need to see student analyses of each exhibit in order to comprehend the context or meaning of the exhibit (Paulson & Paulson, 1994). This was not tested in our study and should be in the future.

Might Inferences From the Scores Lead to Accurate Decisions About Students? Programs? Schools? Herman and Winters (1994) include in this category the diagnosticity, fairness, and consequences of assessments. If diagnosis of individual students is expected from WRP assessments, then any bias in diagnosis becomes an issue of fairness. Despite the encouraging nature of the data just presented, the great differences between schools cautioned us against using portfolio scores to diagnose individual achievement. The number of skills for which there were credible exhibits in each portfolio may have been dependent on the student's school. One school's portfolios showed nearly twice as many skills than did portfolios from most of the

TABLE 14.6
Reliability and Concurrent Validity Statistics for the CRESST Employability Skills Portfolio
Scoring Rubric

Scale	Reliability[a]	Validity[b]
Holistic scales		
General impression of author's skills	.76	.83
Quality of portfolio presentation	.80	.88
Analytic scales		
Total number of skills[c]		.92
Teamwork skills[c]		.88
Leadership	74%	.79
Responsiveness to others	75%	.86
Contributing to a team	79%	.77
Personal management skills[c]		.85
Flexibility and initiative	74%	.83
Responsibility	85%	.76
Communication skills[c]		.76
Writing[d]	89%	n.s.(.26)
Reading	65%	.63
Speaking	76%	.67
Career: career exploration	81%	n.s.(-.05)
Technical skills[c]		.70
Understanding systems[d]	84%	.44
Science and technology	72%	.63
Math	78%	.63
Thinking and problem solving skills[c,d]		.56
Problem solving[d]	82%	.58
Organization and planning[d]	67%	.39

Note. $N = 40$ or 41 due to listwise deletion for reliability and 43 for correlations. For correlations, $p < .05$ unless marked "n.s." for nonsignificant. Data from Troper, Baker, and Smith (in press). Reproduced with permission.
[a]For the holistic scales, Cronbach's alpha for rater pairs. For all other scales, mean percent exact agreement among all pairs of raters.
[b]Simple correlations between Rubric scale score and the mean of employers' judgments of portfolio author hirability.
[c]Each composite variable is the sum of the scores of its constituent skills.
[d]These variables' ranges were restricted. Writing was found in almost all portfolios, whereas other skills were rarely found.

rest of the schools. We do not believe that students from this school had superior skills. The school had poor resources and disadvantaged students; but it had a sophisticated teacher who championed the portfolio process enthusiastically. Perhaps due to effective teaching, these students may become more aware than students elsewhere that their skills can be transferred to the workplace even though their actual workplace skills do not differ.

Preliminary statistics from the recent scoring of Michigan WRPs (Michigan Department of Education, 1996) shed some light on the likelihood that a student will show evidence of a skill depending on what school the student attends. As mentioned earlier, portfolios from the 15 pilot schools contained analyses on exhibits for all skills far more often than portfolios from non-pilot schools. Furthermore, within nonpilot schools, skills less likely to be documented with exhibits typically produced in schools, such as flexibility, team communication, and responsiveness, had no analysis attached for over 90% of the exhibits. It is possible that as schools' portfolio cultures become more mature, even these skills will become documented with analyses by most students as has happened across all skills in a few pilot schools.

Such an influence of school portfolio culture is consistent with evidence from other portfolio assessments. Research has found that students differ in how much support they receive from their teachers on portfolios—across schools, and even within classrooms (Gearhart & Herman, 1995). In another study, underprivileged students may have chosen portfolio exhibits poorly because they did not comprehend the purposes and criteria in the assessment as well as more advantaged students did (Dietel, 1993). If workforce readiness portfolios were used as high-stakes assessments, then, they could be unfair to students in schools with less well-developed portfolio cultures.

How well do the consequences of WRP assessments align with the overarching goal of improving students' readiness for the workplace? We can only speculate at this point. When students are taught to write analyses to explain how each exhibit meets a benchmark, we expect that they will notice how credible or convincing their exhibits are. As a result, the students will become more aware of their skills and their portfolios will become much richer. It is possible that students who have greater awareness of how their skills can be transferred to workplaces may be advantaged in interviews and in applying their skills in new jobs. What skills are explicitly taught in schools may broaden to include those seen as high priority in the workplace, such as responsibility, problem solving, and teamwork. The Michigan experience shows that portfolio exhibits in early phases of implementation tend to be items already found in schools. Skills that require exhibits that are not typically created in schools, that is, skills that do not fall into a common academic discipline, are less likely to be documented. Hence, in the preliminary statistics from recent scoring far more portfolios showed credible exhibits for skills that fit an academic discipline, such as writing (76%) and math (57%), than nondisciplinary skills, such as speaking (15%), and problem solving (7%) (see Table 14.4). In more mature portfolio cultures, portfolios should provide documentation for all skills. Whether this feeds back into improved instruction, and what impact WRP assessment and concomitant alterations in instruction have on students' performance at work, are issues ripe for research.

CONCLUSION: POSSIBLE VALID USES
OF WORKFORCE READINESS PORTFOLIOS
AND FURTHER RESEARCH

Research in portfolio assessment is in its infancy (Herman & Winters, 1994), so our conclusions are suggestive and indicate many open avenues for further research. Those implementing WRP assessment must make many choices. Among those choices are what purposes the assessment will meet, whether to mandate portfolios or allow for voluntary participation, what resources to provide, how to design scoring, and what level of technical soundness to require. There is a high probability that standardized assessment of WRPs will yield scores that confound individual student achievement with the maturity of school portfolio culture. Therefore, we recommend WRP assessment for low-stakes purposes, especially for ongoing, classroom assessment and as one indicator in schoolwide or districtwide evaluation of curriculum and portfolio implementation. By sampling only several portfolios from each school, the cost of such assessment can be made easier for a district or state to bear.

These issues suggest several lines of research on the consequences of portfolio assessment, on the quality of measurement produced by WRP assessments, and on the development of portfolio cultures. To what extent portfolios that show evidence of many skills indicate higher workforce readiness, or whether portfolio cultures that produce such excellent portfolios support the development of stronger workforce readiness, are questions ripe for research. Valuable research could test how well measures of WRPs can be used to evaluate changes in workforce readiness or portfolio curriculum. What individual and school-level variables influence the contents of portfolios and the scores given to portfolios in external assessments is worthy of further research. Some of the variables suggested in this chapter have been school portfolio culture, individual students' workforce readiness skills, and perception of high- versus low-stakes. Because portfolio scores are likely to reflect influences from multiple sources, decisions using portfolio scores are best made based on knowledge of the magnitude of influence from each source. To study relative magnitudes, generalizability studies can be conducted (Linn et al., 1991). Furthermore, researchers can examine what teacher and student preparation, student activities (such as writing reflections or analyses about every exhibit), parental or employer involvement, administrative leadership, and technical support (such as scoring criteria) feed the fires of strong portfolio cultures.

Voluntary participation with resources and leadership provided from educational administration may produce better portfolios. A mandate, however, guarantees that all schools at least try WRPs. Whatever choice is made, resources must be made available to avoid underfunded programs. Given portfolios' potential value for ongoing classroom assessment combined with the multitude of new issues facing teachers implementing portfolios, we recommend major investments in staff development and studies of the impact of staff development on portfolio culture and on students' success in workplaces. To the extent that portfolio scores will be used for decision making, the soundness of those scores must be examined. Reliability is not guaranteed, either in terms of interrater agreement or interexhibit consistency. Many questions remain about the validity of portfolio assessment. Most of the research questions we have raised relate to validity.

Kids are different today. The expectations placed on them in an increasingly competitive economy require alterations in the way they are educated. Developments such as workforce readiness portfolios hold promise for improvements in students' preparation for a changing world. Fulfilling this promise will require time and support as well as restraint from trying to make portfolios do what they cannot.

ACKNOWLEDGMENTS

The work reported herein was supported in part under the Educational Research and Development Center Program cooperative agreement R117G10027 and CFDA catalog number 84.117G as administered by the Office of Educational Research and Improvement, U.S. Department of Education.

The findings and opinions expressed in this chapter do not reflect the position or policies of the Office of Educational Research and Improvement or the U.S. Department of Education.

Special thanks go to Katharine Fry, Carolyn Huie, Harry O'Neil, and Leora Troper for helpful comments on an earlier draft.

REFERENCES

Arter, J. A., & Spandel, V. (1992). Using portfolios of student work in instruction and assessment. *Educational Measurement: Issues and Practice, 11*(1), 36–44.

Baker, E. L. (1993, December). Questioning the technical quality of performance assessment. *The School Administrator, 50*(11), 12–16.

Dietel, R. (1993, March). What works in performance assessment? Proceedings of the 1992 CRESST conference. *Evaluation Comment,* 1–15.

Dweck, C. S., & Leggett, E. L. (1988). A social-cognitive approach to motivation and personality. *Psychological Review, 95*(2), 256–273.

Gearhart, M., & Herman, J. L. (1995, Winter). Portfolio assessment: Whose work is it? Issues in the use of classroom assignments for accountability. *Evaluation Comment,* 1–16.

Gitomer, D. H. (1993). Performance assessment and educational measurement. In R. E. Bennett & W. C. Ward (Eds.), *Construction versus choice in cognitive measurement: Issues in constructed response, performance testing, and portfolio assessment* (pp. 241–263). Hillsdale, NJ: Lawrence Erlbaum Associates.

Herman, J. L., Gearhart, M., & Aschbacher, P. R. (1996). Developing portfolios for classroom assessment: Technical design and implementation issues. In B. Calfee & P. Perfumo (Eds.), *Writing portfolios in the classroom: Policy and practice, promise and peril* (pp. 27–59). Mahwah, NJ: Lawrence Erlbaum Associates.

Herman, J. L., Gearhart, M., & Baker, E. L. (1993). Assessing writing portfolios: Issues in the meaning and validity of scores. *Educational Assessment, 1*(3), 201–224.

Herman, J. L., & Winters, L. (1994). Portfolio research: A slim collection. *Educational Leadership, 52*(2), 48–55.

Kentucky Department of Education. (1994). *Kentucky instructional results information system: 1992–93* (Tech. Rep.). Frankfort, KY: Kentucky Department of Education.

Koretz, D., Stecher, B., Klein, S., & McCaffrey, D. (1994). The Vermont portfolio assessment program: Findings and implications. *Educational Measurement: Issues and Practice, 13*(3), 5–16.

LeMahieu, P., Gitomer, D. H., & Eresh, J. T. (1995). Portfolios in large-scale assessment: Difficult but not impossible. *Educational Measurement: Issues and Practice, 14*(3), 11–16; 25–28.

Linn, R. L., Baker, E. L., & Dunbar, S. B. (1991). Complex, performance-based assessment: Expectations and validation criteria. *Educational Researcher, 20*(8), 15–21.

Mehrens, W. A. (1989). *Michigan employability skills employer survey: Technical report.* East Lansing: Michigan State University.

Messick, S. (1995). Validity of psychological assessment: Validation of inferences from persons' responses and performances as scientific inquiry into score meaning. *American Psychologist, 50*(9), 741–749.

Michigan Department of Education. (1996, January). *Employability skills in student portfolios: 1995 report.* Lansing, MI: Author.

Michigan State Board of Education. (1993, December). *Employability skills assessment kit.* Lansing, MI: Author.

Moss, P. A. (1995). Themes and variations in validity theory. *Educational Measurement: Issues and Practice, 14*(2), 5–13.

Moss, P. A., Beck, J. S., Ebbs, C., Matson, B., Muchmore, J., Steele, D., Taylor, C., & Herter, R. (1992). Portfolios, accountability, and an interpretive approach to validity. *Educational Measurement: Issues and Practice, 11*(3), 12–21.

O'Neil, H. F., Jr., Allred, K., & Baker, E. (1992). *Measurement of workforce readiness: Review of theoretical frameworks* (CSE Tech. Rep. 343). Los Angeles: University of California, Center for Research in Evaluation, Standards, and Student Testing.

Paulson, F. L., & Paulson, P. R. (1994, May). *A guide for judging portfolios.* Portland, OR: Multnomah Education Service District.

Smith, C. (1993). Assessing job readiness through portfolios. *The School Administrator, 50*(11), 26–31.

Stemmer, P., Brown, B., & Smith, C. (1992). The employability skills portfolio. *Educational Leadership, 49*(6), 32–35.

Stemmer, P., & Smith, C. B. (1993, April). *Toward the framing of research for quality student portfolios.* Paper presented at the annual meeting of the American Educational Research Association, Atlanta, GA.

Tennyson, R. D., & Cocchiarella, M. J. (1986). An empirically based instructional design theory for teaching concepts. *Review of Educational Research, 56*(1), 40–71.

Troper, J. D., Baker, E. L., & Smith, C. B. (1994a, June). *Workplace readiness portfolios: An "adequate evidence" strategy shows evidence of high reliability and validity.* Paper presented at the annual Council of Chief State School Officers National Conference on Large-Scale Assessment, Albuquerque, NM.

Troper, J. D., Baker, E. L., & Smith, C. B. (1994b, July). *Workplace readiness assessment: Employability skills portfolio inventory shows high reliability and validity.* Paper presented at the annual meeting of the American Psychological Society, Washington, DC.

Troper, J. D., Baker, E. L., & Smith, C. B. (in press). *Validity and reliability of a low-inference approach to assessing employability skills portfolios* (CSE Tech. Rep.). Los Angeles: University of California, Center for Research on Evaluation, Standards, and Student Testing.

U.S. Department of Labor. (1991, June). *What work requires of schools: A SCANS report for America 2000.* Washington, DC: U.S. Department of Labor, Secretary's Commission on Achieving Necessary Skills.

Valencia, S. W., & Calfee, R. (1991). The development and use of literacy portfolios for students, classes, and teachers. *Applied Measurement in Education, 4*(4), 333–345.

Wolf, D. P. (1989). Portfolio assessment: Sampling student work. *Educational Leadership, 46*(7), 35–39.

Chapter 15

Assessing Teamwork Skills for Workforce Readiness

Phyllis T. H. Grummon
Michigan State University

Teamwork has gained increasing prominence as a necessary workplace skill in recent years (Carnevale, Gainer, & Meltzer, 1990; Commission on the Skills of the American Workforce, 1990). The need for teamwork skills in perspective employees offers schools the opportunity to help their students develop and document their teamwork skills. While schools have always offered students opportunities to work in teams, they have not generally provided students with ways to assess the growth and development of their teamwork skills (Grummon, 1994). Teamwork assessment is not as straightforward as most academic assessments.

What follows is a review of the factors that need to be considered as schools approach the task of developing teamwork assessments. This discussion starts with a review of teamwork definitions and what constitutes effective teamwork. The dimensions of teamwork and teams' growth and development have a significant effect on what can and should be assessed. A review of the purposes of teams in school and work settings follows the discussion of the dimensions of teamwork. The purpose of a team establishes the parameters for how teams conduct their business. As such, teamwork purposes must be considered when developing assessments. Subsequent to the discussion of purposes, the chapter considers general issues in assessing teamwork by describing factors that affect decisions related to who is assessed, what is assessed, and what assessment methods are best suited to teamwork assessments. The chapter closes with an example of a method for assessing teamwork based on the work of the Workplace Readiness Assessment Consortium of the Council of Chief State School Officers.

DIMENSIONS OF TEAMWORK

While the need to work in teams is cited often by employers, there is little agreement in either the research literature or workplaces on what constitutes a team (National Research Council, 1994). This lack of consensus creates problems when considering the specifics of teamwork assessment, particularly in distinguishing teams and teamwork from other forms of work in small groups. It is important, then, to define teams so that the requirements for creating teamwork assessments become clearer (see Cannon-Bowers & Salas, this book, and O'Neil, Baker, & Kazalaurskas, 1992, for a complementary view of teams). Interpersonal interactions to accomplish work can be thought of as falling along a continuum, with groups at one end and true teams on the other. At one end would be such practices as saying that everyone who reports to the accounting manager is part of the accounting work group, regardless of where they work or their job description. Everyone in the same fifth grade reading class could be said to be in a group, even though most of their work is individually driven and focused. While people in groups like these may interact to accomplish a task, it is most likely that they see their responsibilities as individual, and their interactions are focused on helping individuals accomplish individual goals.

Further along the continuum from groups to teams would be sets of individuals who must contribute to some common product or service. Examples of this type of group include all of the workers on a car-door assembly line who put the door together, students who study together for a test, or people who each complete one part of a report and submit it together. While it is likely that more interaction is required in such groups, they do not have all of the characteristics of a team.

A team has a common goal or goals (Dyer, 1987). Each member of a team is committed to contributing to the goals and to supporting other team members in achieving those goals. It is clear to team members that unless they work together, they will not accomplish the team's goals (National Research Council, 1994). Team members are also likely to feel a commitment to one another and to have a sense of belonging to the team (Francis & Young, 1979). Groups given common tasks to accomplish tend to approach them as a series of individual projects to be "glued" together at the end. In contrast, a team has a strong sense of the interdependence of its members in task accomplishment and will make sure that members help each other whenever necessary (Hare, 1992; Zander, 1982).

Effective teams are adept at both product and process skills (West, 1994). The task or product component of teamwork is familiar to most people as the goal the team needs to accomplish. The team's goal may be to produce a report, transplant an organ, build a house, deliver a package on time, or select the site for a new school. All team members need to contribute to the work of accomplishing the team's goal. They need to produce tangible outputs in order for the team to succeed.

In addition to accomplishing a task, teams are also involved in interpersonal or social processes (Zander, 1982). These processes affect how the task is accomplished and the likelihood of the team's success. Team members engage in decision making, negotiation, conflict resolution, and problem solving while working on the team's goal (Hollenbeck et al., 1995). Teams involve people not only in mutually accomplishing a task, but also in social interactions that affect the accomplishment of tasks (Hare, 1992). For example, a team that cannot resolve conflicts about how to organize its resources to produce a product will be unable to accomplish its production goals. The needed resources may be available in abundance, but if team members cannot successfully engage in the interpersonal process of conflict resolution, they will be unable to effectively use those resources.

Teams change the focus of their social interactions over time (Dyer, 1987). Newly formed teams often do not function as well as teams that have lengthier histories. New teams have to focus on working out the division of labor for tasks and the norms for social interaction (Parker, 1990; Stanford, 1977). Members are learning each other's strengths and weaknesses and negotiating how they will make decisions. If there is no designated leader, then either the roles that a leader fills must be allocated among the members or a leader must emerge from the team. Early in their interactions, team members establish the "rules of engagement" for dealing with conflicts. Team members decide how and when feedback will be given to members about their task accomplishment and social interactions. The creation of a team requires a significant investment of time and energy. Whether or not members satisfactorily resolve such issues as leadership, task assignment, and decision-making strategies can determine the success of the team.

The composition of a team affects its ability to perform both product and process tasks. Team members may differ on a number of dimensions. Factors such as gender, ethnicity, and social status may all have an effect on how team members interact (Cohen, 1994; Webb, 1995). The level of knowledge and expertise that members have about the task at hand also affects the team's ability to perform. If the task requires a single right or best answer, then teams with members high in competence will outperform those teams with members of lesser competence. In some teams, knowledge is distributed among the individual members (Hollenbeck et al., 1995; Hutchins, 1990) and must be shared if the team is to function and perform tasks. The ability to share such knowledge is affected by the interpersonal skills of the members, as well as by their status in the team. A cockpit crew is an example of a team with distributed knowledge. Each member is responsible for his or her instrument set, for conveying information from the instruments to the pilot, and for making any necessary adjustments. While the pilot may make the ultimate decision, the pilot cannot do so without the information provided by the other members of the crew.

Another dimension that affects teamwork is the type of task on which the team works. Tasks or problems vary along a continuum from well-structured to ill-structured or ambiguous. The most well-structured tasks are ones with a single right answer or solution. For example, there is only one acceptable solution to what the right door should look like and do on a new car. Somewhat more ambiguous are tasks requiring teams to generate one or more solutions to well-structured problems. How to advertise the junior prom is a relatively well-structured problem with a number of different solutions. The most ambiguous or ill-structured tasks are ones where there are numerous solutions and often few criteria for deciding what represents an acceptable solution. Increasing attendance represents an ill-structured problem (what do we mean by attendance, why is it important, what motivates someone to attend) where there are numerous potential solutions.

Teams that have well-structured tasks interact in different ways than teams with ambiguous tasks. When there is only one correct solution to the problem confronting a team, the team members are more likely to turn to the member(s) perceived as most competent or knowledgeable for the solution. This produces discussions that focus on who has what competence and results in less interaction among members. Status differential becomes particularly salient, and social loafing is more likely (National Research Council, 1991, 1994; Webb, 1995). In contrast, tasks with greater ambiguity encourage more participation by members, reduce status differences, and decrease social loafing (Cohen, 1994). The distribution of knowledge among team members also enhances team interactions (Hollenbeck et al, 1995; Hutchins, 1990). Ambiguous problems and complex tasks, then, appear to increase the need for process skills in order to accomplish the team's goals.

PURPOSES OF TEAMS

Teams vary in their longevity, their ability to manage social processes, their composition, and the types of tasks they must accomplish. The purpose or function of the team also affects team effectiveness. Teams may exist for a variety of purposes in educational and work settings and may have different terms of engagement. In educational settings, teams have three general purposes. One is to produce individual learning. Athletic teams, debate societies, and drama clubs are all examples of teams sponsored by schools that promote the development of the skills of individual members. Such teams tend to exist for extended periods of time and to engage in a variety of activities. Teachers also use teams to aid the learning of academic skills and knowledge. Cooperative learning frequently focuses on using groups or teams to enhance the individual learning of its members (Bossert, 1988–89; Cohen, 1994; National Research Council, 1994). A group or team of children may tutor one another, practice for a test, or otherwise aid one another's mastery of academic material. Teams focused on academic learning tend to exist for short periods of time.

A second purpose of teams in schools is to produce a product or deliver a service. Teachers that ask students to work jointly on a research project or to produce a report are using teams. Extracurricular work, such as producing the school newspaper or running the school store, are other examples of using teams in educational settings to produce a product or deliver a service. The goal of such teams may include individual learning, but the organizing principle of the team is that it must produce or create something for delivery. Again, some of these teams will function throughout an academic year, or beyond, and some will be relatively short-lived.

The third general purpose of teams in schools is to gain acceptance for ideas or changes in behavior. Elementary school teachers frequently have their classes discuss appropriate behavior in school. When they ask a class to agree on rules for behavior, they are using teamwork to help gain acceptance of an idea. Student councils and advisory groups represent another way in which teams are used in educational settings to help solve problems and gain acceptance for ideas or changes in behavior.

The general purposes of teams in the workplace include two of the three purposes seen in educational settings (Parker, 1990). These two purposes are producing a product or service and gaining acceptance of an idea. A third general purpose of workplace teams is to solve problems that affect production or service. Clearly, one of the basic uses of teams at work is in the production of an organization's product or service. Whether one is talking about a case management team at a social service agency or a door assembly team at a manufacturing plant, teams allow organizations to meet their customer's needs. Teams are also used for two additional purposes, to help gain acceptance of ideas and to solve problems. Quality circles, introduced in the early 1980s to many companies, are frequently used to solve production problems. They may also consider the effects of personnel policies and attempt to gain acceptance for changes in behavior at the workplace. For example, the increased use of safety equipment often relies on changes in normative behavior brought about through group interactions (Grummon & Stilwell, 1984).

The function of the team—production, problem solving, gaining acceptance of ideas—relates to the way in which it is organized, including its longevity. Teams may be formed to solve a specific problem or explore an opportunity. Such teams are usually referred to as task forces or project teams. They have a defined objective and generally disband after that objective has been met. These teams are typically used only in problem solving and gaining acceptance of ideas. Process-improvement teams (Wellins, Byham, & Wilson, 1991), a newer phenomena in work settings, also start with a problem or opportunity. However, they differ from task forces or project teams in several ways. First, the problem is likely to be chronic and involve systems that cut across an organization, like purchasing or accounting. Second, process-improvement teams look at the entire process involved and make suggestions about how to improve that process, not just their portion of the process. Finally, they do not usually disband after they arrive at a first solution or suggestion for change. Instead, the job of the process-improvement team is to continue to look for ways to improve a system or to explore an opportunity. Process-improvement teams work best when they are empowered to investigate problems completely and to implement suggested changes (Wellins et al., 1991).

Self-directed or self-managed work teams are a newer phenomenon in the American workplace (Wellins, et al., 1990). In these teams, employees are given the responsibility and authority to plan, conduct, improve, and evaluate their own work. They generally select their own team leader and may have direct contact with customers and suppliers. The goal of these teams is to place important decisions in the hands of those most likely to be affected by those decisions. A self-directed work team would (a) determine what jobs needed to be done in which order; (b) decide who should do what jobs; and (c) address any problems along the way.

The variety of team purposes affects the training and development the teams receive and the ways in which they might be assessed. Teams that focus on individual learning need to have assessment strategies that tap into that learning. Teams whose purpose is problem solving may be assessed on the effectiveness of solutions. Frequently, the implementation of solutions requires the acceptance of new ideas or changed ways of doing business. If these changes are necessary for solution success, then teams may also be evaluated on the process used to develop solutions. If relevant stakeholders are not included in problem solving, or if members have not encouraged sufficient participation within the team, then solutions may not be implemented.

Intertwined with the purpose or function of the team is its longevity. Teams that exist over time, such as athletic teams or production teams, are different from teams with shorter lifetimes. Most of the differences lie in the process side of teamwork effectiveness. For example, teams with longer lifespans often need to replace members. This creates opportunities for either renewal or dysfunction, depending upon how the team deals with the transition. This, in turn, affects how effective a team will be and hence how effective it appears during an assessment activity. Clearly, differences in the lifespans of teams present a challenge to teamwork assessment and require that assessments be sensitive to the issue of a team's history and its projected future.

ISSUES IN ASSESSING TEAMWORK

This brief overview of the dimensions of teamwork and the purposes of teams provides some insight into the complexity of assessing teamwork for workplace readiness. As one develops assessments, or considers what factors to assess, a number of issues arise related to what is assessed, who is assessed, and what assessment methods can be used. While the rest of this section deals with these issues sequentially, the issues interact with one another, and decisions about one affect the options in others.

What Is Assessed

Effective teamwork requires a range of behaviors on the part of members. Team members must have skills that allow them to both produce (teamwork product) and to interact effectively with one another (teamwork process). Teamwork skills have been described by numerous authors (e.g., Cannon-Bowers, Tannenbaum, Salas, & Volpe, 1995; Council of Chief State School Officers, 1995; Francis & Young, 1979; Grummon, 1994; McIntyre & Salas, 1995; Michigan Department of Education, 1993; Schneider & Schmitt, 1990). A few of the skills that have been identified as part of effective teamwork include content knowledge or expertise, negotiation, decision making, task monitoring, goal setting, information seeking, listening, and attending. Assessments can be designed that focus on product and/or process skills and that look at either global teamwork effectiveness or specific teamwork skills.

Product focused assessment looks principally at the dimensions of a team's outputs. This might include ratings on the quality of the product or service, the number of products created or customers served, the timeliness of production or service, or the costs involved. The inference would be made that products that rank high on one or more of these aspects would be related to effective teamwork. For those teams whose purpose is primarily product or service focused, such product focused assessments may be most appropriate. Since effective teamwork requires attention to both product and process, every teamwork assessment should include some determination of the adequacy of any products created.

Most of the research on the use of teams with students focuses on the effects of cooperative learning on the achievement of individual students, or the product side of teamwork effectiveness (Bossert, 1988–89; Cohen, 1994; National Research Council, 1994; Webb, 1995). Research that considers the effectiveness of the team also looks exclusively at measures of group achievement, such as combined scores on tests or the recall of facts (Cohen, 1994). Again, this is a product focus. There is little research with students that assesses their performance, either individually or in teams, on the process side of teamwork (see O'Neil, Chung, & Brown, this book, for research on this topic). More often, process skills, such as encouraging participation, are manipulated to see their effects on achievement. They are not measured as ends in themselves (e.g., how do students naturally differ in their ability to keep a team focused on its task).

An example of a study that did attempt to measure process skills was done by Schneider and Schmitt (1990) for the Michigan Department of Education. In this research, students were involved in assessment center activities that measured their skills on three process dimensions—problem solving, interpersonal, and initiative. Assessment centers are used in business to assess individuals' skills in teamwork, as well as other "soft" management skills like organizing and prioritizing (Thornton & Byham, 1982). Schneider and Schmitt (1990) found that students were engaged by the assessment center methodology, and that raters were able to observe differences among students, although they also tended to rate the same student similarly on each dimension.

The plethora of skills and knowledge that can be described as being a part of effective teamwork means that any teamwork assessment effort must make decisions about what to measure. Beside deciding the mix of product and process skills to be assessed, another initial decision is: Where on the continuum of global to discrete skills should the assessment focus? This is particularly true for process skills. For example, will teams be rated generally on a communication dimension? Or will they be rated specifically on how often each member speaks and how much time is allotted to different members for speaking? Both ratings could yield information related to team effectiveness. Clearly, however, they represent different types of assessment.

The purpose and uses of the assessment will be a primary factor in deciding what skills will be assessed. A teacher wanting to give students specific feedback on their ability to solicit input from all members may well wish to count the frequency and duration of each member's contribution. An employer may be more interested in whether or not someone received a "best team player" award than in information about how often someone asked another team member his or her views. Thus, the uses for a teamwork assessment will determine the types, number, and detail with which skills and knowledge are assessed.

Who Is Assessed

Assessments of teamwork must be clear about whether the focus for feedback is at the individual, team, or interteam level. It is most common, and probably easiest, to assess individual products, such as test scores, that are the result of teamwork. Team products, such as group reports, are also frequently assessed for their quality (Grummon, 1994). The process side of teamwork is more infrequently assessed and presents greater challenges. However, employers are most likely to describe process skills as the ones they want in employees (Carnevale et al., 1990). Assessment centers are an example of an assessment method focused on the process side of teamwork. Rarely are interteam assessments done, even though cooperation among teams is necessary to success in most organizations.

Whether the focus for assessment feedback is individual team members, the team as a whole, or the interaction of the team with other teams, the composition of the team needs to be considered. The rating that any individual team member receives, as well as ratings for the entire team, is affected by who is on the team and the team's purpose. As already described, different purposes and tasks interact with team composition. Team members may differ on general or specific status characteristics (e.g., race, gender, ethnicity, academic competence, peer approval). These status characteristics affect the likelihood that individuals will talk in teams, how much they learn, and how able the team is to reach its goal. For example, a homogeneous team of low achievers will produce different assessment results if it is engaged in a well-structured problem than if it is engaged in an ill-structured problem (Cohen, 1994).

Teams may also differ in the degree of familiarity team members have with one another. Teams may be created just for an assessment, or they may have existed for a while. Team member selection may be random or purposeful. Teams that have a history of interaction behave differently from teams consisting of strangers or of members who rarely interact as a team (Hollenbeck, et al. 1995). Unless they have a history of effective teamwork, past interactions do not necessarily mean that a team will be more effective. For example, a team that has relied on one member to produce results in well-structured situations may be unable to adjust adequately to the demands of an ill-structured problem. Thus, the history of the team members' interactions with one another must be considered when designing teamwork assessments.

Decisions about whether individuals or the entire team should be the focus of an assessment depend primarily on the purpose of the assessment, but also on the resources available to determine the reliability and validity of the assessment. This is particularly true when performance-based assessment methods are used. The fact that individual performance cannot be entirely controlled by the individual places additional demands on any study of the psychometric properties of a teamwork assessment that is to be used for the determination of individual skill levels. Feedback on skill levels at the team level is somewhat less problematic, but still requires multiple measurement opportunities to meet even modest demands for reliability. Again, assessment of process skills is far more difficult than product evaluations.

Assessment Methods

Clearly, decisions about what teamwork skills will be assessed and who will participate in a teamwork assessment affects how assessments are constructed and conducted. The most basic question related to "how" is whether declarative or procedural knowledge will be assessed. That is, does the assessment focus on the knowledge of effective teamwork practices or the actual practice of effective teamwork.

Assessments that focus on knowledge of effective teamwork practices focus on the ability of individuals to identify or generate appropriate responses to different teamwork situations. Knowledge of process skills is generally what is assessed. A written test asking a student to list the stages of team development represents one way to determine if a student can recall a component of team processes. More complex assessments may ask individuals to respond to scenarios or videotaped vignettes of teams by describing what a leader or team member should do next to maximize the functioning of the team (e.g., see O'Neil, Allred, & Baker, chap. 1, this volume, and McLarty & Vansickle, chap. 12, this volume). As with all assessments of knowledge, teamwork assessments may allow the user to make inferences about an individual's likely behavior in a team.

Assessments of teamwork practice may take a number of forms. First, observations can be done of individuals engaged in teamwork as part of the normal course of work or school. These observations may focus on individual members' contributions or on the overall functioning of the group (Grummon, 1994). Observers may rate either individuals or the entire team on either global teamwork skills or specific teamwork behaviors. As with all forms of observational assessment, raters need to be trained in the rating format. When assessment results will be used for decision making, raters need to be checked for reliability among raters and across time.

Teamwork assessments often use simulations or on-demand tasks, rather than the observation of naturally occurring teamwork processes. Simulations are generally more elaborate than on-demand tasks and require participants to play roles other than their own (O'Neil & Robertson, 1992). For example, in a simulation for new product development, participants might be asked to take the roles of production engineer, design engineer, sales person, accountant, or division manager. In contrast, on-demand tasks present the team with a problem to solve, or a product to produce, and ask that participants be themselves. Assessment centers, for example, may present a group of prospective managers with a problem in employee performance and ask the team to develop potential solutions. Team members will then be observed and rated on a number of teamwork dimensions. Manufacturing plants that focus on teamwork may ask perspective employees to participate in a team to build a product. Either the team or individual team members may be rated in an on-demand task, depending upon the purposes of the assessment.

Somewhere in between assessments of knowledge about teamwork and the actual practice of teamwork, lies biographic or biodata assessments. Biodata assessments are objective reviews of candidate information that are statistically related to the desired job behavior (Klimoski, 1993). Businesses have long used biodata to help determine the eligibility of candidates for positions (Cascio, 1987). An applicant's responses to questions on the application form help the business determine if the applicant has the requisite levels of teamwork skills and experience. States such as Michigan (Employability Skills Portfolio, Michigan Department of Education, 1993) and California (Career-Technical Assessment Program, Career Assessment Program, Far West Laboratory for Educational Research and Development, 1995) have developed portfolios to assist students in documenting their workplace skills, including teamwork (see also Troper & Smith, chap. 14, this volume, and Rabinowitz, chap. 13, this volume). These portfolios typically ask students to provide evidence of teamwork skills, the reasons why the evidence represents a successful teamwork interaction, or the results of an on-demand teamwork exercise. When students are rated on the entries in their portfolios, they are experiencing a form of biographic assessment.

PROTOTYPE MODELS FOR TEAMWORK ASSESSMENT

Overview

The work presented in this chapter was sponsored by the Council of Chief State School Officers (CCSSO). CCSSO is a nationwide, nonprofit organization composed of the 57 public officials who head the departments of elementary and secondary education in the 50 states, five U.S. extrastate jurisdictions, the District of Columbia, and the Department of Defense Schools. The Council has served as an independent voice of federal education policy since 1927. In representing the chief education administrators, the Council speaks on behalf of the state education agencies, which have primary authority for education in each state. The Council's members develop a consensus on major education issues, which the Council advocates before the President, federal agencies, Congress, professional and civic associations, and the public.

In 1994, the Council of Chief State School Officers (CCSSO) Workplace Readiness Assessment Consortium developed three prototype models for teamwork assessment. These models were created following extensive consultation with businesses and a review of the literature on teamwork assessment. The goals of the models were to provide educators with options for assessing teamwork in students and a base for educators to develop their own teamwork assessments. The models were intended to be used for only low-stakes purposes—primarily developmental feedback to students on their teamwork skills. It was also hoped that as teachers reviewed the results of the assessments across teams, they would find that information useful in designing a curriculum that encourages the development of a range of teamwork skills.

The prototype models were designed to reflect the three purposes of teams found in schools—to encourage individual learning, to create a product or provide a service, and to gain acceptance of ideas or to solve problems. When the purpose was individual learning, teachers were provided with general instructions on how to adapt a curricular unit for use as a team project. The prototype models that focused on products and the acceptance of an idea included specific prompts, as well as guidance on how to produce prompts locally.

All of the models used the same rubric for observing and rating teamwork process skills (see Fig. 15.1). Observers were instructed to watch each team for 5 minutes at a time and to simply take notes on what they saw during each observation period. These notes would then form the basis for rating teams on each of the seven items in the rubric. Each item had three behaviorally anchored descriptions of team behavior that were rated as either very well, adequate, or poor. These descriptions provided raters with specific information about what constituted performance at each level. If, in reviewing the notes, an observer felt he or she had not seen any behavior related to the item, he or she was instructed not to rate it. The different

Interpersonal Skills

A. How well did the team members reinforce and support others' ideas?
B. How well did the team negotiate agreement and handle conflict?
C. How well did the team encourage open discussion of ideas?
D. How well did the team work for a accept necessary compromises?

Thinking/Problem-Solving Skills

E. How well did the team plan its work and set goals?
F. How well did the team stay on task?
G. How well did the team define and apply a systematic approach to the task?

FIG. 15.1. Items from Team Effectiveness Observation Sheet. Source: Grummon (1994). *Assessing Teamwork Skills for Workplace Readiness Prototype Models*. Washington, DC: Council of Chief State School Officers, Appendix D.

A.   Did the team use correct spelling and grammar in the letter to the Board?
B.   Did the team write so that the letter to the Board could be easily read?
C.   Did the team organize the letter to the Board in a manner that makes the problems
     and rationales easy to follow and understand?
D.   Did the team write the letter persuasively?
E.   Did the team identify problems that are important to students during the merger?

FIG. 15.2. Items from product scoring form for a letter to the board of education. Source: Grummon (1994). *Assessing Teamwork Skills for Workplace Readiness Prototype Models.* Washington, DC: Council of Chief State School Officers, Appendix D.

prototypes required different amounts of time for completion, with the individual learning prototype offering the most opportunities for observations.

In addition to observing teams, teachers were also provided with separate rubrics for rating the outcomes of the product and idea acceptance prompts (see Fig. 15.2). These rubrics focused on the quality, completeness, and effectiveness of the outcomes. While the rubrics rated the team only on the final outcome, they did include instructions to identify how each student in the team had contributed to the outcome. Teachers were encouraged to develop a similar scoring strategy for their individual learning based curricular projects. That is, individual students could identify how they contributed to the outcome, and the entire team would be responsible for specific portions of the project.

Students were provided with the observation rubric (see Fig. 15.1), the product or idea acceptance rubric (see Fig. 15.2), and with a self-assessment of their teamwork skills (see Fig. 15.3). The CCSSO prototypes suggested that students complete the self-assessment before and after their teamwork experience. The self-assessment covered the same seven areas as the observation rubric. Self-assessments could be kept private.

Teachers were encouraged to provide instruction on teamwork on the day prior to the teamwork assessment. The prototypes included a sample lesson plan for introducing the concepts of process and product in team-

How did you help your team plan?

1.   I tried to help the team see that a plan would make things go easier. I volunteered to do the jobs in the group I knew I could do best and for at least one job that no one wanted to do.
2.   I waited for someone else to start planning, but was willing to help once it got going. I volunteered for the jobs I knew I could get done.
3.   I didn't realize we needed a plan and didn't really do anything to help create one. When someone told me what to do, I was willing to try to do it.

FIG. 15.3. Example of student self-assessment item. Source: Grummon (1994). *Assessing Teamwork Skills for Workplace Readiness Prototype Models.* Washington, DC: Council of Chief State School Officers, Appendix D.

work, the types of skills needed to create effective teamwork, and some ground rules for interacting in teams (see Appendix for sample lesson plan). The plan included the distribution of the rubrics on the day prior to the assessment, so that students could review them and better understand how they would be rated. The lesson plan also included suggestions for how to lead a discussion on the teamwork experience on the day following the assessment. Teachers were asked to provide teams with time to talk about what went well during the teamwork exercises, what could have been improved, and to review the observation notes and scores for their team.

Prototypes

The CCSSO prototype models, based on the purposes of teams in schools, were described in a teamwork assessment manual (Grummon, 1994). The first model, individual learning, did not include any specific prompts. Teachers who elected to use a classroom unit for teamwork assessment were provided guidance in how to select an appropriate learning project. The two basic characteristics of an effective teamwork experience included requiring all students to contribute (e.g., providing individual accountability through individual grades or other means) and ensuring that students receive a grade, or other recognition, based on the combined efforts of the team. These characteristics are suggested as necessary for effective collaborative learning (Bossert, 1988–89; Cohen, 1994; Johnson, Johnson, & Holubec, 1987; Webb, 1995).

The other two prototypes provided teachers with prompts for engaging students in teamwork. The first prompt required students to create a promotional campaign for their school by designing and presenting a poster advertising the school and a skit to accompany the poster. Students were rated on the effectiveness and clarity of their poster and skit. This prototype was meant to address the teamwork purpose of creating a product or service. The second prototype focused on gaining acceptance of ideas. It included a prompt that asked students to respond to a merger of two fictitious high schools (see Fig. 15.4). Students were provided with some basic information about the schools. They were told that their team represented a student group that had been asked to provide input to the Board of Education on students' views of the issues involved in the merger. Teams had to brainstorm a list of issues, choose three issues to present to the Board, and write a letter to the Board explaining why these issues needed to be addressed. Letters were rated on neatness, English mechanics, and persuasiveness (see Fig. 15.2).

The selection of these three prototype models was based on a number of considerations. First, the Workplace Readiness Consortium wanted to offer teachers options for each of the purposes of teamwork. The Consortium strongly believed that curriculum, instruction, and assessment should be

The Merger of K and M High Schools

Anywhere Town now has a shrinking population. Because of that, the School Board has decided to join K and M High Schools into one school building. To make the best decisions, the Board has formed a number of teams to discuss the problems and issues that are associated with the merger. You are part of a team made up of students from both K and M High Schools. The Board has asked each team to write the Board a letter with the problems it believes the Board must address to make the merger go smoothly. As a student team, you are expected to list the problems students see as most critical in the merger and why they must be addressed.

One school will house the combined high school, and one will become a middle school for Grades 6-8. M and K High Schools have been intense rivals in sports, academics, and special programs such as art, music, and debate. (A brief description of each school's present programs is included.) Neither school likes the idea of having to merge, but the district can no longer afford to have two high schools. Either school building can accommodate the combined student enrollment of the two high schools. The Board has not decided which building it will use, but both are in similar physical condition and have the same facilities (gyms, swimming pool, athletic field, auditorium, etc.). The computer lab will be in the combined school. Many teachers are eligible for retirement, so no one will lose a job in the merger.

FIG. 15.4. Prompt for acceptance of an ideal teamwork prototype. Source: Grummon (1994). *Assessing Teamwork Skills for Workplace Readiness Prototype Models.* Washington, DC: Council of Chief State School Officers, Appendix D.

integrated and saw that providing teachers with options would encourage them to do that. It is also clear from the research on performance-based assessment that the most accurate pictures of student's abilities emerge when there are multiple assessment opportunities (Dietel, 1993; Linn & Burton, 1994; Shavelson & Baxter, 1992). The businesses consulted by the Consortium also encouraged the use of a range of assessments to acquaint students with the ways in which teams are used in work settings. Finally, the models offered teachers a continuum of time frames for doing the assessment so that they could decide how much time they would devote to it.

The prototypes' development was guided by input from businesses. Prototypes were reviewed by industry representatives to ensure that they represented teamwork tasks that businesses felt were equivalent to those they used for assessment purposes. As prototypes were being developed, they were subjected to review based on a number of criteria identified by the Workplace Readiness Consortium. These criteria included: (a) any classroom should be able to complete the prototype without needing new resources; (b) the prototypes should be able to be used in any educational setting, with all types of students (e.g., rural, urban, limited English-speaking populations, gifted and talented, etc.); (c) the prototypes should not require specialized knowledge about schools or workplaces; and (d) the tasks should be sufficiently unstructured so that they maximized the need for all members to contribute.

Teamwork Observation

The manual accompanying the prototype models offered guidance to assessment professionals in how to prepare observers for rating teams. Since the prototypes were designed to be low-stakes assessments, and the Consortium had no control over how assessments would be administered, the manual did not detail a training protocol that would ensure either intra or interrater reliability. Instead, assessment professionals were encouraged to provide teachers or other observers with an opportunity to use the rating form before participating in the assessment. This might involve viewing a videotape of a team and rating it, or observing other teachers as they did one of the prototype assessments. The goal of training was to emphasize the importance of recording observed behavior and the use of only those observations to rate the team.

The Teamwork Effectiveness Observation Sheet was designed to provide teams with feedback on how their interactions advanced both process and product outcomes (see Fig. 15.1). The Team Effectiveness Observation Sheet is generic and can be used with any of the prototypes or in any situation where a teacher wants to give teams information on their effectiveness. On the sheet two primary dimensions are assessed, interpersonal skills and thinking/problem-solving skills. The interpersonal dimension has four behaviorally anchored items and the thinking/problem-solving dimension has three. Of the four items in the interpersonal dimension, two focus on communication and two on conflict management. Figure 15.1 presents the seven items on the observation sheet. Figure 15.5 provides the behavioral descriptions associated with each ranking for the thinking/problem-solving item on planning and goal setting.

In developing the Team Effectiveness Observation Sheet, the Workplace Readiness Consortium had a number of goals. From its review of the literature, it was clear there were few models available for evaluating the overall effectiveness of teams in educational settings. Research on cooperative learning has focused almost exclusively on the effectiveness of group work in producing academic achievement (Bossert, 1988–89; Cohen, 1994; National Research Council, 1994). Some assessments of specific process skills have been developed, but there were no readily accessible models for determining the effectiveness of teams on a number of process dimensions. The Consortium felt that students would be best served when they could gain a comprehensive view of how team interactions affect team outcomes. Thus, the Consortium wanted to ensure that any assessment looked at the process of teamwork, as well as any products of that work.

Research indicated that most observers are unable to make fine distinctions in behavior when viewing teams (Schneider & Schmitt, 1990). This led the Consortium to limit the number of dimensions assessed in the

How well did the team plan its work and set goals?

Very well   Members discussed the variety of tasks that had to be accomplished to complete the task. Members reached agreement on the order in which tasks needed to be done. Members volunteered for tasks where they had particular expertise to offer. When no one volunteered for a task, the team had a way to equitably assign work. The team monitored its progress to be sure that tasks were done well and in a timely manner

Adequate   Members did not initially discuss the tasks that had to be accomplished, but did eventually recognize that a plan needed to be developed and created one. Members volunteered for tasks that did not appear to be difficult and avoided ones where more work was involved, but eventually all tasks had someone to do them. One or two members assigned tasks without necessarily consulting with the group. Members had an acceptable method for monitoring task completion or quality, with all tasks eventually being completed.

Poor

        The team never really developed a plan for accomplishing the task. Each member just did what he or she wanted to do with recognition of what tasks had to be done to complete the job. One member created a "plan" and made everyone else follow it with little or no discussion and no consensus on the tasks to be included in the plan. The team did not monitor its progress and consequently some tasks were not completed.

FIG. 15.5. Behavioral descriptions for the TEOS item on planning and goal setting. Source: Grummon (1994). *Assessing Teamwork Skills for Workplace Readiness Prototype Models.* Washington, DC: Council of Chief State School Officers, Appendix D.

teamwork observations. However, the Consortium wanted students and teachers to understand that both interpersonal interactions and joint thinking and problem solving were necessary to successful teamwork. Teams must ensure that their members' social interactions advance the work of the team in accomplishing their task. Simply getting along with one another is not sufficient. Neither is achieving the goal at the expense of positive relationships, since that affects future teamwork.

Other Scoring Rubrics

In addition to the Teamwork Effectiveness Observation Sheet, the CCSSO prototypes included rating forms for student self-assessment, the quality of the skit and poster, and the quality of the letter to the Board of Education. The Self-Evaluation for Team Effectiveness paralleled the items on the observation sheet. There were seven items and each had three behavioral descriptions (see Fig. 15.3 and Fig. 15.6). The scoring form for the prototype model that focused on creating a promotional campaign for the

A.   How did you listen to other people's ideas?
B.   How did you present your ideas?
C.   Did you provide feedback to other members on their ideas and behavior?
D.   How did you help your team handle disagreements?
E.   How did you help your team plan?
F.   How did you help your team pay attention to the work that had to be done?
G.   How did you help your team by doing the work that had to be done?

FIG. 15.6. Items from student self-evaluation for team effectiveness. Source: Grummon (1994). *Assessing Teamwork Skills for Workplace Readiness Prototype Models.* Washington, DC: Council of Chief State School Officers, Appendix D.

school included three items assessing the quality and effectiveness of the poster, four items on the quality and effectiveness of the skit, and one item on how well the team integrated the poster and the skit for their campaign. Teachers were instructed to rate students on the scoring form, but it was certainly also possible for other students to provide ratings. The scoring form for the letter to the board of education on the merger of K and M High Schools assessed the letter with two items on mechanics, two items on the effectiveness or persuasiveness of the letter, and one item on the significance of the problems identified by the team (see Fig. 15.3). As with the poster and skit rating form, teachers were asked to rate the teams on the form after reading their letters.

The Workplace Readiness Consortium considered it important for students to be scored on both their teamwork interactions and the outcomes of those interactions. Clearly, businesses are concerned with production and/or service. They see effective teamwork as a means of ensuring the highest quality of production and/or service. Providing students and teachers with a means for assessing team products was important for reinforcing the notion that teams must be concerned with both sides of the teamwork effectiveness equation.

Piloting

During the 1994–95 school year, states that were members of the Workplace Readiness Consortium recruited teachers to pilot the prototype models. Six states participated in the pilots (Louisiana, Michigan, New Mexico, South Carolina, Texas, Wisconsin). A total of nine high schools with eighteen classrooms and over 400 students in the 9th–12th grades took one of the prototype assessments. The piloting took place most often in English classes, but the pilots also included a Science class, a Computer Applications class, and a Journalism class. Students' demographic characteristics varied by school. Students identified as gifted and talented, vocational, Tech Prep, special education, limited English proficient, general education, and college preparatory all participated in the pilot. The ethnic make-up of classrooms varied greatly by school. Included in the pilot were students identified as White, African American, Hispanic, Asian, and Native American.

The principal goal of the piloting was to gain feedback on the feasibility and usefulness of the prototypes from a teacher's perspective. The Workplace Readiness Consortium wanted to learn what worked and what didn't in the prototypes, the scoring rubrics, and the manual. Feedback forms were provided to teachers. Based on the information received from pilot sites, the original manual was revised in 1995. It became two manuals, one for teachers and one for assessment professionals (Grummon, 1995a, 1995b). Other changes were also incorporated to make it easier for teachers to use.

The piloting found that not all of the prototypes were used equally often. Unfortunately, no teacher reported integrating the teamwork assessment into a regular instructional unit so that teams could be observed over an extended time period. The school campaign prototype (poster and skit) was used more often than the school merger scenario (thirteen times versus five times). Some teachers modified the prototypes to fit specific, short-term projects in their classes. For example, one teacher asked the teams to develop a presentation and poster on their ideal world (Utopia Project). The teachers that reported back on the prototypes stated that they were generally engaging for students and that the students enjoyed the exercises.

Teachers also reported that the prototypes worked best when more than one class period could be devoted to the activity. Students, on the other hand, wrote of their frustration when a team member was absent from a meeting. The Consortium had been sensitive to the issue of time for instruction and had designed the assessments so they could be done in 50 minutes. Teachers said that while they could be accomplished in that time frame, it made the assessment too rushed. It was unclear from the teachers' feedback whether or not they also spent the day before the assessment preparing students or the day following the assessment debriefing students on their teamwork experiences.

In reviewing the comments from teachers, and from the classrooms observed by Consortium members, it became clear that the assessment worked better in some situations than others. Those teachers that were already using teams for instruction had the easiest time using the assessments and reported finding them the most useful. The assessment was also viewed more positively when more than one teacher administered it and the teachers helped to observe each other's classrooms. Classrooms where the assessment was simply administered with no prior use of teams or preparation for the assessment were more inclined to report that students had trouble working in teams and were less motivated by the exercises.

The Teamwork Effectiveness Observation Sheets were returned by all pilot teachers. Some teachers included their observation notes and some simply rated the teams. Comments by observers indicated that they were able to discriminate among the items. For example, on one sheet the notes on item D (related to working for compromises) were: "C. discussed a compromise for the lack of music images [for their poster] with K. Both quickly agreed to the words with no images." Student feedback also showed the ability to discriminate among the items. In one classroom, students completed the Team Effectiveness Observation Sheet for their team. This serendipitous circumstance offered fuller insight into how students saw the items. For example, a student's description of the team's behavior on item C—How well did the team encourage open discussion of ideas?—was the following: "Two of our team members rarely provided ideas for the letter. One was too shy, I think, and the other really didn't care. Other team members would ask these team members 'so what do you think?' They would respond with a simple nod or 'yeah.' It was mostly dominated by three people, but we politely discussed issues. We didn't argue." This student rated this item as "Adequate" for her team, an accurate rating given the behavior described.

The primary goal of the Workplace Readiness Consortium—creating options for teamwork assessment in schools—was met through the development and piloting of the prototypes. Teachers were able to use the prototypes both as they were presented and as models for developing other assessment tasks. The Team Effectiveness Observation Sheet was a useful tool for teachers and students. Teachers were able to observe teams and rate them on the items on the Team Effectiveness Observation Sheet. Teachers and students alike were able to discriminate among the various levels of team performance identified on the Team Effectiveness Observation Sheet. The additional scoring rubrics for student self-evaluation and team products were also used consistently by the teachers and the students.

Validity

The principal source of face validity data for the assessment came from the expert review of the assessment by businesses prior to its administration. Members of the Consortium provided copies of the Team Effectiveness Observation Sheet and the prototypes to businesses in their states and requested feedback from them. In addition, a number of businesses, primarily manufacturing, participated in the development of the prototypes. The information received from these businesses was that the assessment did appear to be appropriate and to target teamwork skills they considered important in their businesses. At no point, however, did any businesses administer it to their employees. Thus, face validity is the only area where the assessment has been judged to see if it meets its intended purposes.

Resources have not been available to determine if the assessment is valid in other ways. For example, it would be helpful to know if students who have more or less experience in working on teams score differently on the assessment. Similarly, do students who score high on the assessment perform better on jobs that require teamwork? Since this was designed as a low-stakes assessment and intended to primarily influence curriculum and raise students' awareness of the need for teamwork, there are no plans at this time to conduct any validity studies.

CLOSING THOUGHTS

The teamwork assessment prototype models developed by the Workplace Readiness Consortium were an attempt to fill a gap in workplace readiness assessment. They represent only a start, and there is still considerable work left to be done. While the Team Effectiveness Observation Sheet may be high in face validity and based on behaviors known to contribute to team effectiveness, it has not been subjected to even a moderate review of its validity or generalizability. As a low-stakes assessment, it serves a useful purpose if it increases teachers' and students' awareness of the behaviors that characterize effective teamwork. However, to be most useful, perform-ance-based teamwork assessments must be shown to have a consistent relationship with subsequent performance in teams at work and other settings.

The Consortium hoped that schools would understand that students must be trained and educated in the elements of teamwork if they are to be successful in the workplace. The Consortium believed that while schools do include team or group work in their curricula, they rarely assess students on the process side of teamwork. Without process-based assessments, it is difficult for students to increase their teamwork skills. Process assessment should accompany product assessment, so students can learn how social interactions in a team affect the team's ability to produce. The feedback the Consortium heard from pilot sites indicated that the assessment ac-complished that purpose. A student from an English class had this com-ment on the assessment, "This has taught me that you do need to cooperate with others to make it in this world."

REFERENCES

Bossert, S. T. (1988–89). Cooperative activities in the classroom. In E. Z. Rothkopf (Ed.), *Review of Research in Education, 15*, 225–252.

Cannon-Bowers, J. A., Tannenbaum, S. I., Salas, E., & Volpe, C. E. (1995). Defining competencies and establishing team training requirements. In R. Guzzo, E. Salas, & Associates (Eds.), *Team effectiveness and decision making in organizations* (pp. 333–380). San Francisco, CA: Jossey-Bass.

Carnevale, A. P., Gainer, L. J., & Meltzer, A. S. (1990). *Workplace basics: The essential skills employers want.* San Francisco, CA: Jossey-Bass.

Cascio, W. F. (1987). *Applied psychology in personnel management* (3rd ed.). Englewood Cliffs, NJ: Prentice-Hall.

Cohen, E. G. (1994). Restructuring the classroom: Conditions for productive small groups. *Review of Educational Research, 64,* 1–36.

Commission on the Skills of the American Workforce. (1990). *America's choice: high skills or low wages!.* Rochester, NY: Author.

Council of Chief State School Officers. (1995). *Consensus framework for workplace readiness.* Washington, DC: Author.

Dietel, R. (1993). What works in performance assessment? *Evaluation Comment, Spring 1993,* Center for the Study of Evaluation & The National Center for Research on Evaluation, Standards, and Student Testing.

Dyer, W. G. (1987). *Team building: Issues and alternatives.* Reading, MA: Addison-Wesley.

Far West Laboratory for Educational Research and Development (1995). *Career-Technical Assessment Program and Career Assessment Program.* San Francisco, CA: Author.

Francis, D., & Young, D. (1979). *Improving work groups: A practical manual for team building.* San Diego, CA: University Associates.

Grummon, P. T. H. (1994). *Assessing teamwork skills for workplace readiness: Prototype models.* Washington, DC: Council of Chief State School Officers.

Grummon, P. T. H. (1995a). *Assessing teamwork skills for workplace readiness: Prototype models: Teacher's manual.* Washington, DC: Council of Chief State School Officers.

Grummon, P. T. H. (1995b). *Assessing teamwork skills for workplace readiness: Prototype models: Assessment professional's manual.* Washington, DC: Council of Chief State School Officers.

Grummon, P. T. H., & Stilwell, J. (1984). Unlocking the door to safety: Attitudinal aspects of safe behavior at work. *Professional Safety, 29,* 13–18.

Hare, A. P. (1992). *Groups, teams, and social interaction: Theories and applications.* New York: Praeger.

Hollenbeck, J. R., Ilgen, D. R., Sego, D. J., Hedlund, J., Major, D. A., & Phillips, J. (1995). Multilevel theory of team decision making: Decision performance in teams incorporating distributed expertise. *Journal of Applied Psychology, 80,* 292–316.

Hutchins, H. (1990). The technology of team navigation. In J. Galegher, R. Kraut, & C. Egido (Eds.), *Intellectual teamwork: Social and technological foundations of cooperative work* (pp. 191–220). Hillsdale, NJ: Lawrence Erlbaum Associates.

Johnson, D. W., Johnson, R., & Holubec, E. (1987). *Structuring cooperative learning: Lesson plans for teachers.* Edina, MN: Interaction Book Company.

Klimoski, R. J. (1993). Predictor constructs and their measurement. In N. Schmitt, W. Borman, & Associates (Eds.), *Personnel selection in organizations* (pp. 99–134). San Francisco, CA: Jossey-Bass.

Linn, R. L., & Burton, E. (1994). Performance-based assessment: Implications of task specificity. *Educational Measurement: Issues and Practice, 13,* 5–8.

McIntyre, R. M., & Salas, E. (1995). Measuring and managing for team performance: Emerging principles from complex environments. In R. Guzzo, E. Salas, & Associates (Eds.), *Team effectiveness and decision making in organizations* (pp. 9–45). San Francisco, CA: Jossey-Bass.

Michigan Department of Education. (1993). *Employability skills assessment kit.* Lansing, MI: State Board of Education.

National Research Council. (1991). Enhancing team performance. In D. Druckman & R. Bjork (Eds.), *In the mind's eye: Enhancing human performance* (pp. 247–269). Washington, DC: National Academy Press.

National Research Council. (1994). The performance and development of teams. In D. Druckman & R. Bjork (Eds.), *Learning, remembering, believing: Enhancing human performance* (pp. 113–139). Washington, DC: National Academy Press.

O'Neil, H. F., Baker, E. L., & Kazlauskas, E. J. (1992). Assessment of team performance. In R. Swezey & E. Salas (Eds.), *Teams: Their training and performance.* Norwood, NJ: Ablex Publishing Corporation.

O'Neil, H. F., & Robertson, M. M. (1992). Simulations: Occupationally oriented. In M. C. Alkin (Ed. in Chief), *Encyclopedia of educational research* (6th ed., pp. 1216–1222). New York: Macmillan.

Parker, G. M. (1990). *Team players and teamwork: The new competitive business strategy.* San Francisco, CA: Jossey-Bass.

Schneider, J. R., & Schmitt, N. (1990). *Assessing teamwork skills with assessment center exercises: Final report on employability skills to the Michigan Department of Education.* Unpublished manuscript.

Shavelson, R. J., & Baxter, G. P. (1992). What we've learned about assessing hands-on science. *Educational Leadership, 49,* 20–25.

Standford, G. (1977). *Developing effective classroom groups.* New York: Hart.

Thornton, G. C., & Byham, W. C. (1982). *Assessment centers and managerial performance.* New York: Academic Press.

Webb, N. M. (1995). Group collaboration in assessment: Multiple objectives, processes, and outcomes. *Educational Evaluation and Policy Analysis, 17,* 239–261.

Wellins, R. S., Wilson, J., Katz, A. J., Laughlin, P., Day, C. R., & Price, D. (1990). *Self-directed teams: A study of current practice.* Development Dimensions International, Association for Quality and Participation, and *Industry Week.*

Wellins, R. S., Byham, W. C., & Wilson, J. M. (1991). *Empowered teams: Creating self-directed work groups that improve quality, productivity, and participation.* San Francisco, CA: Jossey-Bass.

West, M. A. (1994). *Effective teamwork.* Leicester, England: The British Psychological Society.

Zander, A. (1982). *Making groups effective.* San Francisco, CA: Jossey-Bass.

APPENDIX: PREPARING STUDENTS FOR
TEAMWORK ASSESSMENTS

Overview

In order to prepare students for the teamwork assessment, it is helpful to spend at least the class period prior to the beginning of the assessment discussing the characteristics of effective teams and the ground rules for team interactions. Teachers should read the tables on *Characteristics of Teams, Teamwork Skills,* and *Stages of Team Development* before presenting teamwork to students, so teachers can better answer students' questions. A list of Ground Rules for Effective Teams is included in the manual. It should be provided to students before they participate in the assessment.

While the directions included in this and subsequent sections are written as a script, it is not necessary to read it verbatim. The script is provided for individuals who feel more confident having explicit instructions. Teachers may review the script and give their students the general instructions. However, **make sure that all procedures are followed.**

Instructions to Teachers

NOTE: Before the start of the class, put the following list on the board. These materials will be referred to later in this section.

- **Ground Rules for Effective Teams**
- **Team Effectiveness Observation Sheet**
- **Product Scoring Sheet**
- **Self-Evaluation for Team Effectiveness**

> Over the next [insert time period for assessment selected], you will be participating in a teamwork assessment. This is not like tests that you are used to taking. You will be working in teams to [insert type of assessment; e.g.; write a report, produce a product, find solutions to a problem, etc.]. The goal of the assessment is to give you feedback on your skills in working in teams. Businesses routinely describe the ability to work in a team as a critical skill for doing well in a job. Virtually all careers require you to work in a team at least part of the time.

> When you come in tomorrow, you will find out who your teammates will be. I will be assigning you to teams by [insert method of assignment; e.g.; random, teams worked in before, teams designed to maximize strengths, etc.]. I will be observing each team and evaluating the results of your teamwork. A little later in this period, I'll show you the forms to be used when I observe your team. I'll also describe how the scoring will work. Right now, I want to tell you about the exercise and discuss effective teamwork.

NOTE: *These instructions are written as if the classroom teacher will be the only person observing the teams. Depending upon the arrangements in a school, another teacher, an outside observer--such as a business person, a counselor, an administrator, or other observer may rate a team. Anyone who observes the teams must follow the instructions on the* **Team Effectiveness Observation Form** *and be trained in using the rating scale. If more than one observation is made per team, the same person would not observe both times.*

Teamwork Product Introduction

The exercise you will be participating in uses skills related to planning, organizing, and [insert specific content of teamwork assessment chosen, see below]

- for classroom projects *the type of project* (researching and writing reports, surveying people, conducting experiments, developing debate topics, etc.
- for product-based tasks *the production of a product*
- for scenarios *the solution of a problem*

NOTE: *Teachers using classroom projects for the assessment should give an explanation of any grading procedures. That is; the team will receive a grade on the entire project and, if the teacher chooses, also on their individual contributions. The teacher may also wish to provide any typical introduction he or she might give to doing projects in that class. Teachers using either the product-based or scenario assessments will need to review the contents of the* **Product Scoring Sheet** *for the assessment they are using. These sheets are included in the sections on those assessments.*

Let's take a few minutes and list some of the things that might need to be done for a team to plan and organize a [insert specific content of teamwork assessment].

NOTE: *Teachers should list ideas on the board and then summarize. Make sure that such key points as the following are included on the list or in the summary; (a) discussing the specific tasks to be completed, (b) making sure that every task has someone to do it, (c) having a way to check on when tasks are completed, and (d) a method for determining the quality of the final product. Teachers may add other items they believe will help the teams plan and organize to complete the assessment.*

Teamwork Process Introduction

As important as the [report, product, solution, etc.] you create, is the process you will be using; that is, working in teams. To get ready for working in teams, we're going to spend some time today talking about what it means to work effectively in a team.

NOTE: *Customize here depending on whether or not the class already uses groups, collaborative learning, project-based learning, etc. in class. Remind students of what they have already done and any ground rules, etc., particularly if this assessment will differ from the typical team practices in the classroom.*

How many of you have ever been on a team or worked with other people to get a job done? It may be on a project at school, in the neighborhood, on a sports team, in a community group, or your family. I'm guessing that sometimes you've enjoyed working on teams and sometimes you've thought the team was a waste of time. First, I'd like to get some impressions about what happened when teams worked well. What made working on a team you liked a good experience?

Write the class's responses on the board under the heading, Teams that Worked Well.

Now, think about a time that a team didn't work well. What happened on that team that made it ineffective?

Write the class's responses on the board under the heading, Teams that Didn't Work Well, without erasing the effective team's responses so the class can compare them.

What do you notice when you look at these two lists?

Note: *Summarize these comments, point out that effective teams have members who pay attention both to what needs to be accomplished by the team (the task of the team) and to each team member's input into the accomplishment of the task (the process of the team). The latter may come out as 'team spirit,' 'everyone cared about everyone else,' 'when we had a disagreement we worked it out,' or 'everyone pulled their own weight.' After the students have commented on their lists, the teacher may wish to distribute a copy of* **Characteristics of Teams.**

As you can see, you've identified many of the same behaviors that are identified by social scientists in their research on effective teams. When you come to class tomorrow to work in a team, think about the characteristics we've discussed today and the part you can play in helping your team work well.

Introduction to Scoring

Remember that the goal of this assessment is to provide you with feedback on how well your team worked together. In order to do that, I'll be observing each team during the assessment. I'll also be scoring the results of your teamwork; that is, the products you produce using teamwork.

NOTE: *Teachers who do the classroom project assessment should be able to observe teams twice for 10 minutes each. Teachers who elect one of the other two assessments will probably only be able to observe teams once for about 5 minutes, depending upon the number of teams in the class. If more observers are available, then teams can be observed for longer periods of times. Teachers should provide students with the most information possible about how long and how often they will be observed. Teachers may wish to give students the opportunity to observe other teams.*

When your team is observed, I'll be taking notes on what I see your team doing. This is a copy of the form I'll be using when I watch each team and the way in which I will be scoring any products you create using teamwork.

NOTE: *Hand out **Team Effectiveness Observation Sheet, Product Scoring Sheet**, and **Self-Evaluation for Team Effectiveness**.*

As you can see, it includes many of the same behaviors we've already talked about as representative of what effective teams do. I'll be rating your team on those behaviors, as well as on your success in creating the required product.

NOTE: *Review the forms with the class, stopping to answer any questions that are raised about definitions, requirements, etc.*

During the assessment, you will also be rating yourself on how well you think you are doing in helping your team work well together and accomplish its task. Look over the form and then I'll answer any questions you have about it. You will see that it is very much like the observation form I will be using.

Note: *Teachers who are doing a classroom project may wish to have students do self-evaluation before they begin working in their teams, half-way through, and at the end. Teachers who are doing product-based task assessments or scenarios may ask students to complete the self-evaluation the day before and the day after the assessment.*

Introduction to Team Ground Rules

As you can see from the Team Effectiveness Observation Sheet, I will be looking at how each of you contributes to the work of the team. In order for you to do well as a team member, each person and the team as a whole need to follow some ground rules. I have some ground rules to show you. We can discuss them and add to them if necessary.

Note: *Hand out Ground Rules for Effective Teams. Discuss with the students and make any modifications agreed to by everyone. Following that discussion, have students complete the Self-Evaluation of Team Effectiveness. That should take to the end of the class period.*

Random Assignment

Randomly assigned teams can be created in the following manner. Prepare materials for random assignment before the assessment period begins. Divide the number of students in each class by 5. If there are leftover student in any class, the teacher may create a team of 4 students or multiple teams of 6 students. Do not have any teams larger than 6 students or smaller than 4 students. A team assignment for each student will be created by writing a single number (e.g., 1, 2, 3, 4, 5, 6) on cards or slips of paper. This number will be the number of the team to which the student is assigned. For example, a class of 28 students would have 5 teams, three of which would have 6 members. The teacher would make five cards with the number 1, five with 2, six with 3, six with 4, and six with 5. Mix up the cards and place them in an opaque container. Once students have settled into their seats, pass the container around and have students select one card and read the number on it out loud. Record each student's team assignment. The students may keep their cards.

Seating Arrangement

Whenever possible, students should move their seats in such a way that teams sit together and are somewhat separated from other teams. If furniture is not movable, then students in the same team should try to sit around a single work space. Allow time at the beginning and end of each assessment period for students to move furniture.

Materials

If possible, have newsprint and markers available for posting team thoughts, decisions, etc. However, this is not necessary for the successful completion of the assessment. Students should have paper and pen or pencil for the assessment.

Chapter 16

Use of Networked Simulations as a Context to Measure Team Competencies

Harold F. O'Neil, Jr.
CRESST/University of Southern California

Gregory K. W. K. Chung
Richard S. Brown
CRESST/University of California, Los Angeles

USE OF NETWORKED SIMULATIONS

In previous reports (O'Neil, Allred, & Dennis, 1992; chap. 10, this volume) we documented several validation studies of our measures of the negotiation subskill of an interpersonal competency (U.S. Department of Labor, 1991, 1992). In this chapter we present results of our initial attempt at measuring teamwork processes using computer simulation.

In the following section we cover four issues: (a) the importance of teams, (b) a taxonomy of teamwork processes, (c) teamwork processes within a negotiation context, and (d) measuring teamwork processes. The focus of our effort is on a simple but fundamental question: How do we assess team processes and outcomes so that the measurement is not only reliable and valid, but also *timely*?

Importance of Teams

There has been an increased interest in team performance over the last 20 years attesting to its importance. Parallel efforts within the educational (e.g., Koffel, 1994; Stasz et al., 1993; Webb & Farivar, 1994; Webb & Palincsar, 1996), industrial (e.g., Druckman & Bjork, 1994; Stasz, Ramsey, Eden, Melamid, & Kaganoff, 1996; Sundstrom, De Meuse, & Futrell, 1990), and military (e.g., Brannick, Prince, Prince, & Salas, 1995; Cannon-Bowers & Salas, this book; Cannon-Bowers, Tannenbaum, Salas, & Volpe, 1995;

Franken & O'Neil, 1994; Swezey & Salas, 1992) sectors have attempted to characterize, measure, and understand the constructs underlying effective teams. Although each sector focuses on different facets of teamwork and employs different paradigms, all three sectors recognize the potential of teams to increase and enhance learning, task performance, work productivity, and product quality. Indeed, one theme that emerges from all three sectors is that under optimal conditions, effective teams often reflect the cliché that the whole is greater than the sum of its parts.

Of particular relevance to our work is how changes in American industry are driving changes in education and training. Numerous studies and commissions have examined the skill requirements for tomorrow's entry level worker. O'Neil, Allred, and Baker (1992, this book) reviewed five major studies on workforce readiness commissioned by federal, state, or private agencies. All studies identified interpersonal and teamwork skills to be essential skills.

The private sector's emphasis on teamwork has gained momentum as global competitive pressures have forced companies to alter management practices. Downsizing, reorganizations, and other structural changes in the workplace have resulted in companies placing greater responsibility on their workforce. A prevailing belief is that teams offer the potential to dramatically improve a company's competitiveness (Cannon-Bowers, Oser, & Flanagan, 1992). Whether employees participate in teams as adjuncts to their regular job (e.g., quality-circle teams) or participate in teams full-time (e.g., negotiating teams), the implication is clear: Increasingly, employees are being required to be part of teams. Given the importance of teams in the American workforce, one of our goals is to develop feasible measures of teamwork processes of high school students. Educators and employers can then use these measures to help them focus on education and training efforts.

A Taxonomy of Team Processes

Morgan, Salas, and Glickman (1993) provide insight into the nature of teams. In their model of team development, Morgan et al. (1993) postulate two tracks of team process, a taskwork track and a teamwork track. The taskwork track accounts for specific activities unique to the task. Taskwork team skills influence how well a team performs on a particular task (e.g., whether or not a team of negotiators reaches an agreement with the other party). Taskwork skills are domain-dependent, task-related activities. The teamwork track or team workskills influence how effective an individual member will be as part of a team and are domain-independent team skills. Team workskills encompass skills such as adaptability, coordination, cooperation, and communication. Effective teams develop competence along both tracks. Members of effective teams possess basic skills required for the task and know how to coordinate their activities, communicate with each other, and respond effectively to changing conditions.

We have drawn on the work of Morgan et al. (1993), Salas, Dickinson, Converse, and Tannenbaum (1992), Burke, Volpe, Cannon-Bowers, and Salas (1993), and others (O'Neil, Baker, & Kazlauskas, 1992; Webb, 1993, 1995; Webb & Palincsar, 1996) to aid our development of teamwork process measures. From the taxonomy of teamwork skills identified by Burke et al. (1993), we have adopted the following six categories: (a) adaptability—recognizing problems and responding appropriately, (b) coordination—organizing team activities to complete a task on time, (c) decision making—using available information to make decisions, (d) interpersonal—interacting cooperatively with other team members, (e) leadership—providing direction for the team, and (f) communication—the overall exchange of clear and accurate information. In the following discussion we first define and then elaborate on each process with examples from the teamwork literature.

Adaptability is the process by which a team is able to monitor the source and nature of problems through an awareness of team activities and factors bearing on the task. Adaptive teams use this information to adjust to situational demands by using compensatory and feedback behaviors.

An important facet of adaptability is the detection and correction of problems. Sometimes this detection process is labeled situational awareness. Members cross-check each other's performance, provide backup when needed, and freely provide feedback to and accept feedback from each other (McIntyre & Salas, in press). For example, Oser, McCallum, Salas, and Morgan (1989) observed moderate correlations between various team behaviors and team effectiveness as measured by a team simulation score. Oser et al. (1989) studied the performance of 13 Navy fire-control crews and found that the top three (more effective) teams, compared to the bottom three (less effective) teams, differed across a range of behaviors. With respect to problem detection and correction, members of more effective teams (a) helped others who were having difficulty with a task ($r = .67$, $p < .05$), (b) assisted others when they had difficult tasks to perform ($r = .49$, $p < .05$), (c) suggested to others to recheck their work to find their mistakes ($r = .54$, $p < .05$), and (d) provided suggestions to others on the best way to find an error ($r = .49$, $p < .05$).

Coordination is the process by which team resources, activities, and responses are organized to ensure that tasks are integrated, synchronized, and completed within established temporal constraints. Our view of coordination is that it primarily involves task accomplishment rather than interpersonal harmony.

One area where coordination has received much attention is in aircrew management. Poor team coordination has been cited as a cause in aircraft disasters (Foushee, 1984) while good team coordination has been associated with better training performance (Brannick, Roach, & Salas, 1993; Kanki, Lozito, & Foushee, 1989a). For example, Brannick et al. (1993) had 52 dyads participate in a low-fidelity (personal computer) flight simulator. Participants were college students participating for extra class credit. Most participants (76%) did not have any personal computer flight simulator experience. The teams' mission was to shoot down as many enemy planes as possible within 30 minutes. Each member had control over different parts of the simulator (e.g., one member would control the piloting and the other would control the fire-control system). Interdependency was built into the task in such a way that the enemy plane could be shot down only with both members working together. With respect to coordination, Brannick et al. (1993) found significant correlations between team coordination and the outcome measure of team performance (as measured by the number of radar "locks" by the enemy plane; $r = .29$, $p < .05$). Brannick et al. defined coordination as properly sequenced behavior and the exchange of useful information. Oser et al. (1989) observed similar results of effective information management. Members of effective teams, compared to less effective teams, showed more coordinated gathering of information ($r = .50$, $p < .05$).

Interestingly, Brannick et al. (1993) also found that experience with computer flight simulators was significantly correlated with coordination ($r = .31$, $p < .05$), as well as with (a) communicating about escaping the enemy plane ($r = -.34$, $p < .05$), and (b) communicating about the meaning of the simulator display ($r = -.58$, $p < .05$). It is interesting that the latter two relationships are negative. One interpretation of their finding is that the familiarity of the situation contributes to how team members coordinate their activities. Team members share an understanding about certain tactical and operational elements of the simulator, which reduces the need to discuss those elements. Such shared mental models have been observed in aircrews familiar with each other compared to aircrews that were working together for the first time (Foushee, Lauber, Baetge, & Acomb, 1986). Foushee et al. (1986) speculated that familiarity aided team performance as crew members were familiar with each other's behavioral and communication styles. This familiarity helped crew members anticipate and respond to each other's needs and actions. Similarly, Kanki, Lozito, and Foushee (1989b) observed that predictable and homogeneous communication patterns distinguished low-error from high-error aircrews using high-fidelity flight simulators. Low-error aircrews exhibited uniform communication patterns with respect to speaking and responding to each other. Conversely, high-error aircrews exhibited widely varying patterns. In effect, these aircrews lacked a standard way of communicating with each other, in contrast to the low-error aircrews.

The nature of these kinds of shared mental models has been the focus of recent models of team performance (Cannon-Bowers, Salas, & Converse,

1993; Rouse, Cannon-Bowers, & Salas, 1992) and is discussed in greater detail in the next section.

Decision making is the ability to integrate information, use logical and sound judgments, identify possible alternatives, select the best solution, and evaluate the consequences.

Shared mental models provide a useful perspective on team decision making (Cannon-Bowers et al., 1993; Rouse et al., 1992). A mental model can be thought of as a person's conception of a system (e.g., a negotiation task). Mental models help people describe the purpose and form of the system, help them explain the system's function and what it is doing, and help them predict what the system will do in the future (Rouse et al., 1992). In terms of team performance, Cannon-Bowers et al. (1993) suggested that four kinds of mental models are pertinent to team performance: (a) models of the task to be performed; (b) models of the operating characteristics of the task-related equipment; (c) models of the roles, responsibilities, and interdependencies of each team member, and of how information flows within the team; and (d) models of each member's skills, knowledge, abilities, and behavior patterns. A shared mental model represents an overlap between individual members' mental models.

Rouse et al. (1992) made several predictions about teams that possess shared mental models. Two are relevant to decision making. First, Rouse et al. (1992) predicted that teams with shared mental models will require less overt planning time for good performance. Essentially, if everyone knows what to do, then they don't need to discuss it. Second, teams with shared mental models will engage in less overt communication but maintain performance. In this case, team members are competent and proficient in what they do and how they interface with each other. Overt requests for information will decrease because team members are able to predict the needs and actions of others. What the shared mental model perspective suggests is that effective decision making is inextricably tied to group coordination. Presumably, the quality of decision making covaries with the quality of coordination. The more coordinated the group, the better the decision-making process and vice versa.

These predictions have important implications for team decision making. Teams with shared mental models will converge on a decision faster because everyone has a comparable understanding of the task demands. Team members capitalize on their shared understanding to minimize nonessential, overt communication. Everyone knows who is responsible for what and whom to communicate with. With a shared understanding, members know whom to provide information to and when to provide it, and the recipient of information knows that the information is relevant and timely.

Support for this view is observed in Brannick et al.'s (1993) study discussed earlier. With respect to time, Brannick et al. observed significant correlations between task time and quality of performance as measured by (a) observers' holistic ratings ($r = -.48$, $p < .05$), and (b) number of locks by the enemy aircraft ($r = .53$, $p < .05$). Teams that took longer to shoot down the enemy plane were (a) rated as performing poorer on the simulation, and (b) had a higher number of radar locks on their plane by the enemy. And as discussed earlier, Brannick et al. (1993) also found significant negative correlations between experience and two communication frequency measures, and found negative correlations (although these correlations were nonsignificant) between experience and five of the remaining six indicators. Further, Brannick et al. (1993) found negative correlations between all communication acts and the outcome measures, suggesting that the more team members "talked it up," the more their task performance suffered. The research conducted by Lahey and Slough (1982) suggests this interpretation in that instructors who graded team training exercises gave lower grades to military teams doing excessive talking.

These negative correlations between communication frequency and team performance are puzzling, given the wealth of findings pointing toward the importance of communicating on teamwork (Foushee et al., 1986; Kanki et al., 1989b; McIntyre & Salas, in press). What may be occurring is that poorly performing teams are "talking it up" in an attempt to improve their condition.

Interpersonal skill is the ability to improve the quality of team member interactions through the resolution of team members' dissent, or the use of cooperative behaviors.

Team reinforcing behaviors have been found to be related to effective team performance (Oser et al., 1989). Interpersonal processes are important not as much to minimize intergroup friction as to foster team interdependence. That is, cooperative behavior reflects a commitment to team performance (Mullen & Copper, 1994). Cooperative behavior reflects a belief that each member is critical to the overall success of the team, and helping others helps the team. Members of effective teams value team performance over individual performance and view themselves as part of a team and not simply individuals working together (McIntyre & Salas, in press).

Oser et al. (1989) found significant relationships between interpersonal behaviors and team performance. Members of more effective teams tended to (a) make positive, motivating statements ($r = .54$, $p < .05$), (b) praise others for a good job ($r = .59$, $p < .05$), and (c) thank others for catching a mistake ($r = .60$, $p < .05$). In addition, Oser et al. (1989) observed negative correlations when team members made negative statements directed at the team or the training ($r = -.70$), or when a member raised his voice when correcting another member ($r = -.60$, $p < .05$). Brannick et al. (1993) found a similar pattern of results. In their study, Brannick et al. (1993) found significant relationships between (a) the amount of suggestions given and team performance ($r = .29$, $p < .05$), and (b) cooperative behavior and team performance ($r = .29$, $p < .05$).

Leadership is the ability to direct and coordinate the activities of other team members, assess team performance, assign tasks, plan and organize, and establish a positive atmosphere.

In their summary of "lessons learned" from their teamwork research, McIntyre and Salas (in press) suggest that team leadership style plays a crucial role in the functioning of a team. In particular, good team leaders exhibit the kinds of behaviors that help team performance in general: the ability to adapt to changing conditions, exchange information, provide and accept feedback, and provide and accept help. A good team leader is a model of teamwork that other team members can emulate.

The Oser et al. (1989) study described earlier provides support to McIntyre and Salas' (in press) analysis that good team leaders reflect good team processes. Oser et al. (1989) observed that leaders in more effective teams, compared to leaders in less effective teams, initiated far more of the critical behaviors (i.e., error identification and resolution, coordinating information gathering, and team encouragement). In their review of the teamwork literature, Macpherson and Perez (1992) concluded that during training, a team leader should be encouraged to solicit recommendations from team members, analyze and comment on those recommendations, and then communicate his or her decisions to the team.

One crucial aspect of good team leadership is the willingness to accept feedback from subordinates. A good team leader establishes an environment where there is a free and bi-directional flow of information between the leader and members. Leaders who constrain this flow of information—by virtue of their leadership style, rank, or personality—impede team performance. Team members in this kind of environment are often inhibited from giving feedback or backup. Foushee (1984) reports that copilots in these kinds of environments often hesitate to voice concerns, even in potentially dangerous situations. For example, in 1979 a commuter plane crashed because the copilot failed to assume command after the pilot became incapacitated. The captain was the company vice president, had a gruff personality, and was noticeably upset on the day of the crash. The copilot was recently hired and still on probation. Foushee (1984) speculates that had the copilot not been intimidated, he would have assumed control and possibly averted the disaster.

Communication is the process by which information is clearly and accurately exchanged between two or more team members in the prescribed manner and by using proper terminology, and the ability to clarify or acknowledge the receipt of information.

Perhaps the most powerful index of team performance is the extent and character with which team members interact with each other. Communication underlies every team process discussed so far. As a process, effective communication couples team members' expectations, actions, responses, and feedback behaviors. Communication facilitates the transmission and reception of support behavior, and the detection and correction of error conditions (McIntyre & Salas, in press; Oser et al., 1989), helps team members synchronize their activities (Foushee, 1984; Kanki et al., 1989a, 1989b; McIntyre & Salas, in press; Mullen & Copper, 1994), influences the quality of decision making (Brannick et al., 1993; Cannon-Bowers et al., 1993; Rouse et al., 1992), influences the character of team cohesion (McIntyre & Salas, in press; Oser et al., 1989), and establishes operational norms between members (Foushee et al., 1986; McIntyre & Salas, in press; Mullen & Copper, 1994). It should be noted that excessive communication is counterproductive (e.g., Lahey & Slough, 1982).

Team Processes in a Negotiation Context

The negotiation context within which we frame our teamwork processes is the mixed-motive interdependence paradigm (Bazerman, 1990; Carnevale & Pruitt, 1992; O'Neil, Allred, & Dennis, 1994). Briefly, this paradigm incorporates two dimensions of negotiation styles, distributive and integrative. Distributive negotiation reflects a view that the issues under negotiation are equally valued by both parties; conceding on an issue necessarily means a loss. Such a view leads to a negotiating strategy of attempting to gain as much as possible across all issues—a zero-sum/win–lose perspective. Integrative negotiation reflects a view that the issues under negotiation have differing values for both parties. A concession on an issue does not necessarily mean a loss. Rather, conceding on an issue of low importance may result in gaining on an issue of high importance. This leads to a negotiation strategy of seeking ways in which both parties can benefit. Integrative negotiation reflects a variable-sum/win–win view of the situation. Expert negotiators adopt an integrative negotiation style. The mixed-motive interdependence paradigm has the potential to elicit many of the team processes already defined. In the following discussion we analyze each team process in the context of this negotiation scenario, outlining the kinds of behaviors we expect to emerge.

Adaptability. In general, a team's adaptive capability affects the nature of the team's response to a given situation. Are team members able to detect problems as they arise and, having detected them, respond effectively to resolve the problems? Do team members exhibit innovative and flexible responses to situations?

Applying our definition of adaptability to the negotiation scenario suggests two kinds of outcomes. First, members exhibiting adaptive behavior should show more willingness to be integrative in their responses. For example, flexible team members would recognize that not all issues are equally important and would thus carry out negotiations accordingly. Issues of low importance could be traded away for gains on issues of high importance. Conversely, members showing little adaptability might not recognize this situation and would hold firm across all issues. We expect integrative behavior to manifest itself as questions or statements that try to uncover the other party's priorities and values.

A second outcome of adaptive behavior is the use of compensatory feedback. That is, members provide information to each other regarding the current situation. We would expect compensatory behavior to manifest itself as members agreeing, disagreeing, asking for ideas, or suggesting possible courses of action.

Coordination. In general, a team's coordination capability affects how that team organizes its resources, activities, and responses. Teams with high coordination will carry out a task that is integrated, synchronized, and completed on time.

Coordination in the context of a negotiation scenario suggests that sequencing would be important. We believe that members of coordinated teams will show a propensity to sequence offers in an attempt to tease out the other team's values and priorities. Conversely, members of less coordinated teams will show a tendency to propose offers that will vary widely from round to round. In addition to sequencing, we also expect that highly coordinated teams will be cognizant of time constraints and will communicate this aspect to other members when appropriate.

Decision Making. A team's decision-making capability affects its ability to capitalize on available information. Effective teams use information about their current situation to help them evaluate the worthiness of potential courses of action.

In the context of a negotiation scenario, we expect to see several kinds of decision-making patterns contingent on how well the teams understand the negotiation situation. First, for teams that enter the negotiation simulation with an existing understanding of how to negotiate, we expect that these teams will quickly reach a resolution and engage in little debate over what should be the proper strategy. In effect, these teams have shared mental models of the negotiation situation, and this understanding helps the team reach an agreement.

For teams with less of an understanding (e.g., novices), one of two things could happen. First, these teams could come to an understanding by reasoned discussion. Such teams would pursue a course of action after discussing alternatives and potential consequences. Thus, we would expect members to provide specific reasons for a particular position, and we would expect these reasons to reflect an integration of information about the specific negotiation situation. For example, a member might suggest bidding lower on an issue because that issue is important to the other party. Alternatively, teams with less of an understanding could simply engage in help-seeking behavior. That is, communication between members would be dominated by appeals for ideas about what to offer, what tactic to take, and the like.

Interpersonal. In general, a team's interpersonal capability affects how well team members work cooperatively. Do team members encourage each other to work together? Do team members help resolve team member dissent?

In terms of our negotiation scenario, we would expect teams with a high interpersonal dimension to encourage member participation, help seeking, and help giving. In addition, we would expect members to freely exchange compliments.

Leadership. In general, a team's leadership capability affects its ability to provide direction for the team. To what extent does a member coordinate the activities of other members? Does the team engage in planning and organization that reflect appropriate priorities?

In our negotiation scenario we expect that teams or team members high in leadership will engage in communication that provides direction and sets priorities. We expect a member strong in leadership to place the current negotiation situation in context of the team goal. For example, such a member would remind other members of the team's priorities, thereby aligning task engagement with task goals. We also suspect that strong leadership would manifest itself in directives. In this case, communication would take the form of directing others to do something. Finally, we believe that good leadership will be reflected in a leader who is receptive to input from other members.

We believe that our earlier efforts using an individual-level negotiation scenario (O'Neil et al., 1994) provide a useful context for studying team-level processes. In the next section we first discuss current approaches to measuring teamwork processes and their limitations. We then describe how we address those limitations in our research.

Measuring Teamwork Processes

Our work focuses on assessing team processes that emerge during the negotiating of a contract. We are interested in the nature of the interaction between team members and how that interaction impacts team performance. Yet a critical measurement issue remains unresolved: How do we assess teamwork processes in such a way that the measurement technique is reliable, valid, and timely?

Our previous work on using computer simulation suggests that our approach is feasible, reliable, and valid when measuring individual negotiation skills (O'Neil, Allred, & Dennis, 1992; O'Neil et al., 1994; chap. 10, this volume). Our simulation technique reliably discriminated between experts and novices, providing strong support for the validity of our approach. Our findings persisted across different negotiation contexts and populations (i.e., high school students negotiating for a job at a movie theater, law students negotiating for a job with a law firm).

However, our preliminary attempts at using teams in the same simulated negotiation context were inconclusive (O'Neil et al., 1994). In O'Neil et al. (1994, Study 2), 3-person teams were compared to 2-person teams. The negotiation scenario was identical to that used by individuals. The primary difference between the scenarios was that the team scenario required the team members to agree verbally on a course of action. Then, one team member would communicate that decision to the computer.

Although O'Neil et al. (1994) found nonsignificant effects on one measure of performance (final counteroffer), the results were in the predicted direction (3-person teams performing better than 2-person teams). On the measure of quality of performance (integrative vs. distributive), O'Neil et al. found no significant differences. Finally, in terms of frequency of agreement versus nonagreement, O'Neil et al. did find significant effects: 3-person teams tended to reach agreement more often than 2-person teams. While the measures used in O'Neil et al. (1994) captured team outcomes, they did not capture the underlying teamwork processes. Thus, the focus of our current work is to develop methods for measuring teamwork processes.

Existing approaches to measuring teamwork processes rely almost exclusively on observational methods (Baker & Salas, 1992). For example, behavioral checklists (e.g., Oser et al., 1989), videotaped and audiotaped observation (e.g., Brannick et al., 1993), and analysis of think-aloud protocols are the most common techniques to measure teamwork processes. These methods are labor intensive and time-consuming (Sanderson & Fisher, 1994). Observations must be transcribed, coded, and analyzed post hoc. Such techniques offer no opportunity for rapid analysis and reporting of team performance. From an assessment perspective, these methods are unappealing because of the lag between test administration and reporting of test results. Further, these methods are neither practical nor cost-effective in large-scale test settings.

Our approach to addressing these limitations was to expand our computer-based negotiation simulation to a new context (union–management negotiation of a contract) and to incorporate real-time measures of team-work processes. To achieve this, we broadened the scope of the computer-based negotiation simulation in the following way. Individuals were assigned to their own computers and collaborated with team members through a custom-developed computer conferencing system (e.g., McCarthy et al., 1993). Our initial system implemented the sending of both predefined messages and free-form messages; more typical conferencing systems allow for sending typed-in, free-form messages only. This approach of linking team members via a network differed from our preliminary investigation of teams (see Study 2 of O'Neil et al., 1994 for differences). In that study, O'Neil et al. (1994) physically grouped one team with one computer (e.g., three participants sharing one computer). Discussion took place face-to-face but was unrecorded. In the present study, each individual operated his or her own computer and interacted with other team members by sending messages.

Our final technique for capturing teamwork processes was to use a predefined taxonomy of teamwork skills discussed earlier and to categorize the expected communication acts team members would engage in. We developed a priori a set of "canned" messages, which members can send to each other, and we categorized these messages as belonging to one of the five teamwork processes (adaptability, coordination, decision making, interpersonal, and leadership). We assume that this taxonomy of teamwork is domain independent and independent of scenarios. By tracking the messages selected and sent (and hence, by definition, the teamwork process category), we get an accurate index of the teamwork process. We assume that each message in a category is as important as any other message, and thus all messages are equally weighed. Using those techniques provides us with a real-time teamwork assessment system. We can administer, score, and interpret in real time.

Our approach of using networked computer simulation is consistent with other computer-based efforts at measuring team performance (see Weaver & Bowers, 1995, for a review). However, what distinguishes our system from other systems is that we measure team processes as they occur in real time. We conducted several pilot studies and a validation study.

PILOT STUDIES

Two pilot studies were conducted to assess the feasibility of our approach. The first represented our initial attempt at assessing the functionality of the computer system and messages. We were interested in feedback from users regarding usage of the system, messages, and task performance. The second reflected several major revisions to the system based on the first pilot study. The two pilot studies are described in greater detail next.

Pilot Study 1

Participants. The first pilot study used three groups of adults. Participants were drawn from available graduate and undergraduate students and other, nonstudent adults in our laboratory. Two groups were tested formally, where participants worked as a team under controlled conditions. In these groups, the programmer of the computer system participated and assumed the role of an advisory member. The third group was tested informally; participants provided comments about the system as they carried out the negotiation task.

Negotiation Task. For our pilot studies we created a scenario where participants collaborated in 3-person teams. Each team consisted of one leader and two advisory members. The team assumed the role of representing a union, and the team's responsibility was to negotiate a contract with a simulated management team. The issues under negotiation were wages, health and welfare, and pension. The simulated management team was programmed to concede on its most important issue when the union (i.e., participant) team did likewise (O'Neil et al., 1994; chapter 10, this volume). Thus, the simulated management team mirrored the negotiating behavior of the participants' team. The team outcome measures, based on previous research (O'Neil et al., 1994; chap. 10, this volume), were (a) whether agreement was reached and (b) whether the agreement was integrative or distributive.

Messages. In order to capture team processes in real time, we developed a set of messages for each teamwork category (e.g., interpersonal skills) previously identified. Traditionally, measurement of teamwork skills is done using human judges either to rate observations (live or videotaped) of a team interacting, or to rate transcripts of team members' think-aloud protocols. In our case, we developed messages a priori to represent team processes. Our intent was to have team members use these messages to communicate with each other regarding the negotiation. For our feasibility study, we included a variety of message types that reflected good and poor team skills. In addition, we included messages to measure communications skills and motivation. The following list gives examples for each category: (a) adaptability—"We should try something different"; (b) communication—"I don't understand, please explain"; (c) coordination—"We need to hurry up"; (d) decision making—"We might want to give a lot on pension"; (e) interpersonal—"I am interested in hearing from you on this";

(f) leadership—"Let's place an offer now"; and (g) motivation—"Our objective is to maximize our interests."

Computer System. The computer platform we used was the Macintosh (1993), and the software was developed with HyperCard 2.2 (1993). We used the built-in networking capabilities of the operating system (System 7, 1992) and HyperCard 2.2 (1993) to implement a rudimentary client-server system. We designed the computer system to enable members to communicate with each other via a common message blackboard. Messages were posted as they were dispatched by members. All members had read and write access to the blackboard.

The message sending capability was organized around teamwork category. To send a message, participants first clicked on a button representing one of the teamwork categories. Then, a window opened that presented a list of predefined messages for that category as well as an option where the participant could type in his or her own message. If the participant chose to send a predefined message, he or she simply clicked on the message and the message was dispatched. Alternatively, the participant could type in a message. Participants were also provided with a paper listing of all the messages sorted by teamwork category. The computer system recorded the number of messages (both predefined and self-generated) sent from each category.

Results. For the first group, the maximum number of rounds was set to 12. (We define one round to consist of an offer from the union team and a counteroffer from management.) The total number of messages sent was 231, and no agreement was reached. The simulation ended after the allotted 12 rounds and ran for 100 minutes. For the second group, the maximum number of rounds was set to 10. The total number of messages sent was 124, and agreement was reached on round 8 after 82 minutes. For the third group, the maximum number of rounds was set to 8. The total number of messages sent was 49, and agreement was reached on round 7 after 70 minutes. These results of the first pilot study point to strengths and limitations of our initial approach. These findings are outlined next.

All groups participated with sustained interest and effort. Despite participants' knowledge that they were negotiating with a simulated management team, all groups approached the task with concerted effort. Participants reported that the task was interesting and that they were genuinely trying to reach an agreement. A review of the message transcripts also suggests a consistent pattern of communication. In general, communication repeatedly took the form of a three-stage process. In the first stage, someone (usually the leader) would suggest a specific offer. The second stage was characterized by a discussion of the pros and cons of that offer, how the offer might impact the team, and potential actions the management team might take in response to the offer. During this stage alternative offers were suggested by all team members. The last stage was characterized by someone (usually the leader) making a suggestion to wrap up the discussion and make a decision. The leaders of all groups sought consensus before sending the offer to management. There was no evidence of a leader acting independently or dictatorially.

The networked computer simulation clearly evoked a strong desire to communicate. Participants wanted to communicate immediately, and using the enter-own-message facility was the most efficient way to do it. In the two formal groups, team members started using the predefined messages in earnest, but quickly resorted to the enter-own-message facility. In the third group, all members immediately resorted to and relied on the enter-own-message facility.

All three groups avoided using the predefined messages primarily because they had difficulty interpreting the message categories. Participants uniformly reported that the message categories were unclear, ambiguous, and cumbersome to use. The presence of message categories impaired rather than facilitated communication. Generating a message was simply easier than searching for an appropriate one. A review of the transcripts suggests that the ease of entering messages was a major reason for not using the predefined categories and messages. However, for many of the self-generated messages, the substance of those messages mapped reasonably well onto our predefined message categories.

Pilot Study 2

The results of the first pilot study provided insight regarding our initial approach and informed us about needed revisions. First, although the presence of the enter-own-message facility clearly enhanced communication and was the preferred method of communication, the number of overlaps between the self-generated messages and the predefined messages suggested that we were able to capture the substance of what participants wanted to communicate. Second, the finding of a stable communication pattern suggested a different way of organizing the messages. Rather than grouping messages by teamwork categories, this finding suggested grouping messages by communication function. We were also able with this scheme to avoid teaching what the categories were and how to use them.

Incorporating these findings in the second pilot study required two major changes to the system. The first change involved eliminating the enter-own-message facility. Our intent was to capture team processes in real time; maximizing ease of communication was of secondary importance. Results from the first pilot study suggested that our predefined messages were able to capture what participants wanted to say. In addition, we refined our message set to reflect what pilot study participants had typed in.

The second change was made to the way the messages were organized for participants. In the first pilot study the organization was done on-screen (i.e., having one button per teamwork category). In the second pilot study we shifted the organization from the computer screen to a paper handout. The handout listed all the messages and was organized functionally. For example, messages that were commonly used were grouped together under the heading "Quick Responses." Each message on the handout corresponded to a button on the screen. To send a message, participants clicked on the button that corresponded to the desired message on their handout.

Participants and Negotiation Task. Six adults (two groups) were drawn from available undergraduates. Participants took part in the same negotiation task as participants in the first pilot study.

Messages. For the first group, we revised the message set from the first pilot study, incorporating new messages based on what participants typed in. And we generated additional messages that we believed to be consistent with our definitions of team categories. Participants received a handout with messages divided into eight areas. The areas were (a) quick responses, (b) input from other members, (c) negotiating priorities, (d) entering offers, (e) discussing offers, (f) wages, (g) health and welfare, and (h) pension. The handout listed 86 messages.

For the second group, we used feedback from the first group to reorganize the handout. We included additional messages that they suggested, as well as new messages we generated. For this group, we partitioned the handout into 11 sections. These sections were (a) a flow-chart depicting the three-stage process we observed during the first pilot study, (b) quick responses, (c) seeking assistance, (d) team spirit, (e) entering offers, (f) adjusting offers, (g) reasons for bidding higher or lower on each issue, (h) time concerns, (i) sending offers, (j) awareness of our team's priorities, and (k) awareness of the management team's performance.

Results. For the first group, the maximum number of negotiation rounds was set to 8. The total number of messages sent was 270, and agreement was reached on round 8 after 60 minutes. For the second group, the maximum number of rounds was also set to 8. The total number of messages sent was 193, and agreement was reached on round 6 after 55 minutes.

Postsession interviews with groups revealed that members felt the system was usable. Members said they were able to communicate with each other adequately (although all participants wanted the capability to enter their own messages). Participants found the message set to be, on the whole, adequate to carry out the negotiation task.

VALIDATION STUDY

Drawing on the taxonomy of teamwork processes outlined by Burke et al. (1993) and our analyses of these processes in the context of the mixed-motive interdependence negotiation, and on our pilot study results, we suggested the following hypotheses for our validation study.

We expect that teams that exhibit higher levels of the team processes (adaptability, coordination, decision making, interpersonal, and leadership) will perform better on the negotiation simulation. Performance is defined in our scenario in four ways: (a) final counteroffer, (b) frequency of agreement, (c) type of agreement (integrated vs. distributed), and (d) time. The widespread finding that effective teams engage in these kinds of processes leads us to believe that team performance on our negotiation simulation will reflect similar patterns.

As suggested by the results of Brannick et al. (1993) and the work on shared mental models (Cannon-Bowers et al., 1993; Rouse et al., 1992), we expect time to be sensitive to team decision making. Specifically, we expect that teams engaging in more communication will take longer to reach an agreement. Our prior results (O'Neil et al., 1994) indicate that good performers (experts) would display less fixed-pie bias (a tendency to see the negotiation as a win–lose proposition; see the discussion of the validation measures that follows).

Finally, our prior results (O'Neil et al., 1994) suggested that more self-regulating activity and less worry distinguished experts from novices. Thus, we expect that higher self-regulating activity would lead to higher team processes and, in turn, better team outcomes.

Methods

Participants. Eighty-one students (34 male, 47 female) participated in this study. All participants were drawn from a high school located in southern California. One teacher provided four classes of students. Students participated as part of that day's class activity. UCLA and USC procedures for protection of human subjects were followed (e.g., parental approval, informed consent).

The overall ethnic composition of the participants was 29 White/Anglos, 3 Black/African Americans, 4 Hispanic/Latinos, 38 Asians, and 7 Middle Eastern/Persians. Participants' ages ranged from 14 to 18 years old (60 sophomores, 21 seniors). Participants came from three 10th-grade classes

(18, 18, and 24 participants respectively) and one 12th-grade class (21 participants). The content of all classes was American history.

Participants drawn from these intact classes covered a range of academic standing, ethnic, and gender compositions. The first class was of normal academic standing (sophomores) and was composed primarily of Asian, Middle Eastern/Persian, and female students (2 Whites, 2 Black/African Americans, 1 Hispanic/Latino, 9 Asians, 4 Middle Eastern/Persians; 6 males, 12 females). The second class was a 10th-grade "resource class" (each student had an individual educational plan). This class was composed primarily of White students (11 Whites, 2 Hispanic/Latinos, 3 Asians, 2 Middle Eastern/Persians; 9 males, 9 females). The third class was of normal academic standing (seniors) and composed primarily of White students (14 Whites, 1 Hispanic/Latino, 5 Asians, 1 Middle Eastern/Persian; 11 males, 10 females). The fourth class was a 10th-grade Advanced Placement class composed primarily of Asian and female students (2 Whites, 1 Black/African American, 21 Asians; 8 males, 16 females).

Negotiation Scenario. The negotiation aspect of the task followed the format established in previous studies (e.g., O'Neil et al., 1994; chapters 9 and 10, this volume). As discussed earlier, the present study extended the negotiation task by adding a teamwork dimension. Table 16.1 presents the general domain specifications embedded in the software.

The scenario presented to participants was that of a union–management labor negotiation situation. The ecological validity of our scenario was facilitated by information from the union and management leadership of the Southern California Airconditioning and Refrigeration Industry. Values for the negotiation issues were taken from the industry's *Master Agreement* (Southern California Airconditioning and Refrigeration Industry, 1990). Our messages were also reviewed by the industry and union leadership. In general, they felt that we had captured the formal negotiation process but that the informal process was not well captured (e.g., business conducted on the golf course, the use negative interpersonal remarks).

Participants were informed that they were part of a 3-person team representing the union or management. (In fact, all teams were representing the union; the computer simulated the management team, and no team was aware that all other teams were also assigned the union role.) Participants were informed that their team's responsibility was to reach an agreement with the other side, and that the team should work together on what to offer the other team. Participants were informed they had eight rounds to complete the negotiations, and they were not told who their teammates were. Further, the negotiation scenario was not an interdependent task. It could be solved by a single student. Thus, in our scenario, the task is very similar to a collaborative learning math environment in a K–12 education setting.

TABLE 16.1
Domain Specifications Embedded in the Software

General Domain Specification	Specific Example
Scenario	Role-play a contract negotiation by exchanging proposals in mixed-motive context
Players	One or more students and one manager (computer software)
Student	Either expert or novice, individual or team
Manager	Computer software (Carnevale & Conlon, 1988; O'Neil, Allred, & Dennis, 1992)
Priorities	Offsetting
Moves	Reciprocal
Rounds	Offer from student and counteroffer from manager
Subcompetencies	Propose options; make reasonable compromises
Negotiation issues	Three in number (e.g., salary) with offsetting priorities
Negotiation measures	Agreement (yes/no), type of agreement (distributive vs. integrative), final counteroffer
Cognitive processes (domain-dependent)	Fixed-pie bias
Cognitive processes (domain-independent)	Metacognitive skills
Affective processes (domain-independent)	Effort, worry

The negotiation was terminated under the following conditions. First, the team could accept the management (computer) team's offer at any time. Second, if the team sent an offer that was acceptable to the management team, then the management team would accept that offer. An offer that was integrative in nature (over 120 points) was accepted. The last way the negotiation could end is if no agreement was reached within the allotted eight rounds. In that case, a final counteroffer was generated by the software.

Table 16.2 gives an example of a group accepting the management team's offer at round 6.

Negotiation Issues and Priorities. The issues under negotiation were (a) wages, (b) health and welfare, and (c) pension. Participants were informed that their team's priorities were (in order of importance) wages, health and welfare, and pension.

To help participants gauge the value of the offers during negotiation, they were provided with an "issue chart." The issue chart was part of the computer display and hence always available. Participants were instructed to use these values to determine the value of their offers to the management team. Participants were instructed that they should try to negotiate an agreement that optimizes their total number of points. Table 16.3 shows

TABLE 16.2
Sample Negotiation With the Union (Participant) Team Accepting the Management
Team's Offer

Round	Management (Computer) Offer	Union (Participant) Offer
1	15-10-5	120-80-35
2	15-0-0	120-70-30
3	30-10-0	105-70-25
4	45-10-5	90-60-20
5	60-20-10	75-50-20
6	60-30-15	The union team accepts the management team's offer.

Note. The management (computer) team always makes the first offer.

the issue chart for both the participants' team and the management team. Note that the issue priorities for the simulated management team are the reciprocal of the union team's priorities. Participants were not informed of the simulated management team's priorities.

The simulated management team (computer) was programmed to reflect priorities that were opposite to the union's team. That is, the computer's highest priority was pension followed by health and welfare, and wages (see Table 16.3). The computer's responses were programmed to reflect an integrative potential. If the union team gave on its value of least importance (pension), the computer would give on its issue of least importance (wages), and the joint outcome would be maximized. Further, the computer was programmed to make offers that mirrored the union

TABLE 16.3
Issue Chart for the Union (Participant) and Management Teams

Wages		Health and Welfare		Pension	
Union	Management	Union	Management	Union	Management
120	0	80	0	40	0
105	5	70	10	35	15
90	10	60	20	30	30
75	15	50	30	25	45
60	20	40	40	20	60
45	25	30	50	15	75
30	30	20	60	10	90
15	35	10	70	5	105
0	40	0	80	0	120

team's offers. If the union team gave on its most important issue (wages), the computer would give on its issue of most importance (pension). For the health and welfare issue (moderate priority to, both the union and management team), the computer responded with a tit-for-tat strategy: A union concession would be mirrored by a equal concession by the computer. In addition, the computer would accept any union offer that totaled 120 points or more (when computed from management's point of view). Thus, in some cases there was an implicit ceiling on the final counteroffer measured.

Computer Conferencing System. Three major components comprised the system: (a) predefined messages, (b) a handout that listed all messages, and (c) software to support the computer conferencing system. Each component is described next.

1. Messages. In order to capture team processes in real time, we developed a priori a set of messages for the five teamwork categories. Team members selected these messages to communicate with each other during the negotiation.

Fifty-one messages were provided to participants for use during the negotiation task. Fifty messages (10 messages per category) belonged to the five teamwork process categories discussed earlier (adaptability, coordination, decision making, interpersonal, and leadership). One message was provided to signal that a previously sent message was mistakenly sent by the participant. This message was excluded from the message categories. A sample message from each category is given below.

- *Adaptability*—recognizing problems and responding appropriately. A message belonging to this category is "What do you think, M1?" (M1 refers to Member 1 of the team.)
- *Coordination*—organizing team activities to complete a task on time. A message belonging to this category is "We only have 5 rounds left."
- *Decision making*—using available information make decisions. A message belonging to this category is "We should go higher on Wages because it's *our* 1st priority."
- *Interpersonal*—interacting cooperatively with other team members. A message belonging to this category is "Help me out."
- *Leadership*—providing direction for the team. A message belonging to this category is "Send the offer."
- *Communication*—the overall activity of the team. No messages were tied to this category; rather, communication represented the aggregate interaction between team members minus the number of error mes-

sages. Our definition of communication was that accurate and timely messages were sent. Because all of the messages were selected and accurately sent via the network, we considered all of the team process messages to be examples of communication.

2. Message handout. Participants received a paper handout with all 51 messages. The organization of the handout followed the three-stage model of communication patterns that emerged during pilot testing. During the first stage, someone (usually the leader) would suggest a specific offer. The second stage was characterized by a discussion of the pros and cons of that offer, how the offer might impact the team, and potential actions the management team might take in response to the offer. During this stage alternative offers were suggested by all team members. The last stage was characterized by someone (usually the leader) making a suggestion to wrap up the discussion and make a decision. (See Appendix A for the actual handout.)

The top section provided general instructions on how to use the handout. This section also contained a flow chart that depicted the three-stage process. The middle and bottom sections contained the actual messages. The middle section contained messages that were unique to each stage, and the bottom section contained messages that were common to all stages.

We subdivided the "discussing offers" area by function into five sections: (a) entering offers, (b) adjusting offers, (c) assessing performance, (d) keeping track of time, and (e) going higher/lower on the issues. Similarly, the "common messages" area was subdivided into two sections: (a) encouraging others and (b) quick responses.

3. Hardware/Software. The school's computer lab was used to run this study. The lab contained 18 computers, and we provided 6 additional computers. The 18 lab computers were Macintosh LC475s, and the remaining computers were 5 PowerBooks and an SE/30. The 18 LC475s were networked with an Ethernet local-area-network (LAN), and the PowerBooks and SE/30 were networked with a LocalTalk LAN. The tutorial used to deliver task instructions was developed with MediaTracks 1.0 (1990), and the software used in the negotiation task was developed with HyperCard 2.2 (1993). The computer operating system used was System 7 (1992), and each computer had at least 4MB of RAM and 40MB of hard disk space. The computer conferencing software tapped the built-in networking and interprocess communication capabilities of HyperCard 2.2 and System 7 to implement message sending between computers.

The user interface is shown in Fig. 16.1. The display was partitioned into three major sections. The top left section contained numbered message buttons. Each button corresponded to a particular message on the participants' message handout. To send a message, participants clicked on

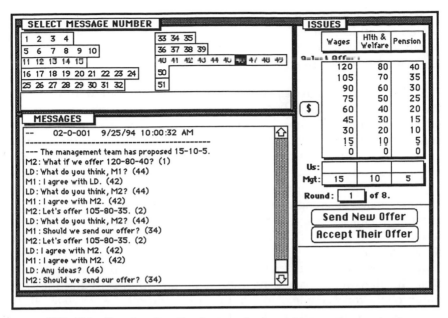

FIG. 16.1. User interface for the team leader. Advisory team members' screens did not have the send and accept offer buttons. Note the three major sections (counterclockwise from the top-left): message buttons, message display, and offer information.

a button and the message corresponding to that button would be sent to the other team members' computers. (See Appendix A for the message handout participants used to map messages to message numbers.) The lower left section of the screen displayed the messages sent by all the members. The messages were listed in the order sent. The right-hand section displayed the issue chart, current union and management offers, and (if the participant was the leader) buttons to send and accept offers. The team's current offer was selected by clicking on one of the issue values. The selection would appear on every team member's computer. Only the leader could select, send, or accept offers.

For certain messages, user input was required. These messages were handled by the use of dialogs that mostly required simple point-and-clicking. For example, the message "What if we offer X-Y-Z?" was handled by the participant simply clicking on the desired issue and selecting the desired value.

A second type of message required users to type in a value. In this case, the value the participant typed in was checked against a list of acceptable values. If the value was not in the list, the user was notified and the value rejected.

Measures. The measures employed in this study focused on team-level outcomes within a negotiation context, process measures of teamwork, metacognitive skills, fixed-pie bias perceptions, and manipulation checks. Each measure is described next.

1. *Team outcome measures.* Our team outcome measures were based on the work of O'Neil et al. (1994) and were designed to measure four aspects of the negotiation scenario: (a) final counteroffer, (b) whether or not an agreement was reached, (c) if an agreement was reached, the type of agreement (i.e., integrative or distributive), and (d) elapsed time on the negotiation task.

The final counteroffer was defined as the sum of the offer levels across all three issues. If an agreement was reached between the management and union teams, the agreed-to offer was used to compute the final counteroffer. For example, if the agreed-to offer was 120-40-0, the final counteroffer was 160.

If an agreement was not reached, the final counteroffer was computed from the offer the management team would have proposed given the last union offer. For example, suppose the last union offer was 120-80-40. In response to this offer, the management team would counterpropose 0-0-0. Thus, the final counteroffer would be 0.

Agreement was defined as whether or not the union team reached an agreement with the management team. No distinction was made for the direction of the agreement. That is, a union team accepting the management team's offer was treated the same as the management team accepting the union team's offer.

Type of agreement was defined as integrative or distributive. An integrative agreement was defined as the union team reaching an agreement worth more than 120 points (see Table 16.3). A distributive agreement was one in which the union team reached an agreement worth 120 points or less.

Elapsed time was defined as the time between the start of the negotiation and the termination of the negotiation. This time does not include the tutorial or question-answering activities.

2. *Individual and team-level teamwork process measures.* The individual and team-level teamwork process measures were designed to capture the degree to which individuals and their team engaged in each of the team processes (i.e., adaptability, cooperation, decision making, interpersonal, leadership, and overall communication). Individual team process measures were computed by counting the number of messages sent in each team process category. For example, if a team member sent 10 messages in the adaptability category, that individual's adaptability score was 10. This method was used for all categories except communication. Communication was measured as the total number of messages sent by an individual minus the number of "I sent the wrong message" messages.

Team-level teamwork processes were measured by summing across individual category counts. For example, if all three team members sent 10 messages each in the adaptability category, then the team-level adaptability performance was 30.

3. *Self-regulation questions.* These questions consisted of measures of metacognition (planning, self-assessment, awareness, and cognitive strategies). In addition, effort and worry were measured. The reliability and validity data for these measures are acceptable and can be found in O'Neil, Sugrue, Abedi, Baker, and Golan (1992) and O'Neil and Abedi (1996).

4. *Fixed-pie bias questions.* Two fixed-pie bias questions, based on the work of O'Neil et al. (1994), were adopted to measure the degree to which participants viewed the negotiation context as either (a) a situation where the issues being negotiated are fixed, and that a gain on one side necessarily means a loss to the other side (a "win–lose" situation), or (b) a situation where the issues being negotiated are not equally important to both sides, thus creating a potential of gaining on an important issue by giving on a less important one (a "win–win" situation).

5. *Manipulation checks.* Nine manipulation-check questions were designed to give us a sense of (a) how well participants understood their team's priorities, (b) how adequate participants found the predefined messages to be (as a communicative device), (c) whether participants knew the identity of their teammates, (d) how easy participants found the system to use, and (e) participants' experience with computers. Finally, participants were given an opportunity to type in messages that they would have liked to have available.

Procedure

Each participant was randomly assigned to a computer, a team, and a role (leader vs. member). The experimenters introduced themselves, informed the participants that they were part of a study examining teamwork and negotiation skills, and gave a brief overview of the scenario and the activities to follow. All participants went through a negotiation tutorial on the computer. The tutorial explained in detail the negotiation scenario, the negotiation issues and priorities, the team scenario, and how to operate the computer.

Team Scenario. Participants were informed by the tutorial that they were part of a 3-person team comprised of a leader and two members. The leader was the only one who could select offers for the union team, and send to and accept offers from the management team. The two members were responsible for advising the leader. Participants were urged to reach consensus on what to offer the management team.

Computer Use. Participants were given computer demonstrations on how to carry out critical functions such as sending messages. This demonstration consisted of playing prerecorded animated sequences that, for example, showed the mouse moving to a button, clicking on it, and showing the computer's response.

Participants who finished the computer tutorial were instructed to study the message handout while waiting for everyone else to finish. Once everyone finished the tutorial, participants began the negotiation task. At the end of the negotiation task, participants were given feedback regarding their performance. Participants were shown their individual performance score (number of messages sent for each message category), team performance score (total number of messages sent by all team members for each message category), negotiation performance score (whether or not an agreement was reached), type of agreement (integrative or distributive if an agreement was reached, not applicable otherwise), and elapsed time for the negotiation task.

Following the performance feedback, participants were presented with 34 self-regulation questions, 2 fixed-pie bias questions, and 9 manipulation check questions. In addition, they had the opportunity to type in any messages they felt the system should have had. Following the questions, the participants were presented with screens describing the purpose of the study, the nature of integrative ("win–win") and distributive ("win–lose") agreements, and the need for effective teamwork skills in today's workplace. Students were escorted back to their classrooms, debriefed, and told that they were all union teams and that the management team was played by the computer. Any questions were then answered.

Data Collection Problems

The first attempt at data collection was marred by catastrophic computer failures. The class was comprised of 18 sophomores (6 groups of 3 participants). Throughout the initial data collection session, computers would crash or freeze. Only one group out of the six completed the entire experimental session. The individuals in this group participated under standard conditions in a new group the following week. Thus, the majority of participants in the first class were not exposed to the negotiation task or subsequent questionnaire-type data collection. These students participated under standard conditions in new groups the following week. The remaining three classes (which did not participate in this first attempted data collection) were rescheduled for the following week.

The computer failures were caused by an interaction of a utility used to extend the HyperCard 2.2 (1993) program's functionality and the network functions of HyperCard. This utility (called an "XCMD"), which provided the window dialog capability, also provided a special kind of window that could be programmatically opened and closed without user intervention (hereafter called a status window). We used status windows to notify the participant that the computer was engaged in a lengthy process. The status window also signaled to the participant that he or she should wait. Immediately after the process was finished the window was closed, signaling to the participant that he or she could continue. However, use of the status window caused the computer to crash when the following conditions occurred simultaneously: (a) the status window was being closed and (b) a message was received from another computer. The problem was eliminated by not using status windows.

A second data collection attempt was made and completed the following week. Three computer failures occurred, of which only one failure was recoverable. During the first session, one computer failed during the negotiation task. Members of this team were immediately escorted back to class and informed that the computer had failed. (These participants were dropped from our analysis.) The failure appeared to be similar to the one that occurred the previous week; however, we were unable to determine the exact cause of failure. During the second session, one computer failed after the participant completed the tutorial. Because there were spare computers, members of this team were moved to those computers. This incident occurred before the class started the negotiation task. The cause of this problem was a HyperCard system crash. This problem did not recur. The third computer failure occurred in the third session during the negotiation task. This team was immediately escorted back to class. As with the first session, this failure appeared to be similar to the one that occurred the previous week, but its cause is unknown.

Another incident occurred in the final session—the 10th-grade resource class (18 participants or 6 groups). This session was far noisier and rowdier than the other three sessions. In particular, some of the teams that finished the negotiation task later in the session were talkative and disruptive while answering the questions. One factor that contributed to the problem was that this group ran nearly to the end of the class period, which was the last class of the school day.

RESULTS

Message Selection Procedure

All of the messages provided in the negotiation scenario were selected from a much larger pool of potential messages derived from student messages in our pilot testing or developed by the authors to represent teamwork processes mentioned previously. The messages were sorted by two independent raters into five categories: adaptability, coordination, decision making, leadership, and interpersonal teamwork processes. Messages could be put in more than one category, with the condition that if a message was sorted into more than one category of team processes, such categories would be ordered according to the rater's perceived strength of fit for that message to each category. Thus, if Message 1 was sorted into both the leadership category and the adaptability category, leadership and adaptability would be ranked for this item according to which category the item fit into better, with 1 indicating the best fitting (primary) category, and 2 indicating the next best fitting (secondary) category. Almost all items were placed in two or fewer categories by each rater.

From this larger pool of potential messages, items were selected that showed strong agreement between the two raters for each category, with the object being to retain ten messages in each category. Agreement between the raters occurred when both raters sorted an item into the same primary category. To obtain the desired quantity of messages in each category, messages with secondary agreement were retained for those categories without ten messages with primary agreement. This practice provided an adequate number of messages for the leadership category. However, in the adaptability, coordination, decision making, and interpersonal teamwork process categories, one message was needed to complete the desired quantity for each category. These were determined by a third rater, who then rated those messages for which the first two raters failed to reach agreement. Thus, at least nine out of ten messages in each category (90% to 100%) resulted from primary or secondary agreement between the two initial raters. In addition, a single error message that did not fit into any of the teamwork process categories was included. The rater agreement for the five teamwork process categories is shown in Table 16.4.

TABLE 16.4
Rater Agreement for Teamwork Process Categories

Teamwork Process	Primary Agreement	Secondary Agreement	Cumulative Agreement for Both Categories
Adaptability	90%	0%	90%
Coordination	40%	50%	90%
Decision making	90%	0%	90%
Leadership	100%	0%	100%
Interpersonal	90%	0%	90%

TABLE 16.5
Frequency Count of Messages

Message #	Frequency	Message #	Frequency	Mesage #	Frequency
1	302	18	35	35	53
2	287	19	9	36	65
3	49	20	62	37	22
4	53	21	14	38	29
5	52	22	26	37	26
6	20	23	19	40	89
7	31	24	17	41	37
8	32	25	62	42	357
9	17	26	3	43	270
10	40	27	16	44	216
11	24	28	20	45	106
12	44	29	40	46	95
13	45	30	14	47	326
14	35	31	21	48	40
15	65	32	49	49	28
16	16	33	551	50	65
17	26	34	129	51	118

Individual-Level Measures

Preliminary analysis of the data revealed that each of the 51 messages provided in the negotiation simulation was used at some point. The observed frequency for each message is provided in Table 16.5. As may be seen in Table 16.5, though all messages were utilized, individual message usage varied from 3 (message #26) to 551 (message #33). The frequency distributions were highly skewed positively. Thus, prior to combining the items to generate team process scales, the message counts were transformed from raw score frequency of usage counts for each subject to percent of overall item usage for that subject relative to all subjects. That is, if Subject A had a frequency count of 6 on item X that had a total usage of 50, Subject A's score on item X was transformed from 6 to .12 (6/50). This put each item on a common metric (0–1) and preserved the underlying distribution because each item's frequency count transformed by division of a constant value.

As was mentioned earlier, during the initial class periods of data collection, software system malfunctions made it necessary to reschedule subjects to the following week, once the malfunction problems had been mostly resolved. This resulted in one of the classes having some prior exposure to the negotiation software, since the malfunctions occurred after subjects had been introduced to the system and navigated through the tutorial, but before they started the negotiation scenario.

Separate statistical analyses were conducted for those subjects who were not exposed previously to the program, for those who were, and for the group as a whole. The first session had 6 groups with varying levels of prior exposure to the program. The other three sessions had a total of 20 groups used in the analysis. Comparing those groups with exposure ($N = 6$) to those without exposure ($N = 20$) on each of the outcome measures— score ($M = 115$, $SD = 16.8$ vs. $M = 124$, $SD = 32.7$); time ($M = 26.5$, $SD = 9.0$ vs. $M = 28.9$, $SD = 9.8$); agreement (83% vs. 85%); agreement type (both at exactly 50% integrative, 50% distributive)—did not result in significantly different scores. Similar results were found for comparisons on all team process categories and self-regulation measures. No differences between the groups were found. In summary, separate analyses did not result in any substantive differences in performance on any of the measures or conclusions drawn therefrom. Thus, data from all four class sessions were retained for analysis and are reported here.

For each subject, individual scores were calculated for the adaptability, coordination, decision making, leadership, and interpersonal teamwork processes by counting the messages from each category. In addition, a total communication score was computed by summing each of the five teamwork process scores for each individual less the total number of error messages for that individual. An individual's scale score was not adjusted for error messages. Means and standard deviations for each scale are presented in Table 16.6.

As may be seen in Table 16.6 of the observed frequency for each process, the average participant "sent" approximately 10% of the messages in each category. Our metric does not allow comparison between categories.

Teamwork Process Reliabilities

The five team process scales were subjected to reliability analysis using item raw scores (counts of times utilized), transformed item scores (proportion scores), and reduced scale proportion scores. There is an issue of

TABLE 16.6
Individual-Level Teamwork Process Scales ($N = 81$)

Process	Mean	SD	Minimum	Maximum
Adaptability	.11	.09	.00	.38
Coordination	.10	.12	.00	.66
Decision making	.10	.11	.00	.43
Leadership	.10	.10	.00	.46
Interpersonal	.10	.15	.00	1.07
Communication	.49	.38	.01	2.15

independence of the item scores because a message could be selected by a participant more than once. In most examples of testing, an item can be answered only once. Thus, we explored a variety of reliability measures. We also looked at both parametric and nonparametric statistics, two of which we report here.

Improvements for each scale in interitem consistency were made by eliminating from one to three items from each message category, based on low item-total correlations (see Table 16.7). These exclusions included eliminating item #6 from the adaptability scale; items #16, #17, and #19 from the coordination scale; items #26 and #27 from the decision making scale; items #12 and #21 from the leadership scale; and items #48 and #49 from the interpersonal scale (see Appendix A for the final item set). Cronbach's alpha coefficients ranged from .51 for adaptability to .76 for interpersonal in the reduced scales, indicating moderate interitem consistency, with the interpersonal scale showing the greatest internal consistency.

Additionally, nonparametric statistical tests indicate moderate but significant agreement between the items on each scale. The reduced version of each scale was analyzed using Kendall's coefficient of concordance (W statistic; see Hays, 1973). The coefficient of concordance indicates the extent to which judges (or items) agree in their ranking of subjects. It has a range from 0 to 1, with zero indicating no agreement whatsoever, and one indicating total agreement. The statistic follows a chi-square distribution with $(N-1)$ degrees of freedom, where N = number of subjects being judged. The null hypothesis that there is no agreement among the items

TABLE 16.7
Summary of Reliability Analysis for Teamwork Process Scales

Teamwork Process Scale	Items	Adjustments	Cronbach's Alpha
Adaptability	4,6,7,22,42,43,44,45,46,50	Use raw values	.49
	4,6,7,22,42,43,44,45,46,50	Use proportion scores	.43
	4,7,22,42,43,44,45,46,50	Remove #6	.51
Coordination	9,11,13,14,15,16,17,18,19,24	Use raw values	.62
	9,11,13,14,15,16,17,18,19,24	Use proportion scores	.57
	9,11,13,14,15,18,24	Remove #16,#17,#19	.63
Decision making	1,3,25,26,27,28,29,30,31,32	Use raw values	.44
	1,3,25,26,27,28,29,30,31,32	Use proportion scores	.61
	1,3,25,28,29,30,31,32	Remove #26,#27	.59
Leadership	2,5,8,10,12,20,21,23,33,34	Use raw values	.34
	2,5,8,10,12,20,21,23,33,34	Use proportion scores	.57
	2,5,8,10,20,23,33,34	Remove #12,#21	.63
Interpersonal	35,36,37,38,39,40,41,47,48,49	Use raw values	.57
	35,36,37,38,39,40,41,47,48,49	Use proportion scores	.75
	35,36,37,38,39,40,41,47	Remove #48,#49	.76

TABLE 16.8
Kendall's Coefficients of Concordance for Teamwork Scales

Process	W Statistic	Chi-Square	df	p <
Adaptability	.24	173.51	80	.001
Coordination	.31	170.92	80	.001
Decision making	.29	184.05	80	.001
Leadership	.24	154.94	80	.001
Interpersonal	.30	191.98	80	.001

can thus be tested, provided the number of items is roughly equal to or greater than 8 (Hays, 1973, p. 803), which is satisfied for all but one of these scales. The results of this analysis are presented in Table 16.8. The coefficients ranged from .24 to .31. In each case, the coefficient of concordance is significantly different from zero at $p < .001$, indicating agreement among the items. Thus, both parametric and nonparametric procedures show moderate consistency and reliability for each of the five teamwork process scales.

Self-Regulation Measures

In addition to teamwork process scores, measures of self-regulation activity were computed for each subject. These included measures of effort, worry, cognitive strategy, self-checking, planning, and awareness. The values for each of these scales were calculated by summing the items from that scale. The means, standard deviations, number of items, and coefficient alpha reliabilities for each metacognitive scale are presented in Table 16.9. In general, reliabilities for the metacognitive measures are acceptable, with a low of .64 for the awareness scale to a high of .84 for the effort scale.

Additional Measures

Two questions presented to each subject dealt with the issue of fixed-pie bias, or a tendency to see the negotiation as a win–lose proposition. On the first question, regarding the management team's most important issue in the negotiation scenario, 76.5% of the subjects showed a fixed-pie bias. An even greater percentage (84.0%) indicated a fixed-pie bias on the second question, dealing with the management team's least important issue. These results are consistent with our prior data on novices (O'Neil et al., 1994).

We were also concerned with whether subjects understood what the priorities of their own negotiating team were. Of the 81 subjects in the study,

TABLE 16.9
Descriptive Statistics and Reliabilities for Metacognitive Scales

Metacognitive Process	Mean	SD	# of Items	Chronbach's Alpha
Effort	20.78	3.46	6	.84
Cognitive strategy	17.15	3.59	6	.70
Worry	9.81	3.52	5	.74
Awareness	15.78	2.82	5	.64
Planning	18.78	3.66	6	.79
Self-checking	13.43	3.33	5	.74

71 (86.4%) indicated that they accurately comprehended their team's priorities. Moreover, since the negotiation within teams was intended to be among anonymous participants, we inquired regarding subjects' knowledge of the identity of the other members in their team. In general, subjects were mostly unaware of the identity of their other team members (Table 16.10). The overall mean was 1.79 indicating average knowledge somewhere between 1 "Not at all" and 2 "Somewhat."

Finally, information regarding subjects' perceptions of the simulation program, message effectiveness, and computer literacy was solicited. Generally, subjects found the program to be easy to use ($M = 3.40$, $SD = .83$; [82.7% of subjects indicated either "often" or "almost always" as opposed to stating "almost never" or "sometimes"]). Though the messages didn't always capture what the subjects wanted to say ($M = 2.63$, $SD = .90$ [46.9%]), they were considered to be effective ($M = 3.27$, $SD = .82$ [81.5%]) and coherent ($M = 3.52$, $SD = .73$ [86.4%]). Additionally, the subjects indicated high levels of computer use ($M = 3.37$, $SD = .87$ [81.5%]).

TABLE 16.10
Means and Standard Deviations for Teammate Identity

Variable	Mean	SD
Knowledge of leader	1.83	1.22
Knowledge of member 2	.70	1.14
Knowledge of member 2	1.83	1.19
Overall	1.79	1.02

Note. These values were computed excluding self-knowledge (i.e., Leaders were excluded from the Knowledge of Leader question, M1s were excluded from the Knowledge of Member 1 question, and M2s were excluded from the Knowledge of Member 2 question).

Team-Level Measures

Each of the individual-level teamwork process measures was used to calculate comparable team-level scores. The individual scores for adaptability, coordination, decision making, leadership, interpersonal, and error messages were summed among the three members of each team to generate a team score for each process. A total team communication score was created by subtracting the total team error messages from the sum of the five team-level teamwork process scores. Means, standard deviations, and intercorrelations for each of the team-level teamwork process measures are presented in Table 16.11. Communication is a composite (simple sum) of the other five team process measures, which are restricted to a 0–1 range. However, the range for communication is from .22 to 3.05. Unlike the other team process measures, communication is a composite, rather than a proportion.

Outcome Measures

The negotiation scenario generated a total of four team-level outcome measures: team performance score, whether an agreement was reached, type of agreement reached, and time spent on the negotiation. Team performance scores were calculated for teams reaching agreement as the total point value of the agreed-upon offer. For teams not reaching an agreement, a team performance score was computed by taking the total point value of the inverse of the last offer submitted by the negotiating team. Upon inspection, we found one team that completely misunderstood the concept of the scenario, which resulted in a team score of zero. Through discussions with the members of this team, we realized that they mistakenly believed that a score of 0 was their ideal result, as opposed to the instructions indicating the objective was to maximize the points for

TABLE 16.11
Descriptive Statistics for Team-Level Teamwork Process Measures

Process	Mean	SD	2	3	4	5	6
1. Adaptability	.33	.20	.69***	.82***	.49**	.61***	.89***
2. Coordination	.28	.21		.72***	.65***	.45*	.88***
3. Decision making	.29	.20			.66***	.41*	.89***
4. Leadership	.28	.18				.37	.77***
5. Interpersonal	.24	.15					.66***
6. Communication	1.38	.76					

Note. *p < .05. **p < .01. ***p < .001.

TABLE 16.12
Descriptive Statistics for Team-Level Outcome Measures (N = 26)

Outcome Measure	Mean	SD	Minimum	Maximum
Performance score	121.73	21.07	50	150
Agreement	.85	.37	0	1
Agreement type	.50	.51	0	1
Time in negotiation (min)	28.31	9.02	13.00	44.00

their team according to their stated priorities. As a result, this team was excluded, leaving data for 26 teams for all team analysis.

The outcome measure of agreement was coded dichotomously (1 = agreement; 0 = no agreement). The great majority of teams reached an agreement within the 8-round negotiating session (85%). Agreement type was determined by the value of the overall score measure. Teams with overall scores greater than 120 received a score of 1, indicating an integrative agreement. Teams with overall scores less than or equal to 120 received a score of 0, indicating a distributive agreement. Time spent in the negotiation scenario was captured by the program for each team and is reported in minutes. Descriptive statistics for each of these measures can be found in Table 16.12.

Tests of Hypotheses

We hypothesized several relationships between the teamwork processes and outcome measures, as well as some relationships among the outcome measures themselves. Our unit of analysis was the team (N = 26), not the individual student. Our first hypothesis was that team processes would be positively associated with team performance. This hypothesis was generally supported by the direction and moderate magnitude of the correlations between these measures; however, few are statistically significant. Failure to find statistical significance for these correlations is not surprising considering the relatively low power of this test given a sample size of only 26. According to Cohen (1992), for a correlation coefficient test of significance with a power of .80 to detect a small effect size (defined by Cohen to be > .10) at the .05 level of significance would require a sample size of N = 783. To detect a medium effect size (e.g., > .30), a sample size of N = 85 would be required. In the current study, only large effect sizes (e.g., > .50) could be expected to be detected (necessary sample size, N = 28).

Table 16.13 provides the correlations between the teamwork processes and outcome measures (Hypothesis 1). Correlation coefficients for team performance and adaptability, coordination, and leadership range from .29

TABLE 16.13
Correlations Between Team Processes and Outcome Measures ($N = 26$)

Team Process	Performance Score	Agreement	Agreement Type	Time in Negotiation
Adaptability	.31	-.24	.24	.72***
Coordination	.29	-.08	.15	.65***
Decision making	.41*	-.04	.41*	.69***
Leadership	.31	-.02	.19	.65***
Interpersonal	.03	-.52**	-.05	.56***
Communication	.35	-.20	.24	.80***

Note. *$p < .05$. **$p < .01$. ***$p < .001$.

to .31. The correlation between team performance and team communication is higher ($r = .35$) and approaches significance ($p = .07$). Only team decision making was significantly related to team performance ($r = .41$; $p < .04$). The interpersonal teamwork process showed no relationship to team performance ($r = .03$; $p > .88$); however, it did show a strong negative relationship with reaching an agreement ($r = -.52$; $p < .01$). Apparently, though most teams reached an agreement, those teams that failed to reach an agreement (4 of 26 teams, or 15%) utilized more interpersonal messages than did teams that reached an agreement. This statement is implied by the negative correlation between agreement and the interpersonal team process measure (-.52). The mean values and standard deviations of the interpersonal measure for each of these groups were as follows: The agreement group ($n = 22$) had a mean of 0.20 and a standard deviation of 0.13, while the non-agreement group ($n = 4$) had a mean of 0.42 and a standard deviation of 0.16. These differences are significant at the $p < .01$ level ($t = 3.00$, $df = 24$).

Similar results are found for our additional hypothesis that teamwork processes would be positively associated with quality of agreement. Again, moderate positive but nonsignificant correlations were found between type of agreement and adaptability, coordination, leadership, and communication (r's = .24, .15, .19, and .24, respectively). Decision making showed a significant positive relationship with agreement type ($r = .41$; $p < .04$), but as with team performance, the interpersonal teamwork process showed no relationship to quality of agreement ($r = -.04$; $p > .81$).

There was more support for our prediction of a positive relationship between the teamwork processes and time involved in negotiation. All of the correlation coefficients between time involved in negotiation and the teamwork processes are positive and significant beyond the $p < .001$ level.

They range from a low of .56 for the interpersonal scale to a high of .80 for the communication measure. Thus, not surprisingly, more communication of all types among the team members is related to longer time involved in the negotiation process.

Our second hypothesis predicted that team performance would be negatively related to time in negotiations. Results for this prediction are mixed. Although time in the negotiation scenario was negatively related to reaching agreement ($r = -.40$; $p < .05$), the correlation between time and team performance was positive but not significant ($r = .28$; $p > .16$). Similarly, results indicate no relationship between time and agreement type ($r = .10$; $p > .64$). Thus, teams that reached agreement took less time in the negotiations, as we predicted. However, less time did not relate to significantly higher team performance values or better agreements.

The third hypothesis proposed a negative relationship between team performance and the degree of fixed-pie bias exhibited by the team. Team-level fixed-pie bias was calculated as the sum of bias exhibited by the three members of the team on two questions; thus, the possible range for this team-level variable is from 0 (no bias by any member on either of the two questions) to 6 (all three members exhibit bias on both of the fixed-pie questions). However, due to the large amount of bias shown at the individual level (76.5% on the first question and 84% on the second question), the team bias score ranged from 3 to 6 ($M = 4.81$; $SD = 1.02$). Correlations between team performance measures and team-level bias did not support our hypotheses. Contrary to our prediction, the relationship between team performance and bias ($r = .29$; $p > .15$) was positive but not significant. Likewise, no significant relationship was found between bias and type of agreement ($r = .04$; $p > .85$).

We also proposed relationships between the self-regulation measures (effort, cognitive strategies, awareness, planning, and self-checking), measures of worry, and team performance measures (Hypothesis 4). All but worry were predicted to show positive correlations with the team performance measures (Table 16.14). Though not all of the proposed relationships were exhibited, we found support for the measures of effort and cognitive strategies. Teams exhibiting more effort on the task showed higher team performance ($r = .57$; $p < .005$) and better agreements ($r = .57$; $p < .005$). Likewise, teams higher on the cognitive strategy measure showed higher performance ($r = .38$; $p = .05$) and better agreements ($r = .36$; $p < .07$), though this relationship only approached statistical significance. As seen in Table 16.14, none of the other relationships is significant, though many are in the proposed directions and are of low to moderate magnitudes, magnitudes that are consistent with earlier research in this area (e.g., Brannick et al., 1993). As noted earlier, one of the limitations of sample sizes as small as the one used in these group-level analyses ($N = 26$) is that

TABLE 16.14
Correlations Between Metacognition and Outcome Measures ($N = 26$)

Metacognitive Scale	Performance score	Agreement	Agreement Type	Time in Negotiation
Effort	.57**	.48*	.57**	.08
Cognitive strategy	.38	.23	.36	.15
Worry	-.26	-.21	-.33	.25
Awareness	.22	.23	.16	-.03
Planning	.30	.26	.28	.26
Self-checking	-.15	.08	-.17	.23

Note. *$p < .05$. **$p < .01$. ***$p < .001$.

only very large effect sizes are likely to be detected and recognized as statistically significant. Clearly, additional investigations with more groups are desirable to ferret out small to moderate effects of teamwork processes and metacognitive activities on team performance outcomes.

DISCUSSION

The results of this study indicate that the measurement of teamwork processes can be accomplished in a reasonably reliable and much more time-efficient manner than that of earlier approaches. The entire negotiation procedure, including individual and team process and outcome scoring and interpretation, took place in less than one extended class period. This kind of improvement in the timeliness of data collection and analysis, without a loss in the reliability and validity of such measures, greatly enhances the ability of researchers to address pressing issues, and it speeds up the accumulation of knowledge in the area.

In addition, as an initial small-scale investigation, this study provides promise that larger scale experimentation may provide more stable and statistically significant estimates of the relationships between teamwork measures, such as adaptability, coordination, decision making, leadership, and interpersonal processes, and outcome measures, such as whether negotiated agreements are reached, the type and quality of those agreements, and the time involved in reaching such agreements.

Moreover, we have shown that the use of technology, specifically computerized networked simulations, can be effectively incorporated in assessing the relationships among these relevant issues. Students found the simulation to be interesting, easy to use, effective, and enjoyable. Few other testing procedures enjoy such enthusiastic, positive appraisal from the test-taking populace. Acceptance of and support for the procedure by students can facilitate additional investigations in the area and may lead to additional improvements in the way teamwork processes and other relevant constructs are measured.

Improving the ability to assess teamwork skills and their relationships to performance outcomes will serve not only the interests of educational researchers, but also the interests of training and development specialists, and American industry as well. With the increasing importance placed on teamwork skills in the private sector, and the desire to encourage and develop such abilities in young adults entering the workforce, the development of efficient teamwork taxonomies and assessment procedures is of critical importance. This study provides direction in pursuing this objective and moves us closer to accomplishing that goal.

In conclusion, existing measures of team processes are labor intensive (think-aloud protocols or ratings of videotaped team sessions) and thus are not timely. Our approach suggests the feasibility of computer-scoring and reporting of team processes in real time. We plan to conduct more statistical analysis to understand more clearly the relationship between team processes and outcomes. Each student, when finished with the simulation, would have a score for each team process (e.g., coordination or leadership) as well as a score of the team outcome. Our software approach has been designed to be domain independent and thus should transfer to other computer-based team environments. Such a reliable and valid measure would offer the capability of assessing quickly team processes and outcomes in collaborative learning environments (K–12) or team training in industry or the military.

ACKNOWLEDGMENTS

The work reported herein was supported under the Educational Research and Development Center Program cooperative agreement R117G10027 and CFDA catalog number 84.117G as administered by the Office of Educational Research and Improvement, U.S. Department of Education. The findings and opinions expressed in this report do not reflect the position or policies of the Office of Educational Research and Improvement or the U.S. Department of Education.

We wish to thank Dr. Eva Baker, Mr. Robert Dennis, Dr. Howard Herl, and Dr. Joan Herman of CRESST/UCLA and Dr. Keith Allred, now at Teachers College, Columbia University, for their assistance. We also wish to thank the leadership of both the Airconditioning and Refrigeration Contractors Association of Southern California, Inc., and the United Association of Journeymen and Apprentices of the Plumbing and Pipefitting

Industry, Airconditioning and Refrigeration Fitters Division, Local No. 250 (AFL-CIO). In particular, we wish to acknowledge the assistance of union member Mr. Jack Ferrara. Finally, we also wish to thank Dr. John Brady, Ms. Susan Hall, Ms. Michele Jansen, and their students for their assistance and participation.

REFERENCES

Baker, D. P., & Salas, E. (1992). Principles for measuring teamwork skills. *Human Factors, 34*, 469–475.

Bazerman, M. H. (1990). *Judgments in managerial decision making* (2nd ed.). New York: Wiley.

Brannick, M. T., Prince, A., Prince, C., & Salas, E. (1995). The measurement of team process. *Human Factors, 37*, 641–451.

Brannick, M. T., Roach, R. M., & Salas, E. (1993). Understanding team performance: A multimethod study. *Human Performance, 6*, 287–308.

Burke, C. S., Volpe, C., Cannon-Bowers, J. A., & Salas, E. (1993, March). *So what is teamwork anyway? A synthesis of the team process literature.* Paper presented at the 39th annual meeting of the Southeastern Psychological Association, Atlanta, GA.

Cannon-Bowers, J. A., Oser, R., & Flanagan, D. L. (1992). Work teams in industry: A selected review and proposed framework. In R. W. Swezey & E. Salas (Eds.), *Teams: Their training and performance* (pp. 355–377). Norwood, NJ: Ablex.

Cannon-Bowers, J. A., Salas, E., & Converse, S. (1993). Shared mental models in expert team decision making. In N. J. Castellan (Ed.), *Individual and group decision making: Current issues* (pp. 221–246). Hillsdale, NJ: Lawrence Erlbaum Associates.

Cannon-Bowers, J. A., Tannenbaum, S. I., Salas, E., & Volpe, C. E. (1995). Defining competencies and establishing team training requirements. In R. A. Guzzo, E. Salas, & Associates (Eds.), *Team effectiveness and decision making in organizations* (pp. 333–380). San Francisco, CA: Jossey-Bass.

Carnevale, P. J., & Conlon, D. E. (1988). Time pressure and strategic choice in mediation. *Organizational Behavior and Human Decision Processes, 42*, 111–133.

Carnevale, P. J., & Pruitt, D. G. (1992). Negotiation and mediation. *Annual Review of Psychology, 43*, 531–582.

Cohen, J. (1992). A power primer. *Psychological Bulletin, 112*, 155–159.

Druckman, D., & Bjork, R. A. (1994). *Learning, remembering, believing. Enhancing human performance.* Washington, DC: National Academy Press.

Foushee, H. C. (1984). Dyads and triads at 35,000 feet: Factors affecting group processes and aircrew performance. *American Psychologist, 39*, 885–893.

Foushee, H. C., Lauber, J. K., Baetge, M. M., & Acomb, D. B. (1986). *Crew factors in flight operations III: The operational significance of exposure to short-haul air transport operations* (NASA Technical Memorandum 88322). Moffett Field, CA: Ames Research Center.

Franken, J., & O'Neil, H. F., Jr. (1994). Stress induced anxiety of individuals and teams in a simulator environment. In H. F. O'Neil, Jr. & M. Drillings (Eds.), *Motivation: Theory and research* (pp. 201–218). Hillsdale, NJ: Lawrence Erlbaum Associates.

Hays, W. L. (1973). *Statistics for the social sciences* (2d ed.). New York: Holt, Rinehart & Winston.

HyperCard 2.2 [Computer software]. (1993). Cupertino, CA: Apple Computer.

Kanki, B. G., Lozito, S., & Foushee, H. C. (1989a). Communication as a group process mediator of aircrew performance. *Aviation, Space, and Environmental Medicine, 60*, 402–409.

Kanki, B. G., Lozito, S., & Foushee, H. C. (1989b). Communication indices of crew coordination. *Aviation, Space, and Environmental Medicine, 60,* 56–60.

Koffel, L. (1994). *Teaching workplace skills.* Houston, TX: Gulf Publishing Company.

Lahey, G. F., & Slough, D. A. (1982, January). *Relationships between communication variables and scores in team training exercises* (NPRDC-TR 82-25). San Diego, CA: Navy Personnel Research and Development Center.

Macintosh [Computer hardware]. (1993). Cupertino, CA: Apple Computer.

Macpherson, D., & Perez, R. (1992). *Review of team/crew unit training research: Implications for team training development and research.* Alexandria, VA: Army Research Institute for the Behavioral and Social Sciences.

McCarthy, J. C., Miles, V. C., Monk, A. F., Harrison, M. D., Dix, A. J., & Wright, P. C. (1993). Text-based on-line conferencing: A conceptual and empirical analysis using a minimal prototype. *Human–Computer Interaction, 8,* 147–183.

McIntyre, R. M., & Salas, E. (in press). Team performance in complex environments: What we have learned so far. In R. Guzzo & E. Salas (Eds.), *Team effectiveness and decision making in organizations.* San Francisco: Jossey-Bass.

MediaTracks 1.0 [Computer software]. (1990). Emeryville, CA: Farallon Computing.

Morgan, B. B., Jr., Salas, E., & Glickman, A. S. (1993). An analysis of team evolution and maturation. *Journal of General Psychology, 120,* 277–291.

Mullen, B., & Copper, C. (1994). The relation between group cohesiveness and performance: An integration. *Psychological Bulletin, 115,* 210–227.

O'Neil, H. F., Jr., & Abedi, J. (1996). Reliability and validity of a state metacognitive inventory: Potential for alternative assessment. *Journal of Educational Research, 89,* 234–245.

O'Neil, H. F., Jr., Allred, K., & Baker, E. L. (1992). *Measurement of workforce readiness: Review of theoretical frameworks* (CSE Tech. Rep. No. 343). Los Angeles: University of California, Center for Research on Evaluation, Standards, and Student Testing (CRESST).

O'Neil, H. F., Jr., Allred, K., & Dennis, R. A. (1992). *Simulation as a performance assessment technique for the interpersonal skill of negotiation* (Deliverable to OERI, Contract No. R117G10027). Los Angeles: University of California, Center for Research on Evaluation, Standards, and Student Testing (CRESST).

O'Neil, H. F., Jr., Allred, K., & Dennis, R. A. (1994). *Assessment issues in the validation of a computer simulation of negotiation skills* (CSE Tech. Rep. No. 374). Los Angeles: University of California, Center for Research on Evaluation, Standards, and Student Testing (CRESST).

O'Neil, H. F., Jr., Baker, E. L., & Kazlauskas, E. J. (1992). Assessment of team performance. In R. W. Swezey & E. Salas (Eds.), *Teams: Their training and performance* (pp. 153–175). Norwood, NJ: Ablex.

O'Neil, H. F., Jr., Sugrue, B., Abedi, J., Baker, E. L., & Golan, S. (1992). *Final report of experimental studies on motivation and NAEP test performance* (Report to NCES, Grant # RS90159001). Los Angeles: University of California, Center for Research on Evaluation, Standards, and Student Testing (CRESST).

Oser, R. L., McCallum, G. A., Salas, E., & Morgan, B. B., Jr. (1989, March). *Toward a definition of teamwork: An analysis of critical team behaviors* (NTSC Tech. Rep. 89-004). Orlando, FL: Naval Training Systems Center, Human Factors Division.

Rouse, W. B., Cannon-Bowers, J. A., & Salas, E. (1992). The role of mental models in team performance in complex systems. *IEEE Transactions on Systems, Man, and Cybernetics, 22,* 1296–1308.

Salas, E., Dickinson, T. L., Converse, S. A., & Tannenbaum, S. I. (1992). Toward an understanding of team performance and training. In R. W. Swezey & E. Salas (Eds.), *Teams: Their training and performance* (pp. 3–30). Norwood, NJ: Ablex.

Sanderson, P. M., & Fisher, C. (1994). Exploratory sequential data analysis: Foundations. *Human–Computer Interaction, 9,* 251–317.

Southern California Airconditioning and Refrigeration Industry. (1990). *Master agreement. Los Angeles and Orange Counties. 1991–1994.* Los Angeles: Airconditioning and Refrigeration Contractors Association of Southern California, Inc. and United Association of Journeymen and Apprentices of the Plumbing and Pipefitting Industry, Airconditioning and Refrigeration Fitters Division.

Stasz, C., Ramsey, K., Eden, R. A., DaVanzo, J., Farris, H., & Lewis, M. (1993). *Classrooms that work: Teaching generic skills in academic and vocational settings* (MR-169-NCRVE/UCB). Santa Monica, CA: Rand.

Stasz, C., Ramsey, K., Eden, R. A., Melamid, E., & Kaganoff, T. (1996). *Workplace skills in practice* (MR-722-NCRVE/UCB). Santa Monica, CA: Rand.

Sundstrom, E., De Meuse, K. P., & Futrell, D. (1990). Work teams: Applications and effectiveness. *American Psychologist, 45,* 120–133.

Swezey, R. W., & Salas, E. (1992). Guidelines for use in team training development. In R. W. Swezey & E. Salas (Eds.), *Teams: Their training and performance* (pp. 219–245). Norwood, NJ: Ablex.

System 7 [Computer software]. (1992). Cupertino, CA: Apple Computer.

U.S. Department of Labor. (1991, June). *What work requires of schools: A SCANS report for America 2000.* Washington, DC: U.S. Department of Labor, Secretary's Commission on Achieving Necessary Skills.

U.S. Department of Labor. (1992). *Skills and tasks for jobs: A SCANS report for America 2000.* Washington, DC: U.S. Department of Labor, Secretary's Commission on Achieving Necessary Skills.

Weaver, J. L., & Bowers, C. A. (1995). Networked simulations: New paradigms for team performance research. *Behavior Research Methods, Instruments, and Computers, 27,* 12–24.

Webb, N. (1993). *Collaborative group versus individual assessment in mathematics: Processes and outcomes* (CSE Tech. Rep. No. 352). Los Angeles: Center for Research on Evaluation, Standards, and Student Testing (CRESST).

Webb, N. (1995). Group collaboration in assessment: Multiple objectives, processes, and outcomes. *Educational Evaluation and Policy Analysis, 17,* 239–261.

Webb, N., & Farivar, S. (1994). Promoting helping behavior in cooperative small groups in middle school mathematics. *American Educational Research Journal, 31,* 369–395.

Webb, N., & Palincsar, A. S. (1996). Group processes in the classroom. In D. Berliner & R. Calfee (Eds.), *Handbook of research in educational psychology* (pp. 841–873). New York: Macmillan.

Author Index

Subject Index

I

Immigration
wages and, 31
Intelligence tests
alpha/beta testing, 259, 260, 263,
276
cognition and, 256, 257
see Military
Interdependency,
compatibility, 162
mental models, 162
shared/compatible knowledge, 163,
172
team skills, 160
International competition, 29
jobs and, 27
Interpersonal competency, computer-
based simulation, 205-225
defining negotiation skills, 206, 207
distributive/integrative negotiations,
209, 211
feasibility, 224, 225
methodology, 214
mixed-motive interdependence, 211,
214
Interpersonal competency, computer-
based simulation of, 219-220
assessment context 209-212
results, 220-223
SCANS study, 206-209, 224
self-serving bias, 213, 223
validation study (expert/novice), 213
Interpersonal skills, computer simulation, 205, 229-251
discussion, 250, 251
expert/novice criterion group
approach, 232
fixed-pie bias, 232, 233
integrative/distributive dimensions,
231
participants, 234, 235
performance measures, 237-245
pilot study for team assessment, 248-
250
procedure, 235, 236
SCANS, negotiation simulation, 229-
231
self-regulation results, 246
self-serving bias, 233, 234, 247, 248
validation study, 234
Interpersonal skills

competency, 189-197
see Teamwork skills

J

Job behavior study
SCANS, 82, 83
Job contract negotiation
see Computer simulation; Interpersonal skills
Job performance
predicting, 258, 264
entry-level, 4
see Military
Job predictability, 277-289
military, 277-281
see Career-Technical Assessment
Program (C-TAP)
Job profiling
see Work Keys System

L

Learning opportunities
for work, 79
Literacy
as continuum, 256
cognitive competence, 256, 257
gap, 255
law of supply and demand, 255-257
model of, 260-262, 273, 275
Organization for Economic Co-
Operation and Development
(OECD), 225
SCANS and, 255, 256
see Cognition; Predictability; Jobs;
Intelligence; Military; Testing

M

Management
new directions in, 4
Marketplace
globalization of, 78
Maryland, educational reform,
Skills for Success (SFS) program
context/expectations for
assessment, 70-74
crafting SFS, 51-55
curriculum transformation, 66-68
defining, SFS, 51-55